Development, Properties, and Industrial Applications of 3D Printed Polymer Composites

R. Keshavamurthy
Dayananda Sagar College of Engineering, India

Vijay Tambrallimath
Manipal Institute of Technology, Manipal Academy of Higher Education, Manipal, India

J. Paulo Davim
University of Aveiro, Portugal

IGI Global
PUBLISHER of TIMELY KNOWLEDGE

A volume in the Advances in Chemical and Materials Engineering (ACME) Book Series

Published in the United States of America by
IGI Global
Engineering Science Reference (an imprint of IGI Global)
701 E. Chocolate Avenue
Hershey PA, USA 17033
Tel: 717-533-8845
Fax: 717-533-8661
E-mail: cust@igi-global.com
Web site: http://www.igi-global.com

Library of Congress Cataloging-in-Publication Data

Names: Keshavamurthy, R., 1983- editor. | Tambrallimath, Vijay, 1989- editor. | Davim, J. Paulo, editor.
Title: Development, properties, and industrial applications of 3D printed polymer composites / R. Keshavamurthy, Vijay Tambrallimath, J. Paulo Davim, editors.
Description: Hershey, PA : Engineering Science Reference, [2023] | Includes bibliographical references and index. | Summary: "The book introduces readers to the concepts of Additive Manufacturing, as well as the use of various materials in polymer Additive Manufacturing, optimization of process parameters for composite material fabrication, case studies, and industrial applications"-- Provided by publisher.
Identifiers: LCCN 2022034395 (print) | LCCN 2022034396 (ebook) | ISBN 9781668460092 (h/c) | ISBN 9781668460108 (s/c) | ISBN 9781668460115 (eISBN)
Subjects: LCSH: Polymeric composites--Industrial applications. | Additive manufacturing. | Three-dimensional printing--Materials.
Classification: LCC TA418.9.C6 D458 2023 (print) | LCC TA418.9.C6 (ebook) | DDC 620.1/18--dc23/eng/20220909
LC record available at https://lccn.loc.gov/2022034395
LC ebook record available at https://lccn.loc.gov/2022034396

This book is published in the IGI Global book series Advances in Chemical and Materials Engineering (ACME) (ISSN: 2327-5448; eISSN: 2327-5456)

British Cataloguing in Publication Data
A Cataloguing in Publication record for this book is available from the British Library.

For electronic access to this publication, please contact: eresources@igi-global.com.

Table of Contents

Detailed Table of Contents

There is a huge scope for natural fiber-reinforced composite materials, which can be used in 3D printing technology. Non-degradable materials used in the additive manufacturing field can be replaced by the innovative application of natural fiber-reinforced composite materials. This chapter introduces the advantages of 3D-printed natural fiber composite materials compared to polymer materials. The homogeneity and thermal stability of fiber reinforcement are the critical characteristics of the composite material. Also, the standardization of material testing has considerable space for research. Continuous natural fiber-printed composite and short natural fiber-printed composite are the major focuses for researchers in the field of additive manufacturing. Various fiber materials yield diverse results in the final product manufactured. Various additive manufacturing techniques can be applied and compared for the same reinforcements and matrix material.

In the coming years, researchers and manufacturers will be more concerned with the demand for products that are easy to construct and can be rendered seamlessly, even at complicated geometries, with the touch of a button. These requirements will be met by the advent of additive manufacturing. This will serve as the catalyst for a revolution. Additive layer manufacturing (ALM), also known as additive manufacturing (AM) or more commonly referred to as 3D printing, is a modern fabrication technology that uses a variety of raw materials to produce items such as medical implants and aircraft wing components by printing layers of material based on 3D digital models. In this chapter, the authors provide a comprehensive overview of the mechanical and tribological behaviour of ceramic, metallic, and fiber-reinforced polymer composites that are made by a range of additive manufacturing (AM) techniques. These composites can be used in a variety of applications, including aerospace, automotive, and medical.

This chapter highlights recent developments and provides an overview of the rapid application of fused deposition modelling (FDM) for polymeric smart and composites. The review is divided into sections that describe the processing conditions and characteristics of FDM components made of polymer and its composites as well as shape memory polymers/composites. The chapter covers a wide range of applications, including PVDF structures and components. While FDM adoption has been rapid in this field, more coordinated efforts in the areas of smart polymer feedstock synthesis, process tuning, and testing are required. This chapter provides an overview of 3D printed smart polymer materials and composites as well as their properties, performance, and potential applications. Additionally, this chapter discusses the motivation for future 3D printing research.

3D printing uses fused plastics and composite filaments to layer components. This method produces parts for household tasks, industrial applications, and rapid prototyping. Developing a printing model requires determining extruder speed, extrusion height, and bed temperature (slicing). Poorly defined parameters can result in poor dimension precision, surface integrity, mechanical properties, and finally printing flaws in plastic and composite sections, limiting or prohibiting their use. During the first stage of 3D printing, a component's first layer is created. This chapter primarily reviews research on the use of image processing methods at different phases of the fused modeling process. Additionally, a review of the broad literature on the junction of FDM and image processing has been presented. The advantages and disadvantages of image processing for the FDM process, as well as existing barriers, have been discussed.

Polymers are macromolecules with a very high molecular weight that are formed as a result of one or more small molecules bonding to each other, typically by covalent bonds. The concept of macromolecules consisting of many repeating units are published in the article titled "Uber Polymerization," which was

Preface

A wide range of industrial and academic players have shown a very keen interest in additive manufacturing technology. The opportunity for exploration and research from materials delivery to final moment has expanded due to this technology's versatility.

In recent years, additive manufacturing has grown in popularity for the production of composite materials, owing to the fact that 3D printing technology allows designers and engineers to create products that can be fabricated at low volumes in an economical manner which are not able or difficult to fabricate using traditional processes.

There is a lot of potential for polymer composites in product designs, as they were used in the early development of AM. Furthermore, because these materials are diverse and have distinct properties, research and development are advancing rapidly. It's become an interesting topic to explore in the additive manufacturing industry since many researchers have done studies on polymer composites. In addition, the characteristics of prevalent materials can be enhanced by the reinforced polymer composites even while performing well in finished products.

Additive manufacturing (AM) of composites using various methods has begun to grow in prominence as a result of its benefits for creating complicated structures. Certain polymer AM processes can now produce dense parts for a wide range of materials, including plastics-based composites, with high reliability. High-performance materials will most likely expand the AM material variety in the near future. A reference book that examines the latest advancements and technological developments in additive manufacturing and additive manufacturing applications for polymer composites is an important reference source. The feedstock materials selection, the synthesis principle and behaviour while manufacturing process, the resultant properties, as well as the relationships between all these elements receive special study when designing materials for additive manufacturing parts.

Additive manufacturing of polymer composites has seen significant developments in recent years. The book discusses these developments and their prospective future. The proposed book, as the title suggests, focuses on advanced technologies, advanced materials, processing methods, and applications. A summary of 3D printed materials' life cycle and performance parameters, as well as information on advances and applications in aerospace, medical, and automobile fields, has been included.

The examination of case studies, approaches to industrial problems, process issues, difficulties of integrating reinforcements and matrices, industrial applications, optimizing using machine learning and artificial intelligence so on, re all addressed comprehensively in this book. The challenges faced by practitioners, researchers, or readers in terms of selecting composite materials, applying optimal process parameters, or applying composite materials developed by additive manufacturing are extensively discussed throughout the book, paving the way for a smooth journey.

In addition, machine learning concepts and algorithms are introduced and applied on the process of additive manufacturing. This book contains contributions from leading scientists and worldwide researchers from business, university, government, and corporate academic institutions all over the world. This highly application-oriented book in polymer composites will benefit academics, researchers, industrialists, scientists, engineers, and students in the automotive, medical, and aerospace domains.

Students in additive manufacturing courses, researchers in the field, polymer composite materials enthusiasts, medical practitioners and engineering professors interested in additive manufacturing will find the book useful.

Further, the book serves to help enthusiasts explore their interests and prepares them for tackling a wide variety of new materials and processes within the field of engineering. It would be extremely beneficial to medical practitioners to gain a deeper understanding of all the different technological approaches related to additive manufacturing and materials.

Chapter 1, "A Look at 3D Printing in the 21st Century," discusses design aspects, challenges, and opportunities in AM. AM technology's process parameters and classification. This chapter also discusses a sustainable approach to AM process technology in comparison to conventional forms. Chapter 2, "A Bibliometric Approach: What, Where, and Who Is Researching This Topic," provides a review of polymer composites and their use in the AM process. Technologies used in AM for the fabrication of polymer composites and the use of materials in various processes. The third chapter, "Properties and Applications of Natural Fiber Reinforced 3D Printed Polymer Composites," provides an overview of the AM design concept for natural fibre reinforced polymer composites. View of functional requirements and design possibilities. Chapter 4, "Mechanical and Tribological Properties of FDM-Developed Polymer Composites: Physical Parameter Studies, Dimensional Stability, and Surface Roughness Studies," is a comparison of mechanical property variation with reinforcement variation for AM parts. AM developed tribological studies of various polymer matrix compositions. "Fused Deposition Modelling of Polylactic Acid (PLA)-Based Polymer Composites" (Chapter 5) discusses various aspects of PLA usage in FDM technology. The chapter also includes case studies on the use of PLA. Chapter 6, "Additive Manufacturing Developments in the Medical Engineering Field," focuses on medical applications of AM technology. Chapter 7 investigates the physical and mechanical properties of polymer biomaterials for the advancement of biodegradable polymer composite materials for uses in biomedical additive manufacturing. The processing method, material selection and customization, and application are all important considerations. Chapter 8, "An Overview of 3D Printed Smart Polymers and Composites," focuses on polymer composites and their use in AM processes, technologies involved in AM for polymer composite fabrication, and material usage in various processes. Chapter 9, "Image Processing Technique Applications in Fused Deposition Modeling," connects electronics applications using AM technology. The chapter discusses various image processing processes that can be used with the AM process. Chapter 10, "The Importance of Polymers in Medicine and Their FTIR and Raman Spectroscopic Investigations," discusses FTIR and Raman approaches for polymer composites in detail. Chapter 11, "Industrial Robot Integrated Fused Deposition Modelling for 3D Printing Process," discusses industrial applications of 3D printing technology by robots in detail. In the current scenario, the use of robotic arms for 3D printing is becoming more prevalent. Chapter 12, "Applications of Fused Deposition Modeling in Dentistry," covers dental AM processes, material selection, current applications, case studies, and challenges. Chapter 13, "Additive Manufacturing Development of High Performance Polymer Composites," discusses the behaviour of high performance polymer composites in extreme conditions which includes high pressure, high temperature, and corrosive chemicals developed by AM. Chapter 14, "Machine Learning Applications for 3D Printed Polymers and Composites," aids in the identification of predictive and actionable recommendations to improve part quality and process design.

The advantages of machine learning for AM data types, data sources, potential variability in experimental and simulation data, and the applicability of these data in ML algorithms are all discussed. Predictive modelling is useful in providing intelligent, efficient, customizable, high-quality, and sustainable production processes. The use of various algorithms and their applications in making AM smart.

This book fills gaps in existing literature by allowing readers to fully comprehend the interactions between polymer composite materials, additive manufacturing applications, and machine learning approaches. This would be the first publication of its kind, benefiting a wide range of readers.

R. Keshavamurthy
Dayananda Sagar College of Engineering, India

Vijay Tambrallimath
Manipal Institute of Technology, Manipal Academy of Higher Education, Manipal, India

J. Paulo Davim
University of Aveiro, Portugal

Chapter 1
Glimpses of 3D Printing in the 21st Century

Vijay Tambrallimath
https://orcid.org/0000-0001-5153-9143
Manipal Institute of Technology, Manipal Academy of Higher Education, Manipal, India

R. Keshavamurthy
Dayananda Sagar College of Engineering, India

ABSTRACT

As the ability to fabricate highly complicated computerized geometry makes additive manufacturing a significant technology in the present day, it has a significant impact on both the manufacturing and design industries. As a result of their intrinsic mechanical and anisotropic properties, pure polymer parts which are 3D printed completely have a limited application potential. Their capabilities have been improved by the filler inclusion for more versatility and application. Innovative benefits of 3D printing include simplicity of material modification, low production cost, less waste, and customised geometry. The overview describes the advancement of matrix composites printing technologies for 3D printing materials with improved qualities including its uses in the electronics, automotive, aerospace, and biomedical. In conclusion, the technology of 3D printing is analysed, highlighting its capabilities and shortcomings.

INTRODUCTION

Additive manufacturing has developed rapidly and ceaselessly ever since being brought into the spotlight at the dawn of the 21st century. In the industrial landscape of the world, the influence of 3D printing has been felt in every corner. It is strongly rooted inside Industry 4.0's expanding presence as an essentially futuristic technology that addresses the flaws that conventional manufacturing processes have been battling for decades. Manufacturing and construction are changing at an alarming pace, and additive manufacturing plays an essential role in this transformation. In addition to innovative and sustainable processes, additive manufacturing has led to smarter machines, more efficient management, and more automated operations. With additive manufacturing, multiple layers of materials are sequentially deposited

DOI: 10.4018/978-1-6684-6009-2.ch001

in layers to create three-dimensional objects (R.Keshavamurthy et. al. (2018)). The machine functions in accordance with the planned design that is fed in the form of an extensive digital layout. Rapid prototyping and forging intricate geometric things are the two main applications of additive manufacturing. Additive layer manufacturing and 3D printing are two more names for additive manufacturing.

The use of additive manufacturing technology for producing useful parts and goods has increased significantly during the past 20 years. Some standard thermoplastic materials such as polylactic acid (PLA) (Melnikova R et.al. (2014), Tran P et. al. (2017) Al., Tymrak B et.al. (2014)), acrylonitrile butadiene styrene (ABS) (Sun Q et al. (2008)), polycarbonate (PC) (Garcia CR et.al. (2012)), and polyamide (PA) (Caulfield B et. al. (2007)), related to certain thermosetting materials utilised in additive manufacturing. Epoxy resins serve as thermosetting polymers, require heat or ultraviolet radiation to cure properly, they are not recommended for 3D printing. (Gu H et. al., (2016)). Different industries benefit from additive manufacturing in numerous ways. There is considerable potential for lightweight and complex cross-section areas like honeycomb cells to be manufactured using this technology in aerospace and automotive industries. Structural optimization and form finding of light weight structures. There are different parts in composite structures that contain cavities and angles for managing weight to strength ratios. To study the feasibility and functionality of complex structures, architects find it very convenient to model them (Wong KV- et.al. (2012)). Using the technology for 3D printing in museum exhibitions, artifact education, and artifact restoration can facilitate physical concept and visualization learning (Short DB (2015)). An evaluation of the mechanical properties and 3D printed scaffolds' biological properties made from tricalcium phosphate doped with silica and zinc oxide. The lack of strength and load bearing ability of polymer materials limits the application of 3D printing to conceptual prototype design and modelling. Pure polymers' practical applicability is limited by these drawbacks. Matrix and fillers have been added to 3D-printed polymers to improve their structural and functional properties, which is unachievable as a single component. Particles, fibres, and nanomaterials are incorporated into polymer matrix composites because they exhibit high mechanical and physical properties. Casting and moulding, as well as machining, were traditionally used to manufacture polymers and composites with complex geometries (Huang SH et. al. (2013)). Despite being well-organized, these procedures and approaches are unable to govern intricate inner systems. Contrarily, because the design is managed by software, additive manufacturing offers a platform for fabricating intricate geometries with little wastage of materials and excellent accuracy. By making manufacturing convenient and precise, AM technologies revolutionized manufacturing. 3D printing has attracted massive interest from researchers over the last three decades, which has led to a growing interest in new polymer printing techniques and methods. Polymer composites with enhanced desirable properties have achieved remarkable developments in 3D printing in recent years (Vijay Tambrallimath et. al. (2021)).

It is with rapid prototyping that 3D printing has demonstrated its capability to fulfil its original purpose, which was to demonstrate just how useful it is. As machine technology advanced and 3D printer compatible materials became more accessible, the number and quality of producible parts increased. In the past, this technique was mainly used to create temporary artefacts, including such early developmental stages or visual representations of thought processes. Now that AM goods are available individually on store shelves or integrated into larger structures, they can be considered finalized products in and of themselves. Thermoplastics were historically used for AM operations because of their low costs. Material development, however, has become increasingly important to the advancement of 3D printing over the past few years. Additive manufacturing offers qualitative advantages over traditional manufacturing with end-use applications reaching feasibility at an increasing rate.

Three-dimensional printing has earned a solid reputation over the years due to several properties specific to the technology. The results of additive manufacturing have consistently been associated with high performance polymers, for example. With the ability to form into complex geometric shapes, it is highly resistant to intense temperatures and extreme stress, and their adaptability to a wide range of uses had given them an unsurpassed popularity. Aside from this, AM is also employed frequently in the manufacturing of industrial parts using metals such as titanium and steel. AM can also produce extremely strong and lightweight ceramic materials like alumina and zirconia, but traditional manufacturing is challenging, giving AM a competitive advantage. Environmental considerations also resonate with the drive to advance material development. The AM industry has considerable room for growth, despite the possibility of producing environmentally friendly products through 3D printing.

3D PRINTING TECHNOLOGY OVERVIEW

Rapid prototyping techniques includes three-dimensional printing involve the material deposition layer by layer in accordance with the CAD model. Polymer composites can be manufactured using a variety of AM technologies. Processes for 3D printing with various parameters and materials includes Stereolithography, Fused Deposition Modelling, Selective Laser Sintering, and Powder-Liquid 3D Printing. Other methodologies-based techniques are currently in the early stages of development. Material, parameters, or the functionality of the composite product determine the advantages and disadvantages of each technique. As a result, these former attributes need to be taken into account when choosing any AM technology. In Table 1, an overview of various methods is presented.

Table 1. Commonly used AM techniques (Saroia, J. et. al. (2020))

Technology	Approach	Material Used	Form of material
FDM	Extrusion and Deposition	Thermoplastic polymers	
PLP	Drop on demand binder	Any material in powder form	Powder
SLS	Scanning with laser	PCL and polyamide powder	Powder
SLA	Scanning with UV	Photocurable resin	Liquid photopolymer
Robocasting	UV assisted curing	PCL, PLA, hydrogel	Liquid or paste

3D PRINTING TECHNIQUES FOR MATRIX POLYMER COMPOSITES

Since polymers have high ductility and low melting points, they are widely used in 3D printing. They are easily available due to their inexpensive cost, versatility for usage in diverse operations and also has an excellent strength to weight ratio. The products of Pure polymer, on the other hand, have modest functional features and mechanical properties, making them ideal for conceptual prototypes and instructional uses. Pure polymers have undergone substantial improvements in their structural strength and fabrication techniques in recent years. Filler particles added to polymer matrix enhance the mechanical and functional properties while improving printing accuracy. MPC (Matrix polymer composites) are the name for these materials. Table 2 shows the details of these classifications.

The new composite materials' development with nanomaterials ceramics, and fiber has been the focus of many types of research aiming for high-performance additive manufacturing of such materials (Wang X et. al. (2017)) (Takezawa A et. al. (2017)).

Table 2.Various reinforced polymer composites

Matrix polymer composites with particle reinforcement	Pure polymers that can be 3D printed are very important since they are readily accessible and affordable. Polymers have improved structurally and functionally because of matrix polymer composites. SLA, SLS, and FDM printing technologies are ideal for printing powders, thermoplastics, or liquids with different types of particles. Glass beads have been added to polymers to improve their tensile strength and storage modulus (Chung H et. al. (2006)). A study conducted by some researchers found an increase in tensile strength when copper and iron particles were added. Aluminium and aluminium oxide were added to nylon-6 to overcome some surface roughness as well as wear concerns. Whereas adding tungsten to PC and Al2O3 to UV-cured resin (Kurimoto M et.al. (2015) has increased the polymer composites' dielectric permittivity. Cuboidal or cylindrical shapes characterized all test samples. A variety of techniques were utilized, and improvements in properties were analysed, but no other structural geometries were manufactured and tested.
Matrix polymer composites with fiber reinforcement	The properties such as stiffness, corrosion resistance and high strength-to-weight ratio of fiber-reinforced structures have made them extremely important and applicable. For enhancing the properties of pure polymers, these fibers have been used as fillers in 3D printing technologies. Fiber-reinforced matrix composites are fabricated primarily using FDM and direct write techniques. The filaments of FDM are made by passing a polymer blending with fibers through extruders. Filaments such as these are used to print 3D models of desired shapes. Short glass fibres and carbon fibres (Tekinalp HL (2014)) are the most common discontinuous fibres found in fiber-reinforced matrix composites. Increased thermal, electrical, and mechanical properties, displaying unique advancements. As a result of fiber reinforcements, tape swelling can be reduced during deposition procedure, resulting in a greater stiffness of the material during printing. Van Der et al. (2015) using a MarkOne printer with dual printing heads for nylon and CFRP, studied the mechanical properties of 3D printed composites with continuous fiber reinforced polymer reinforcements. On top and bottom, the continuous fibres were jammed between thermoplastic nylon polymers. Improvements of mechanical properties were found in the results. To improve the printing technique, detailed studies were conducted on the process parameters.
Matrix polymer composites with nano-particles reinforcement	The use of nanotechnology in the present world has a huge and wonderful impact on our daily lives, particularly in the fields of electronics, medicine, sports, automobile and among others. Researchers have also developed nanoparticle reinforced matrix polymer composites because of nanotechnology's influence on additive manufacturing. Adding nanoparticles to polymers has been researched recently to improve their mechanical properties. Tensile strength improvement of 39% and 7.5% was observed for printed composite parts with 10% carbon nanofiber and 10% multi-walled carbon nanotube combined with polymers (Hector Sandoval J (2006)). However, the nanoparticles present in these composites decrease the elongation and increase the brittleness of the material. Lin et al. (2015) used SLA to explore the mechanical characteristics of a composite made of graphene oxide and photopolymer. The printed composite's tensile strength and ductility were improved by 62.2% and 12.8% when graphene oxide was added at 0.2%. Increased crystallinity of graphene oxide contributed to the enhanced ductility of the reinforced polymer. The electrical characteristics of polymer composites were improved by adding carbon-based nanoparticles like graphene, carbon nanotubes, carbon-black and carbon nanofiber. Wei et al. (2015) a composite filament was printed using FDM and graphene was added to increase the electrical conductivity of ABS material.
Ceramics polymer composites	The use of 3D printing in ceramics has been revolutionary since it allows for the creation of complex designs and facilitates increased manufacturing speeds and improved fabrication methods. Using polymers as binders during 3D printing, polymers can enhance mechanical and functional properties. There are numerous medical applications for natural hydrophilic polymers and ceramic powders, which includes scaffold printing (Zhou Z (2018)). As the ceramic particles cure, the binder solutions of these polymers help them become wet and cement them together. Zein et al. (2022) used FDM to create innovative scaffolds with a honeycomb pattern. For additive manufacturing, ceramic powder particle size has a great deal of significance since it determines how the particles adhere to each other while curing and drying. Ceramic small particles have a smaller surface volume and contain larger pores, enabling binders to keep the particles firmly together during printing, whereas large ceramic particles create thick layers during printing that give a fine surface finish and increase the printing capacity. With the ability to make extremely large micro and macro components from ceramics, ceramics have a multi-dimensional application. The technology lends itself well to the design and manufacturing of bioceramic scaffolds. Ceramic-filled mixtures can be printed using SLA, the best AM technique. Zanchetta et al. (2016) shown how the 3D printed ceramic structures of high-performance are made utilizing photosensitive resin. As a result, the SLA technology makes it simple to build ceramic structures with enhanced characteristics, such as crack-free, multifunctionality, cost-effectiveness and high density (Schwentenwein M et.al. (2015)).

APPLICATION

Wohler's research revealed this in recently that the industry of aerospace has contributed 18.2% in the market of AM technology and the percentage is expected to continue increasing in the future Wohlers T ((2014)). Lightweight structures with enhanced mechanical and structural properties can be manufactured using various 3D printing techniques. Particularly, the electron beam manufacturing and selective laser sintering processes have opened new avenues for fabricating complex shapes and designs out of pre-alloyed metal powder. The manufacture is simple and efficient owing to these processes. The researchers are also able to create complex CAD models and high-accuracy airfoils and propellers using carbon fiber reinforced polymer composites. As a revolutionizing technology in aviation, additive manufacturing has shown great promise. The weight and production time of different functional components are being decreased by several aeroplane manufacturing businesses employing 3D printing techniques. Wang et al. (2018) examined if 3D printing is practical for the aviation sector. By using the processes of fuzzy analytical hierarchy and fuzzy geometric mean, the different aspects of 3D printing techniques were aggregated and studied. The discussion included the current AM techniques as well as other crucial elements such as simplicity in assembly, lightweight components, and cost effectiveness. Powder bed fusion has recently gained popularity for producing small parts with high precision and accuracy. With this approach, the manufacturer will have a great deal of freedom to optimise the design and its practical features. Powder bed fusion was used to manufacture the brackets for the Airbus A350 XMB. A comparison of the printed parts to the traditional manufacturing method showed that they were 30% lighter and taken 75% less time to manufacture. A variety of biomedical applications are made possible by 3D printing by using computerized designs in the manufacturing process. Natural and synthetic materials can both be printed using AM technology. In addition to demonstrating many functional attributes and applications, different biomedical structures can be visualized in 3D using MRIs and their 3D printed duplicates. The AM technology market is estimated to be dominated by biomedical applications, which represent about 11% of the market. Nowadays, 3D printing technology is integrated with electrical component manufacturing techniques, which allows for rapid prototyping of a wide variety of products. The development of rapid prototyping machines and techniques to improve existing electronics production methods is of interest to many researchers (Espalin D et. al., (2014)).

FUTURE POTENTIAL AND CHALLENGES

Researchers have developed new methods for producing products with better mechanical properties and customized features using additive manufacturing. Various industries require high strength and lightweight products, including aerospace and biomedical. The modern world pays great attention to matrix composite materials because of their excellent thermoelectric properties, durability, and mechanical properties. The fillers inclusions is however restricted to a predetermined quantity known as the threshold point. Following that, the particles of fillers begin to cluster and enhance the formation of void, which lowers the mechanical characteristics. Combining materials with differing physical properties or particle sizes can cause the printer to have difficulty extruding correctly, resulting in reduced bonding forces between the materials, which again results in void formation.

AM techniques face a variety of challenges depending on the process involved. The most popular, user-friendly, and economical 3D printing method is called FDM. A simultaneous heat and cooling process occurs during layer deposition and FDM extrusion, which results in a high level of void formation. Anisotropic structures are the result, which decrease the mechanical properties. Powders or alloys are cured layer by layer in 3D printing processes using heat fusion. Temperature gradients during curing can cause variations in vertical and horizontal directed properties. The printed part may exhibit anisotropic behaviour. Anisotropic properties and processing speed have also been identified as problems with other manufacturing techniques. However, AM technology has proven to be extremely versatile and has enhanced its utilization despite these many challenges and drawbacks. Researchers are attempting to enhance 3D printing techniques in all the work stated above to lessen its limits when compared to conventional manufacturing techniques. Researchers nowadays have complete flexibility to evaluate and manage the characteristics and functionality of products connected to their functional utilisation due to the polymer composites development in recent years, which has also created a brand-new field for the development of the material.

REFERENCES

Caulfield, B., McHugh, P., & Lohfeld, S. (2007). analyzing the effect of build parameters on polyamide components' mechanical properties. *Journal of Materials Processing Technology*, *182*(1-3), 477–488.

Chung, H., & Das, S. (2006). Processing and properties of glass bead particulate-filled functionally graded Nylon-11 composites produced by selective laser sintering. *Materials Science and Engineering A*, *437*(2), 226–234.

Espalin, D., Muse, D. W., MacDonald, E., & Wicker, R. B. (2014). 3D Printing multifunctionality: Structures with electronics. *International Journal of Advanced Manufacturing Technology*, *72*(5-8), 963–978.

Garcia, C. R., Correa, J., Espalin, D., Barton, J. H., Rumpf, R. C., Wicker, R., & Gonzalez, V. (2012). Anisotropic metamaterials' 3D printing. *Prog Electromagn Res*, *34*, 75–82.

Gu, H., Ma, C., Gu, J., Guo, J., Yan, X., Huang, J., Zhang, Q., & Guo, Z. (2016). Multifunctional epoxy nanocomposites: A brief overview. *Journal of Materials Chemistry. C, Materials for Optical and Electronic Devices*, *4*(25), 5890–5906.

Hector Sandoval, J., & Wicker, R. B. (2006). Functionalizing stereolithography resins: Effects of dispersed multi-walled carbon nanotubes on physical properties. *Rapid Prototyping Journal*, *12*(5), 292–303.

Huang, S. H., Liu, P., Mokasdar, A., & Hou, L. (2013). Additive manufacturing and its societal impact: A literature review. *International Journal of Advanced Manufacturing Technology*, *67*(5-8), 1191–1203.

Keshavamurthy, Tambrallimath, Prabhakar, & Sekhar. (2019). Additive Manufacturing Processes and their Applications for Green Technology. In Handbook of Research on Green Engineering Techniques for Modern Manufacturing. IGI Global. doi:10.4018/978-1-5225-5445-5

Kurimoto, M., Yamashita, Y., Ozaki, H., Kato, T., & Funabashi, T. (2015). 3D printing of conical insulating spacer using alumina/ UV-cured-resin composite. In *Electrical insulation and dielectric phenomena (CEIDP), IEEE Conference on, 2015*. IEEE.

Lin, D., Jin, S., Zhang, F., Wang, C., Wang, Y., Zhou, C., & Cheng, G. J. (2015). 3D stereolithography printing of graphene oxide reinforced complex architectures. *Nanotechnology*, *26*(43), 434003.

Melnikova, R., & Ehrmann, A. (2014). Finsterbusch K 3D printing of textile-based structures by Fused Deposition Modelling (FDM) with different polymer materials. In *IOP Conference Series: Materials Science and Engineering* (vol. 1). IOP Publishing.

Schwentenwein, M., & Homa, J. (2015). Additive manufacturing of dense alumina ceramics. *International Journal of Applied Ceramic Technology*, *12*(1), 1–7.

Short, D. B. (2015). 3D. *Print Addit Manuf*, *2*(4), 209–215.

Sun, Q., Rizvi, G., Bellehumeur, C., & Gu, P. (2008). FDM polymer filament bonding quality influenced by processing conditions. *Rapid Prototyping Journal*, *14*(2), 72–80.

Takezawa, A., & Kobashi, M. (2017). An optimization-based multimaterial topology and additive manufacturing method for making porous composites with variable thermal expansion. *Composites. Part B, Engineering*, *131*, 21–29.

Tambrallimath, V., & Keshavamurthy, R. (2021). Mechanical Characterization of PC-ABS based Graphene reinforced Polymer Nanocomposites fabricated by FDM Process. *Journal of Polymers*, *13*(17). Advance online publication. doi:10.3390/polym13172951

Tekinalp, H. L., Kunc, V., Velez-Garcia, G. M., Duty, C. E., Love, L. J., Naskar, A. K., Blue, C. A., & Ozcan, S. (2014). Highly oriented carbon fiber–polymer composites via additive manufacturing. *Composites Science and Technology*, *105*, 144–150.

Tran, P., Ngo, T. D., Ghazlan, A., & Hui, D. (2017). computational study of bio-inspired composite structures using bimaterial 3D printing under transverse and in-plane loads. *Composites. Part B, Engineering*, *108*, 210–223.

Tymrak, B., Kreiger, M., & Pearce, J. M. (2014). by employing open-source 3-D printers, the components of Mechanical characteristics produced in real world environments. *Materials & Design*, *58*, 242–246.

Van Der Klift, F., Koga, Y., Todoroki, A., Ueda, M., Hirano, Y., & Matsuzaki, R. (2015). 3D printing of continuous carbon fibre reinforced thermo-plastic (CFRTP) tensile test specimens. *Open J Compos Mater*, *6*(01), 18.

Wang, X., Jiang, M., Zhou, Z., Gou, J., & Hui, D. (2017). 3D printing of polymer matrix composites: A review and prospective. *Composites. Part B, Engineering*, *110*, 442–458.

Wang, Y., & Wei, Q. (2020). A review on 3D printed matrix polymer composites: Its potential and future challenges. *International Journal of Advanced Manufacturing Technology*, *106*, 1695–1721. https://doi.org/10.1007/s00170-019-04534-z

Wang, Y.-C., Chen, T., & Yeh, Y.-L. (2018). Advanced 3D printing technologies for the aircraft industry: A fuzzy systematic approach for assessing the critical factors. *International Journal of Advanced Manufacturing Technology*, 1–11.

Wei, X., Li, D., Jiang, W., Gu, Z., Wang, X., Zhang, Z., & Sun, Z. (2015). 3D printable graphene composite. *Scientific Reports*, *5*, 11181.

Wohlers, T. (2014). Tracking global growth in industrial-scale additive manufacturing. 3D. *Print Addit Manuf*, *1*(1), 2–3.

Wong, K. V., & Hernandez, A. (2012). A review of additive manufacturing. *ISRN Mechanical Engineering*.

Zanchetta, E., Cattaldo, M., Franchin, G., Schwentenwein, M., Homa, J., Brusatin, G., & Colombo, P. (2016). Stereolithography of SiOC ceramic microcomponents. *Advanced Materials*, *28*(2), 370–376.

Zein, I., Hutmacher, D. W., Tan, K. C., & Teoh, S. H. (2002). Fused deposition modeling of novel scaffold architectures for tissue engineering applications. *Biomaterials*, *23*(4), 1169–1185.

Zhou, Z., Cunningham, E., Lennon, A., McCarthy, H. O., Buchanan, F., & Dunne, N. (2018). Development of three-dimensional printing polymer-ceramic scaffolds with enhanced compressive properties and tuneable resorption. *Materials Science and Engineering C*, *93*, 975–986.

Chapter 2
3D-Printed Polymer Composite: A Bibliometric Approach - What, Where, and Who Is Doing Research on This Topic

Issam Hanafi
https://orcid.org/0000-0001-8765-2631
Université Abdelmalek Essaâdi, Morocco

Francisco Mata Cabrera
Universidad de Castilla-La Mancha, Spain

Jalal El Bahaoui
Abdelmalek Essaâdi University, Morocco

Juan Carlos Campos Rubio
https://orcid.org/0000-0001-6187-6442
Universidade Federal de Minas Gerais, Brazil

Wanderson de Oliveirade Leite
https://orcid.org/0000-0002-5723-8525
Instituto Federal de Minas Gerais, Brazil

ABSTRACT

Due to the recent innovations in 3D-printed polymers research, this study presents a systematic overview of the area, exposing gaps and interesting directions for future research. The current study investigated the trend of research growth using 2558 research papers with 97131 references data collected from the Web of Science Core Collection database (WOS), over the period from 2005 to May 2022, using bibliographic coupling and keyword co-occurrence. The research results allow the authors to conclude that the number of publications in this field's importance has grown tremendously over the last 20 years, with the United States, China, United Kingdom, and India emerging as the countries that publish the most. The top five researchers in 3D printed polymer composite were also identified.

DOI: 10.4018/978-1-6684-6009-2.ch002

INTRODUCTION

Additive manufacturing, widely termed as three-dimensional printing, has progressed and completely altered fast in recent decades. Various technological aspects of 3D printing are now frequently employed in many areas (Ngo, 2018).

Production cost and time is a crucial differentiator between 3D printing and traditional manufacturing. This dynamic is altered by 3D printing. It removes the need for tooling and allows manufacturing to begin immediately after design. This implies a shorter time to market, which gives a competitive advantage (Steenhuis, 2016), (León-Calero, 2021). Each new unit generated by 3D printing incurs no further expenditures. This implies it can create a single part as well as hundreds of pieces at almost the same cost per part (Attaran, 2017). On the other hand, 3D printing operates on the premise of additive manufacturing. The product is designed from the ground up to reduce waste (Gomes, 2022). Topology optimization, and generative design may all be used to minimize material utilization and product weight.

One of the additive industrial technologies is extrusion-based 3D printing that encompasses depositing melted thermoplastic material layer by layer sideways a specified processing pathway after extruding it from a heated nozzle. Since the 1990s, this method has been the furthermost current 3D printing procedure for thermoplastic supplies (Parandoush, 2017), (Shahrubudin, 2019).

The thermosets, thermoplastics, polymers, blends and biological systems are among the polymers employed. Polymer design, processing parameters and additives are discussed in relation to increasing build speed and improving surface polish, accuracy, functionality and mechanical qualities (Ligon, 2017).

Previous research on material extrusion 3D printing has resulted in substantial advancements in material, equipment, and production method (Blok, 2018). Various studies strive to emphasize the variety of 3D printed metal-metal, ceramic-metal and polymer-based applications while exploring the benefits and limitations of additively made multi-material structures (Bandyopadhyay, 2018) The number of studies on 3d printing of multi-material components has increased remarkably. Some multi-material additive manufacturing surveys are already available in the literature. In response to the increased attention in this field, the current article attempts to give a more thorough and up to date assessment of 3D printing of multi-material components by assembling a full collection of additive manufacturing using multiple materials technologies that have been published in peer-reviewed journals (Rafiee, 2020). The functional features of the 3D-printed complex structures are intriguing, and showed how a 3D-printed elastomer like this may be used and opened the door to the future creation of revolutionary soft robotics (Kuang, 2018).

Many researchers have created novel compositions and substances for use in 3D printing processes, the majority of polymer-based biocompatible are 3D printed, the objective is to create functioning and viable organs, and in order to do so, numerous biomaterials and production processes must be investigated (Jammalamadaka, 2018). Some research, on the other hand, focus on the temperature history of printed in 3D polyphenylene sulfide (PPS) parts and assess the thermal processing parameters' impacts and circumstances of heat treatment on mechanical properties. According to the findings, forced air cooling in fused deposition modeling can improve the precision of 3D printing (PPS) samples (Geng, 2018).

Related to the mechanics of helicoidal composites, the adaptive conduct in helicoidal designs is demonstrated through completing a mechanistic investigation of the variations happening in the cuticle of the mutabilis over its life cycle, was exanimated with 3D printing sample testing and a comprehensive examination of the role of pitch angle in the intrinsic mechanics of helicoidal designs (Zaheri, 2019).

Thermoplastics may be processed using extrusion-based additive manufacturing methods because they become soft and processable above their melting point on a regular basis (Shafranek, 2019). Because of their restricted mechanical qualities, they aren't constantly acceptable for structural applications in tidy condition, and additional additions such as reinforcing fillers are usually required to modify the 3D printed characteristics objects (Tian, 2016). Despite the lack of covalent connections, thermoplastics 3D printed can be readily recycled and re-melted.

Light exposure selectively cures a liquid photocurable resin in vat photopolymerization 3D printing, and the structure is generated point by point or layer by layer. Functional monomers employed in these vat photopolymerization additive manufacturing technologies include phenolic polymers, methacrylates, unsaturated polyesters, epoxy resins, organosilicons, and acrylates (Bîrcă, 2019). Similar to dental resins, such systems may be additively built to yield soft materials as well as photopolymers with high mechanical strength and stiffness (Lin,2020).

Mechanical qualities of 3D printed items are impacted by intrinsic material characteristics, build orientation, and the selected manufacturing process among many other aspects. As a result, it is critical to maintain a consistent focus on enhancing the inherent characteristics of polymer materials at all levels, from molecular to macroscopic (Valino, 2019).

Other works analyze the outcome in terms of fiber volume content, layer thickness besides mechanical design orientation performance of 3D printed continuous fiber reinforced composites components manufactured with 3D printer. The mechanical response of the printed specimens is determined by three-point bending and tensile experiments. The impacts of process factors on failure modes are determined by analyzing SEM pictures of cracked surfaces.

The outcomes reveal that strength and stiffness improve by way of fiber volume content rises in most situations, however the amount of augmentation in mechanical performance is low by means of sustained increase in fiber content, notably for glass fibers (Chacón, 2019).

Making tension in fiber, feed rate, printing temperature and fiber surface preparation are all difficulties along the path to producing a high-quality composite product. The experimental findings reveal that the tensile and bending characteristics of the composite and reinforced composite with carbon fiber have significantly improved (Heidari-Rarani, 2019).

Focusing on the properties of 3D printed samples with diverse processing circumstances, similarly fill speed, density, infiltrations and various materials, as well as morphologic, thermal and mechanical. The results revealed that raising the infill density increased the printed components' Young's modulus (Abeykoon, 2020).

The amount of interfacial adhesion among pathways, discrete layers, print head and the intrinsic characteristics of the polymer causes anisotropy in mechanical properties. Raster angle, air gap, component design, post-processing and build orientation are all process characteristics that might help with anisotropy study (Wu, 2020).

Also, others are only interested in multiple material via vat photopolymerization and its capability to spatially modify mechanical characteristics utilizing an admixture of different reaction chemistries and materials inside 3D printed designs. Cortés (2020) reported that 3D printing was frequently used to increase the stress sensing and mechanical capabilities of components made of carbon nanotubes. The impacts of post-curing treatment were studied, in addition the composition of the CNT. The treatment had an effect on mechanical characteristics by increasing glass transition temperatures and Young's modulus, but it had less of an effect on electrical properties.

Other researchers examined at 3D printing, a usual approach for creating digital solid structures in 3D. To manufacture high quality 3D printed elements, the best printing settings must be employed.

In spite of, this type of printing is conducted at a fixed width, which neither enables design flexibility nor an ideal arrangement of carbon fiber reinforced polymer filaments (Kumekawa, 2022). This work experimentally validated a three-dimensional printing approach with a high DOF in which path width and layer thickness were modified by adjusting the nozzle height during CFRP printing. The range of thickness changes achievable was calculated based on the fundamental confirmation of modifying the layer thickness. The outcomes demonstrated that the optimized width model and fiber path successfully lowered weight while increasing strength.

The most influential factor on fabric polymer matrix composite space utilization is the bead aspect ratio, whereas the most influential factor on composite strength is the fiber volume fraction. However, the effectiveness of 3D printed parts may be enhanced thru minimizing the distance among tows, reducing fabric thickness besides using a plain fabric technique rather than the traditional resin transfer molding. (Fu, 2022).

Isotropic mechanical characteristics were observed in composites produced with poor fiber alignment. Aligning the fibers in the composite, on the other hand, produced direction-dependent characteristics with dramatically increased modulus and strength in the alignment direction (Wright, 2022).

Furthermore, high-performance 3D printing technologies and multi-material have gained popularity. In recent years, high-performance polymer and composites have been described for extruded 3D printed material. Because of the large changes in mechanical, tribological characteristics and thermal of reinforced Carbon/Glass fiber functionally graded materials are interesting materials for 3D printing. The objective of this study is to offer a unique approach for producing functionally graded materials utilizing continuous welding filaments in order to enhance the overall performance of 3D printed reinforced Carbon/Glass fiber polymer (Wang, 2022).

It was possible to obtain experimental thermoplastic filaments with a high fiber volume percentage of continuous carbon fibers. The fiber volume reached as for 30-50%, well above those previously investigated and commercially accessible. The primary goal of this work was to identify potential barriers to the usage of these experimental filaments for structural applications in aerospace and beyond. The produced microstructure by 3D, besides the material's mechanical characteristics and fundamental thermal, were studied and compared across a wide variety of fiber volume fractions (Parker, 2022).

As a breakthrough 3D-printing process, in situ pin-assisted melt impregnation in the printing head, where melt impregnation happens precisely before printing, was established (An, 2022). Various printing parameters were used to assess the printability and impregnation of carbon fiber. Furthermore, the tensile strength of printed samples made by means of the novel production by 3D method was related to that of samples made using traditional printing methods, indicating that evenly infected carbon fiber reinforced thermoplastics and carbon fibers had superior mechanical properties.

Another method for combining their qualities is to blend two or more polymers. Because of their high processing temperatures and high melt viscosities, adapting these high-performance polymers to additive manufacturing is extremely difficult. These materials require unique printing conditions in order to be 3D printed. The materials produced here will contribute in the production of functional prototypes built of high-performance composites exploiting the fused filament fabrication additive manufacturing technique for diverse applications (Diouf-Lewis, 2022). Other study revealed that by altering the weight fraction and type of the polymer binder system, the morphology and mechanical characteristics of composites incorporating hybrid reinforcement may be adjusted (Gackowski, 2022). Previous study, however, has revealed that the composites 3D printed using this method exhibited weak bonding inside the HBPs and in the interlayer areas (Gackowski, 2022).

Unfortunately, bibliometric analysis of 3d printed polymer composite is rarely investigated, so, there is a specific requirement of objective and systematic analysis, A quantitative assessment of total published data in order to increase a profounder overview of the research centers also possible research tendencies.

Bibliometric analysis by Pritchard is an effective method for quantitatively assessing the trend of scientific research and the current state via the use of visualization and mathematical statistics methodologies based on the vast mainstream of published materials. The statistics information may be used to analyze the number of interconnected research, which increases from year to year, significant worldwide output performance and international collaboration, recognize hot spots in the field, besides lastly recommend forthcoming research objectives, with an attention on type and the number of created documents, the nations that have made the largest contributions to this evolution, as well as the primary researchers that have concentrated on the subject. The work presented here, more particularly, intends to address 3 critical issues, the responses to which are intended to lead to more effective and a greater comprehension of the reality experienced in relations of 3D printing for polymer composites, notably:

- What is the observed reality in terms of 3d printing for polymers composite?
- Which countries contributed the most to the research, and where?
- Who has made the most significant contribution to 3D printing- Who?

METHODOLOGY

The purpose of this study is to illustrate and evaluate research development in the field of 3D printing from 2005 until the fifth month of 2022 (May). Based on a content and descriptive analysis, the bibliometric analysis technique will be used in this study. When used to measure and study information and texts, it not only enables the monitoring and analysis of large volumes of data., but it is also highly verified by strong theories.

Bibliometrics refers to the use of statistical tools to evaluate bibliometric publication information such as reviews, peer-reviewed journal papers, chapters, books and conference proceedings (Rousseau, 2014). It is often used to show the connections between quantitative methodologies and study domains.

It is a method for monitoring, quantifying and evaluating the number of scientific articles published by organizations and persons during a certain time window and region, and moreover the links between these publications.

A review of a collection of research papers on 3D printed polymer composites considering quantitative variables such as temporal progression, furthermost prolific researcher, furthermost productive country, number of citations, etc. The statistical examination of outcomes released by institutions or individuals, in addition to their time window and location, as well as the connecting points or linkages among these publications. Is an approach to measure researchers or the academic standing of journals quantitatively using statistical methodologies for example TC index and citation rates. Using publication patterns to identify technical breakthroughs and gaps in the domain.

The data for the research papers obtained from the Web of Science Core Collection database (WOS). It is a famous and widely chosen a research database that is easily accessible to researchers and serves as the basis for the effective combination besides interface of data offered by it. In this regard, highlighting the latent presented by the database and conducting a bibliographic review, with topic choice of "3d printed" in addition "polymer composites," which was the subject of the study conducted along with the scope of bibliometric investigation in 603 sources.

As shown in figure 1, Articles take dominance here, with a weight of 82 percent (2107 papers) of the types of documents published, followed by review papers (14% —355 papers) and article early access (2 percent —63 papers).

As a result, the database is regularly updated, besides all systematic searches were undertaken on a different daytime to avoid any possible bias. Web of Science Core Collection database "export" function was used to send "cited references and suitable documentation" from returned records to bibliometric tools for additional processing. Then, detailed data on general information, such as authors, institutions, locations, source journals, annual publications, number of citations, research areas, funding sources and H-index, was acquired. The aforementioned method was used independently by 2 investigators, and any discrepancies were handled through discussions.

Figure 1. Document type

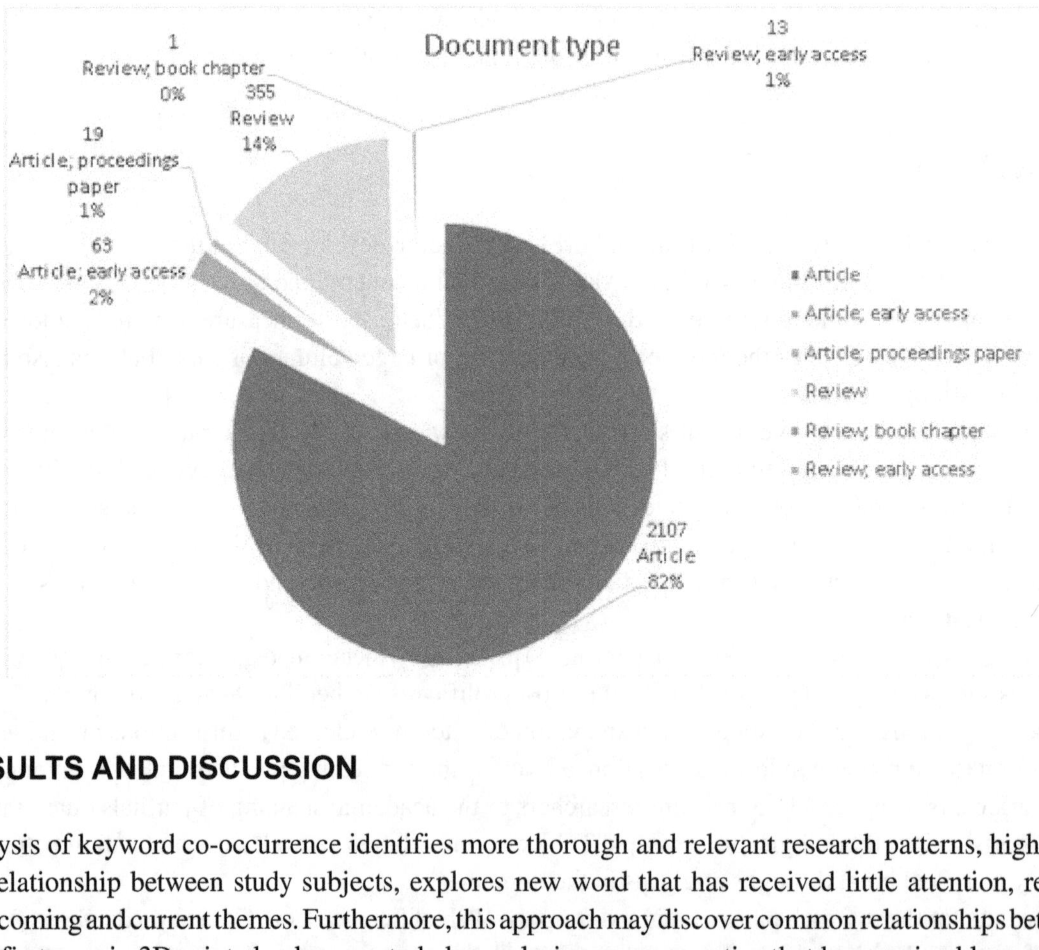

RESULTS AND DISCUSSION

Analysis of keyword co-occurrence identifies more thorough and relevant research patterns, highlights the relationship between study subjects, explores new word that has received little attention, reveals forthcoming and current themes. Furthermore, this approach may discover common relationships between specific terms in 3D printed polymers study by exploring new semantics that has received less consideration and disclosing present and forthcoming issues. In this study, 9142 keywords were found from 2558 publications, including keywords extracted from the title and abstract as well as author keywords.

The co-occurrence network knowledge mapping was displayed in figure 2, with word repetition limited to 250 times. A set of terms with higher research value was chosen, resulting in a network representation of co-occurrences of the top 268 keywords.

As a result, each node represented a separate term. The circle's area represented the number of times each phrase occurred. The co-occurrence was represented by the curve connecting the circles, and its thickness was proportional to the intensity of the association.

The color denoted keyword clusters, near keywords were frequently more strongly related than distant terms. The top ten most popular keywords were "3d printing", "additive manufacturing", "mechanical properties", "composites", "fused deposition modeling", "tissue engineering", "fused filament fabrication", "4d printing", "polymers", "nano composites", "biomaterials" and "polymer matrix composites". It was clear that polymer matrix composites were the principal focus of the investigation.

The keywords were then categorized into 13 major clusters, which were reflected by the colors yellow, blue, green, and red. The green cluster with the most items (42 items) was emphasized on 3D bio printing, such as "bio compatibility", "bio materials", etc. The green category is mostly interested with polymer composites and nano polymer composites (33 items), such as "PEEK", "thermoplastic polyurethane", and "rheological characteristics", in addition to the others. The red category focused mostly on 3D printing, methods, and process factors counting "fabrication", "digital light processing", "sensors", "surface treatment", and "properties". Finally, the yellow cluster includes terms associated with polymer matrix composites such as "thermal analysis," "electrical characteristics," and "finite element analysis," among others.

Figure 2. Network map of the top 268 keywords based in co-occurrence

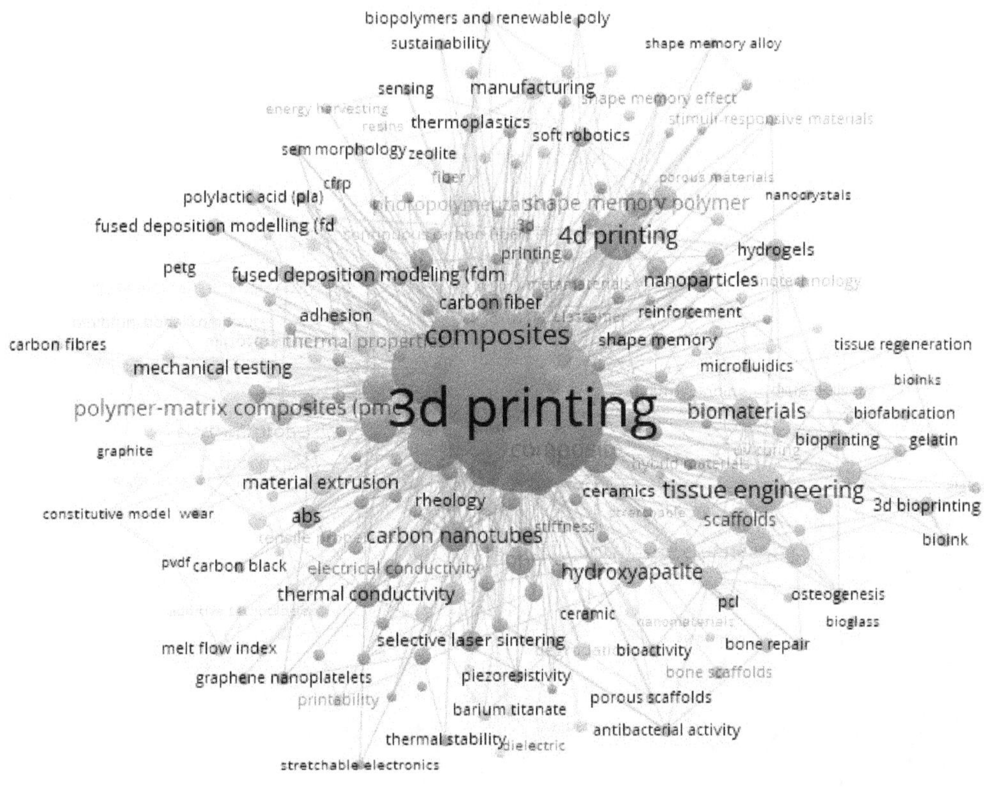

Figure 3. Selection of terms in the bibliometric map: (a) 3d printing (b) polymer matrix composites

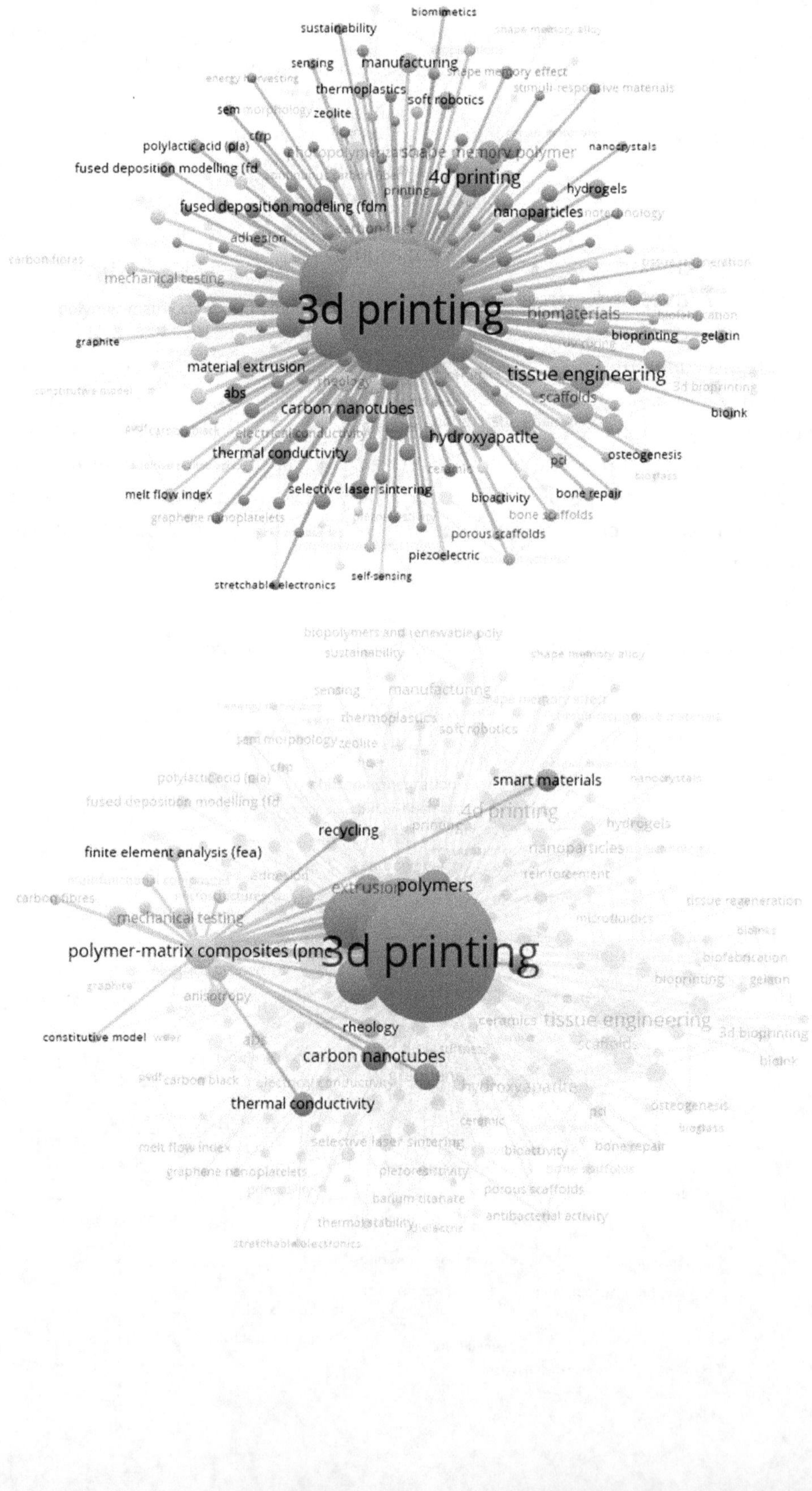

Figure 4. Publications growth trend research from 2005 to 2022

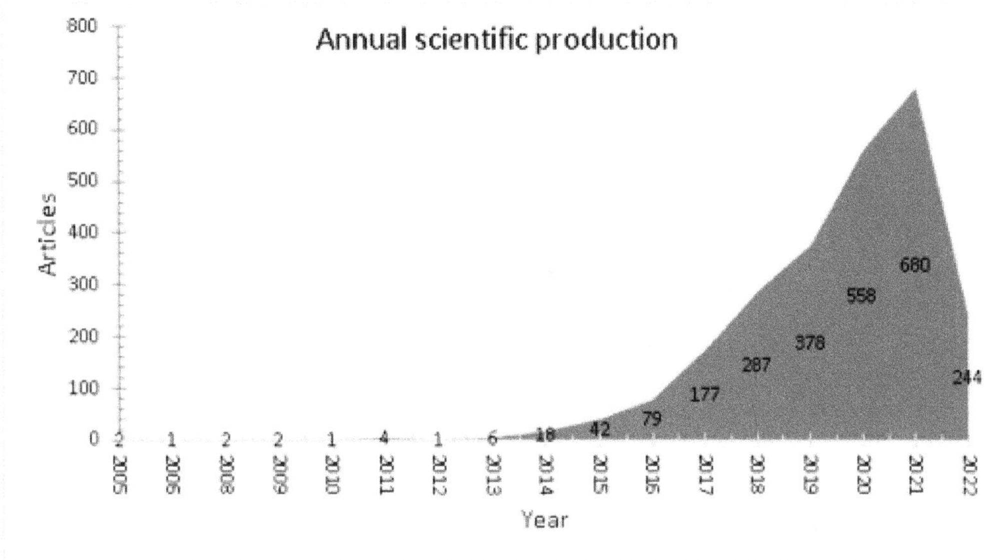

The statistical dispersion of 3D printing-related publications was investigated, and the yearly distribution is shown in Fig. 4. Articles regarding 3D printing began to appear in 2005, and the number of papers has steadily increased with small adjustments over time. It had only lately attracted significant academic attention.

The entire era was divided into two key stages, based on the article's growth trend: the emerging stage (2005-2014) and the affluent phase (2014–2021). In the early phases, the number of published publications was comparatively low, with not as much of than one per year, besides the yearly growth rate was slow. Following then, the quantity of papers in 3D printing increased from 18 in 2014 to 680 in 2021.

This growth can be due to both increased scientific interest and the fact that as the various additive processes progressed, it became clear that composite removal would no longer be the typical treatment working method conducted by a tool moving through a 3D work envelope, converting a mass of materials with a chosen shape layer by layer, contributing in a 37.7 time increase in publications compared to 2014. It is interesting to note that as technology has advanced and printer prices have reduced, this rising trend has been reinforced.

When we concentrate on published papers, we see that there are 5 scientific journals that have made more data existing in this this subject of study during the years under consideration. The topmost twenty productive journals are presented in Figure 5, which are Polymers (153), Materials (104), Additive manufacturing (103), Composite part B engineering (78), and Acs applied materials and interfaces (76). The last five years have clearly highlighted the importance that the issue of 3D printing has taken on, with a steadily rising trend as shown in Figure 6. (39 documents in 2017; 80 during 2018; 121 in 2019; 165 in the course of 2020; and 212 over 2021; with May 2022 already existing 79 documents published).

These journals with the most papers published have had a continuous increase in the number of local citations in the previous 5 years, with a notable increase in the last 5 years. with distinct implication to Advanced materials, through a number of local citations (LC) 4253, followed by Composites part B

Figure 5. Documents by year

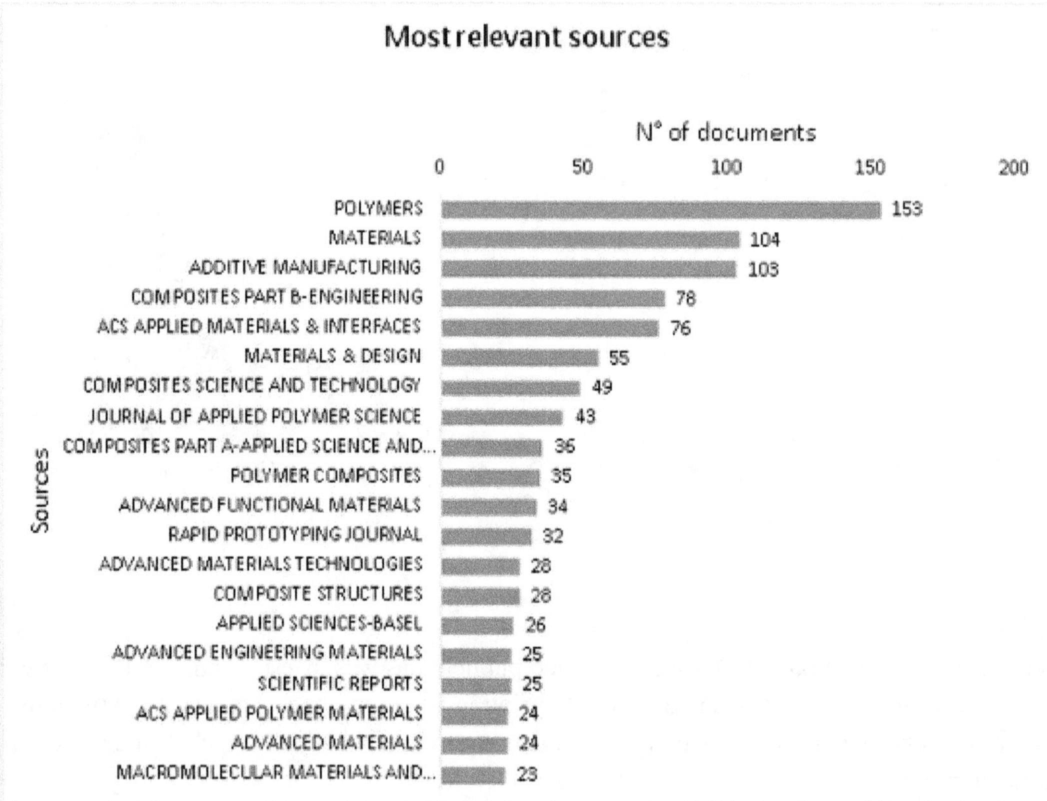

Figure 6. Source documents by year

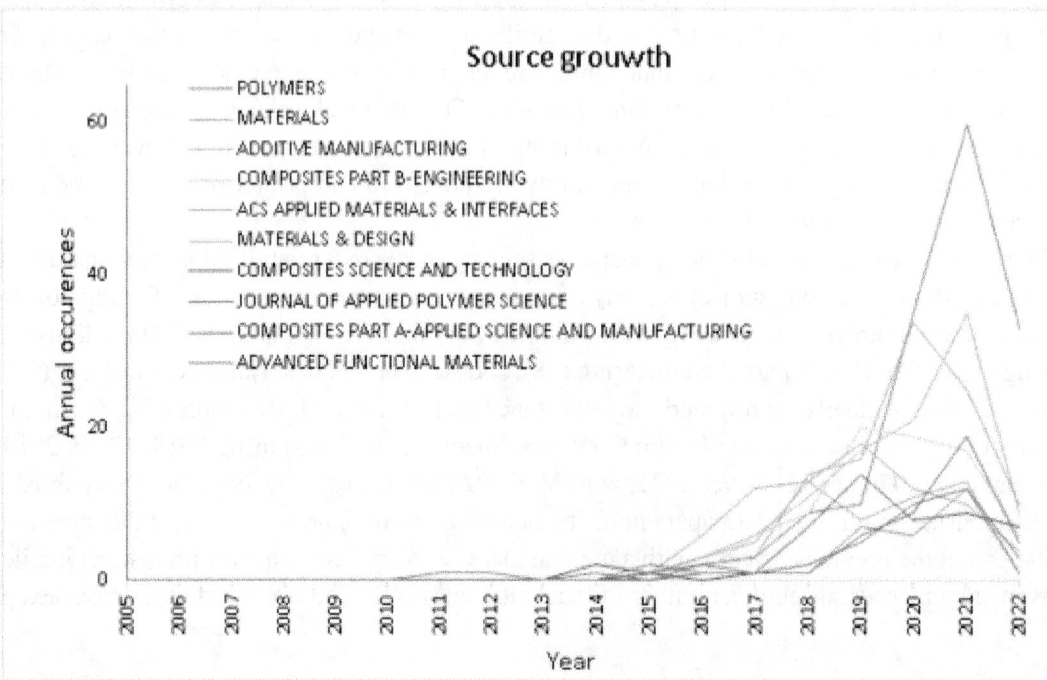

Figure 7. Source citations

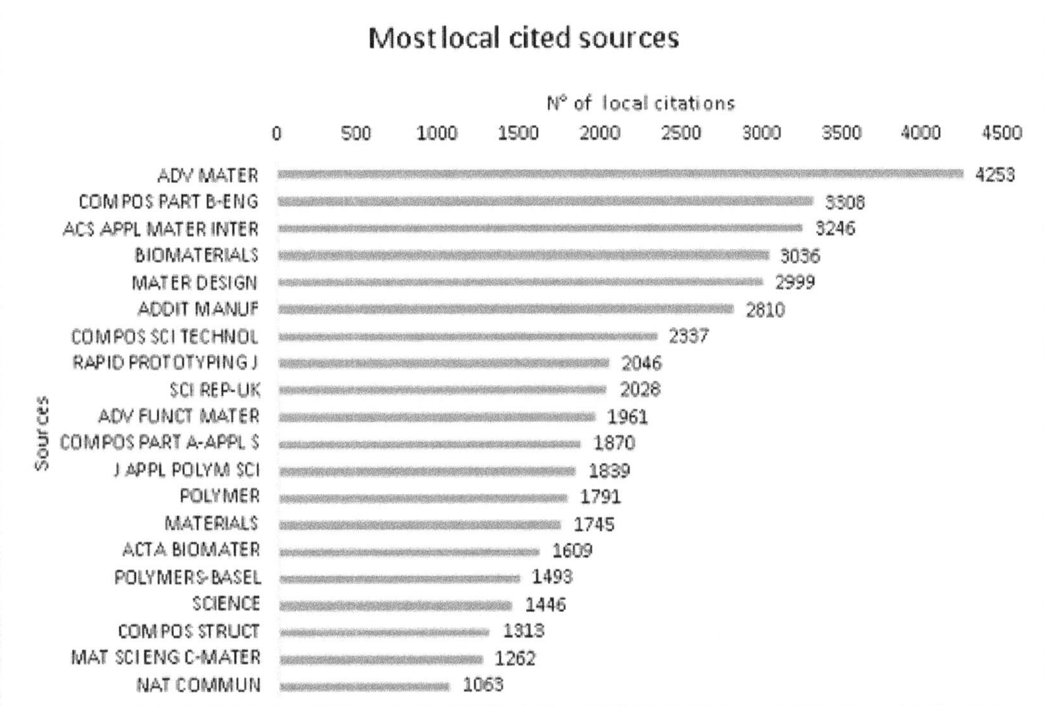

engineering with number of LC 3308, followed by Acs applied materials and interfaces with a number of LC 3246, as shown in figure 7.

Looking at the source local effect by Tc index (Figure 7), one can infer that although Composites part B engineering continues to take the lead (with 6006), Acs applied materials and interfaces comes in second with 3086, followed by Advanced materials with 2844.

The total production was produced by 9405 authors. Thus, the average number of authors per paper is 3,68, collaboration index equals to 3,71, with average number of co-authors per publication is 5,36. Qi H.J. from United State surpassed the list with 28 documents, tracked by 23 documents for Zhang J., Lalevee J. (20), Wang P. (19), Wang Y. (19), Singh R. (18), Lee J. (16), Leng S.J. (16), Chen Y. (15), and Kumar R. (15) are among the most important researchers, as presented in Figure 9. As previously said, 3D printing is thought to be a crucial driver in the development, promotion of working procedures, with a substantial many of researchers have focused on the issue in recent years. In his works, and focusing exclusively on the most recent of Qi H.J., active composites, which are made up of components that react differently to external stimuli, have the ability to change shape. By combining 4D printing and active composites, the printed frame can have a pre-designed complicated property distribution or material on many small voxels, providing immense design freedom. However, this vast design area presents a barrier in effectively locating relevant designs to achieve a desired form change. To assist the design process, a unique evolutionary algorithm-based technique and machine learning is provided here. For forward shape-change prediction, a recurrent neural network-based machine learning model inspired by beam deformation characteristics is constructed, with training dataset obtained from finite element simulations (Sun, 2022).

Figure 8. Source local impact by TC index

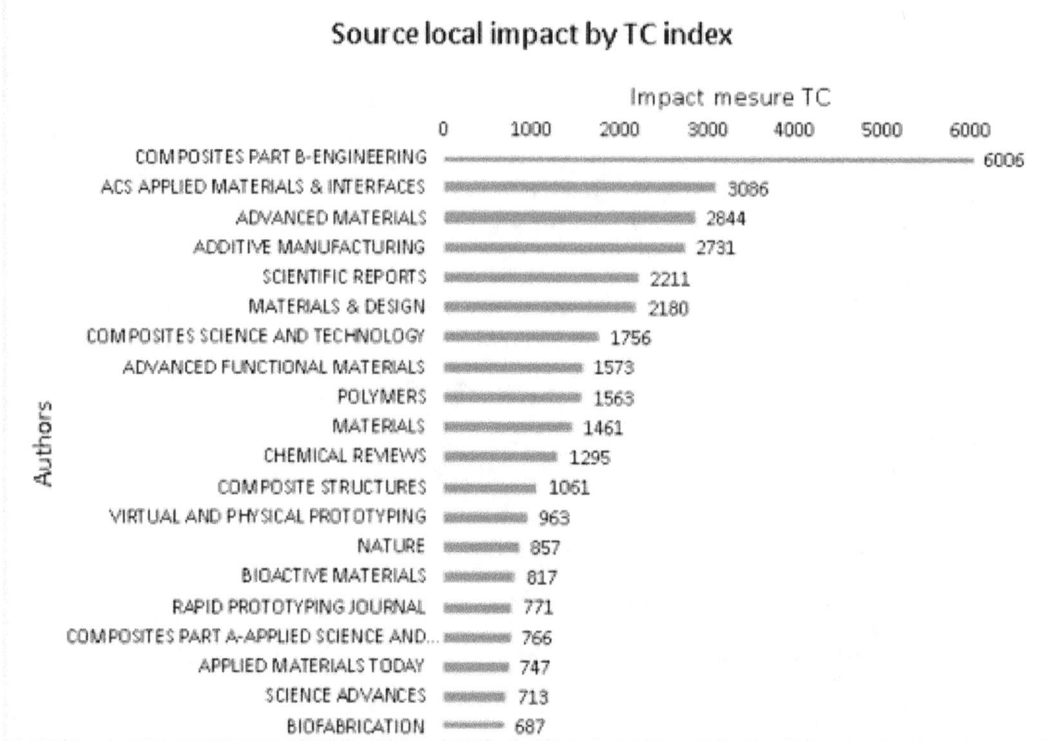

Evolutionary algorithm using machine learning is then utilized to address the inverse issue of identifying the appropriate design. The machine learning-evolutionary algorithm technique is highly efficient for numerous target shapes of varying complexity. A breakthrough paradigm is presented by merging machine learning-evolutionary algorithms with computer vision algorithms, allowing active straight beams to be produced using hand-drawn lines besides 4D printed that morph into the drawn outlines under stimulation. As a result, the approach offers an extremely efficient tool for creating 4D-printed active composites, gives particular emphasis to themes such as 3D printing of anisotropic polymer materials using various processes like as material extrusion, sheet lamination and powder bed fusion (Chen, 2022).

Concerning Zhang J, the second author (Feng, 2020), (Xiao, 2022), with the most published papers, it is noteworthy that this author tails a typical the subjects investigated by the above-mentioned author.

Figure 10 depicts highly referenced researchers who were cited more than 250 times, with the furthermost exceedingly mentioned author, who obtained more than 650 citations, being from the United States.

Qi H.J. is the most often cited author (660 citations), headed by Hui D. (648 citations), Dunn M.L. (548 citations), Wang X. (405 citations), Gou J.H. (400 citations), Jiang M. (400 citations), Zhou Z.W. (400 citations), Ozcan S. (330 citations), Dunn C.K. (324), and Kunc V. (304).

As a result, the time period for which citations were recorded, or the citation window, has an influence on the number of citations received; cumulative citation per year was employed to manage the volatility in citations.

Figure 9. Source documents

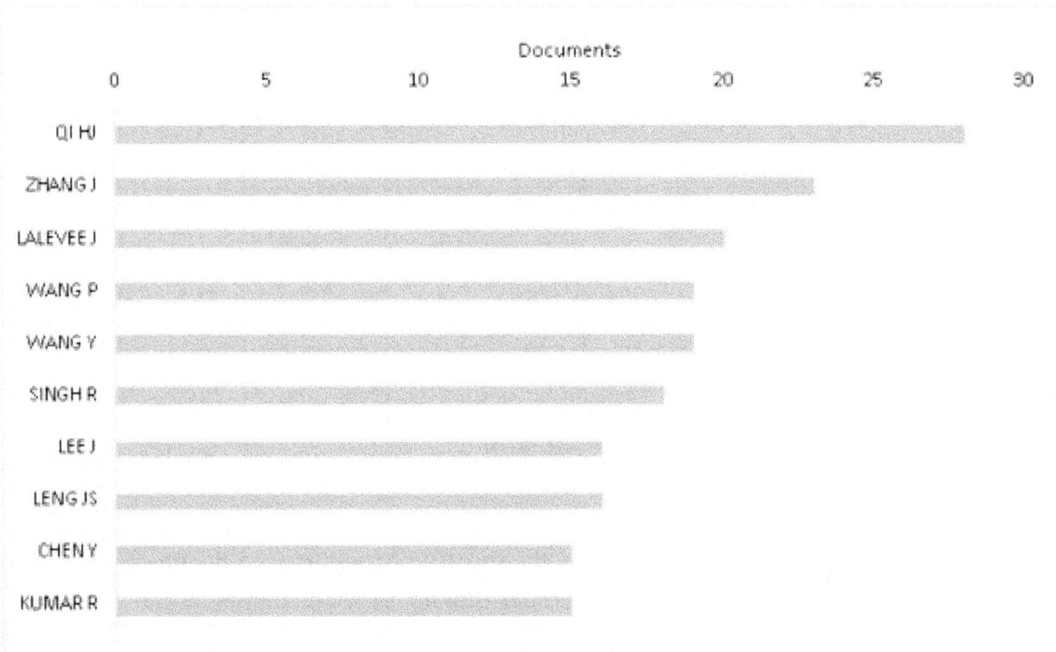

Figure 10. Most local cited authors

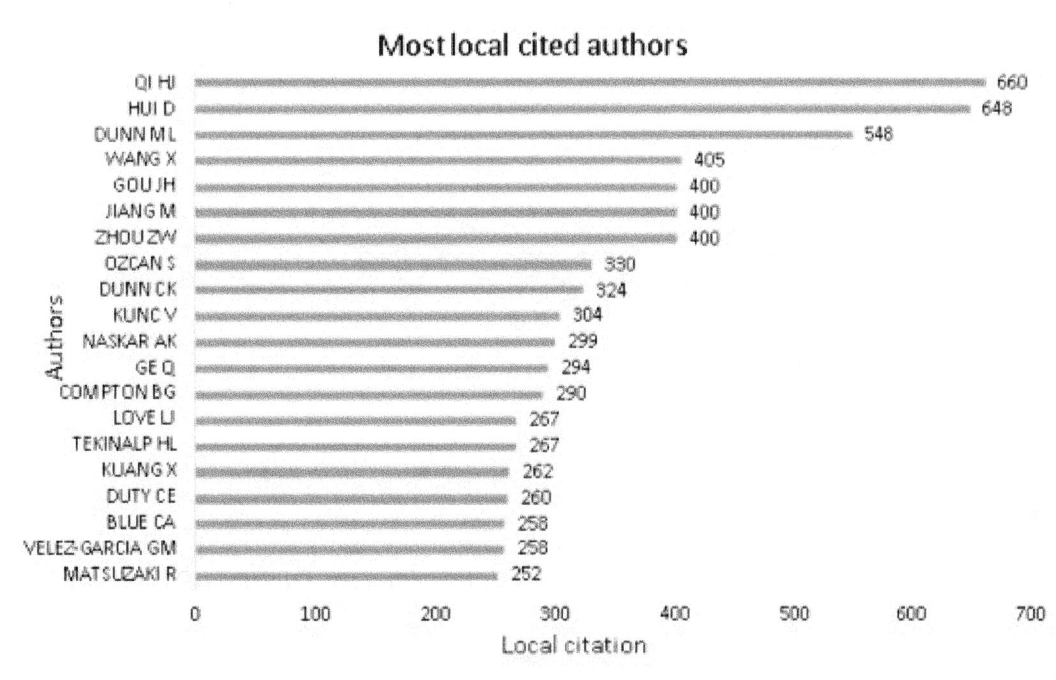

Table 1. Most top ten cited papers

Paper	DOI	Total Citations	TC per Year	Year
Ngo TD	10.1016/j.compositesb.2018.02.012	2125	425	2018
Kim Y	10.1038/s41586-018-0185-0	711	142.2	2018
Turnbull G	10.1016/j.bioactmat.2017.10.001	430	86	2018
Ligon SC	10.1021/acs.chemrev.7b00074	1295	215.833	2017
Wang X	10.1016/j.compositesb.2016.11.034	1290	215	2017
Ge Q	10.1038/srep31110	490	70	2016
Matsuzaki R	10.1038/srep23058	444	63.429	2016
Khoo ZX	10.1080/17452759.2015.1097054	436	54.5	2015
Compton BG	10.1002/adma.201401804	897	99.667	2014
Tekinalp HL	10.1016/j.compscitech.2014.10.009	631	70,111	2014

Four distinctive publishers were identified as the top 10 most referenced documents based on our findings, as presented in Table 1. Elsevier, National Library of Medicine, ACS publishing, Nature, Taylor & Francis, and Wiley are among the publishers. Elsevier published the majority of the journals, four in total. In contrast to the ACS publication, Nature, Taylor & Francis, and Wiley each published one. Another significant publisher ranked in the top ten was the National Library of Medicine, which had one document in total.

Figure 11. Most relevant keywords

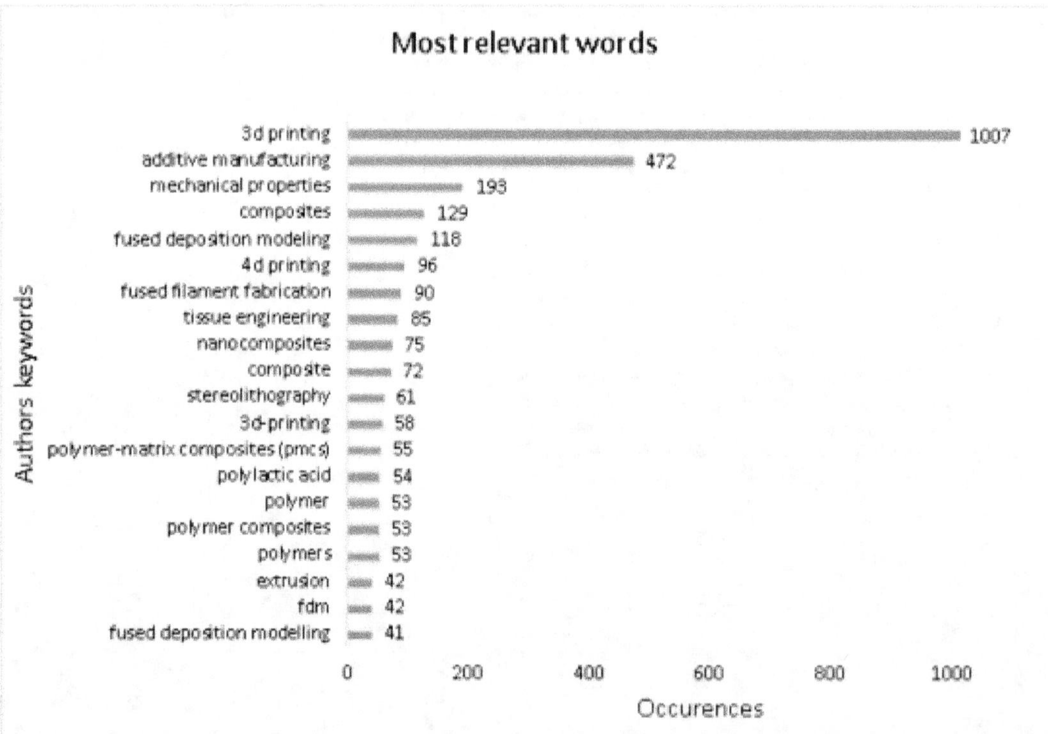

Figure 11 demonstrates the furthermost important author keywords. The bulk of the terms (35%), additive manufacturing (17%), and mechanical features (7%), are associated with 3D printing. Author keywords revealed similar research patterns in general; the keywords indicated the concentration of fabrication associated studies. As indicated, author keywords such as "composites", "nano composites", "polymers", "polymer matrix composites", and "additive manufacturing" focused on study materials, methodologies and procedures.

Figure 12. Growth of the top 10 keywords

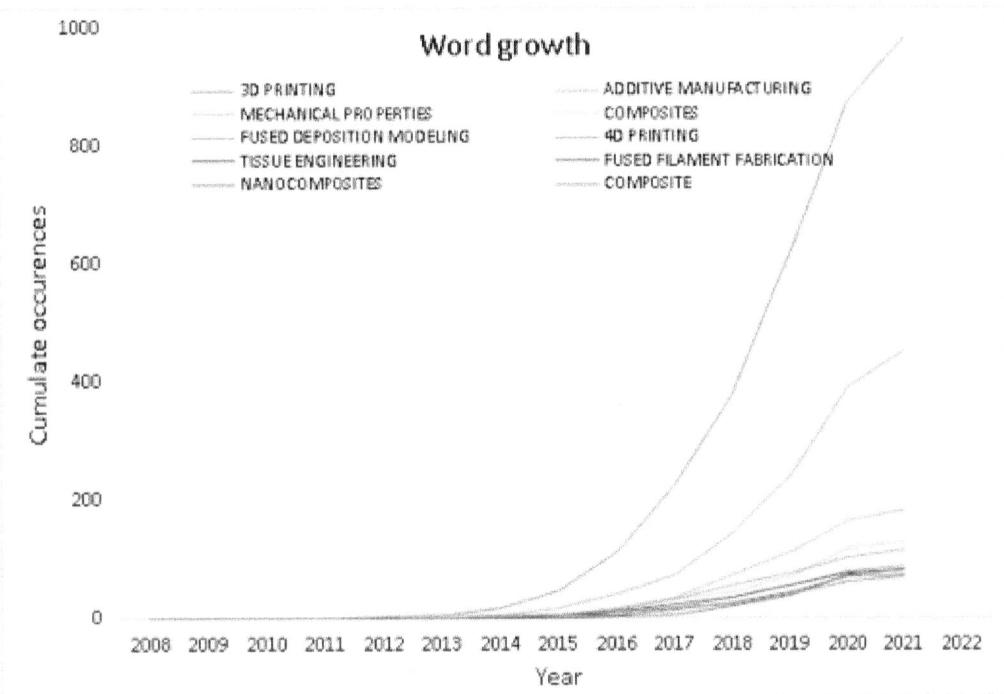

Another important aspect to investigate is the twelve-monthly growth of the top 10 keywords. Inclusive, every term increased year after year, illustrating Buildings' ongoing and consistent growth, and clearly depicts the trend from 2013 to 2021 as shown in figure 12. The facts show that "3D printing" and "additive manufacturing" have maintained the top two occurrences across the past decade. Since 2017, the third the furthermost frequent occurrence has been "mechanical characteristics", with a significant rise in the curve. "Composites" has grown massively meanwhile 2019, and it will be the fourth most prevalent term in 2021. Moreover "fused deposition modeling" has revealed remarkable growth since 2017, we may expect it to continue to expand in the future.

Another fascinating topic is to examine the hot themes in 3D printing. Figure 13 illustrates the top subjects based on keyword occurrences that have been constantly examined since 2016. The blue-filled circle on the left side of the graphic shows the subject frequency from 250 to 1000, and the bigger the diameter of the circle, the greater the occurrence. In the early years, the one popular theme was "bio-fabrication", "self-assembly", and "rapid prototyping."

Then, between 2017 and 2019, additional issues including “3D printing,” “additive manufacturing,” “polymer matrix composites,” and “composites” began to get scientific interest.

In 2019, the term “3D printing” received more attention than others. Following that, in 2020, “carbon fibers”, “tensile testing,” and “pvdf” came on the list of hot subjects and continued to get scientific interest. It is clear that the keyword “3D printing” with approximately 615 cumulate occurrences has surpassed “mechanical properties” with 112 cumulate occurrences and 242 cumulate occurrences for “additive manufacturing” as the most important issue in 2019. However, from 2020 to 2021, the terms “carbon fiber”, “sem”, and “tensile testing” became prominent. Furthermore, we anticipate that these newly trendy subjects will continue to gain traction.

Figure 13. Growth of the top 10 keywords

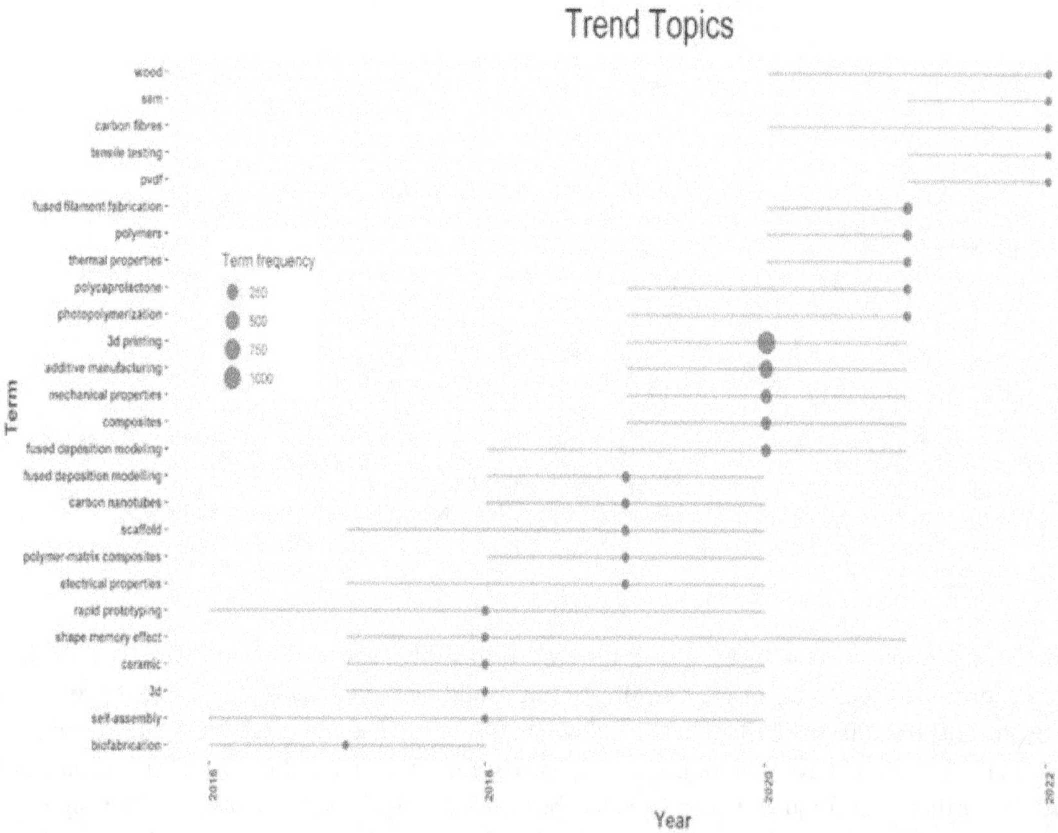

A thematic chart depicts quatern distinct typologies of themes built on two sizes: density and centrality. The upper-right quadrant contains the motor themes, contain crucial and well-developed clusters in 3D printing. Tissue engineering, biomaterials, and hydroxy apatite are found in Cluster 1, whereas 3d printing, composites, and nanocomposites are found in Cluster 2.

Figure 14. Thematic plot of keywords

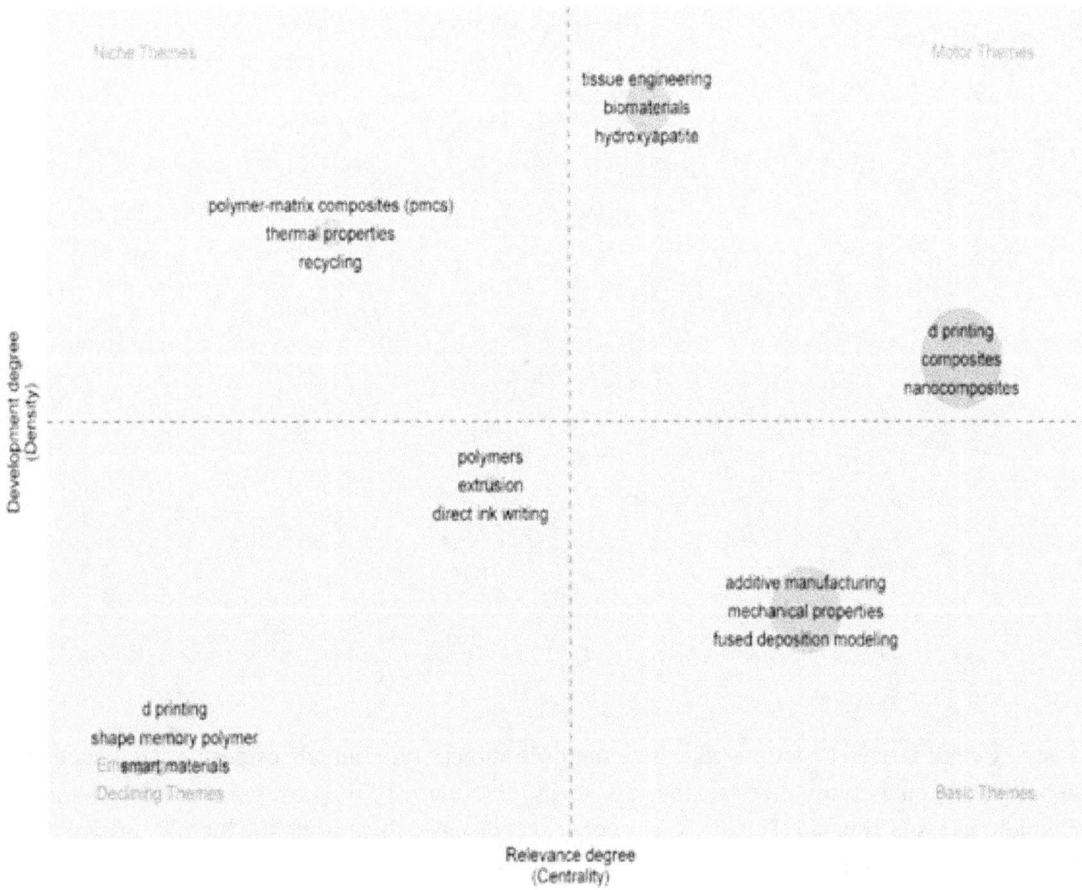

As seen in figure 14, Cluster first position indicates its largest density between these terms in addition its above-average significance. Clusters in the upper-left quadrant's specialized themes quadrant are well-developed but not vital to the subject.

Cluster one is made up of "polymer-matrix composites", "thermal characteristics" and "recycling" from this quadrant. At 3D printing, clusters with inferior density and centrality can be highlighted in the bottom-left quadrant labelled developing or declining subjects. Cluster 1 is made up of "polymers", "extrusion" and "direct ink writing" whereas Cluster 2 is made up of "3D printing", "shape memory polymer" and "smart materials".

Also, the bottom-right quadrant labelled core themes comprises clusters that are crucial for the subject nevertheless are undeveloped, as well as clusters that are domain-specific. This quadrant's cluster one contains "additive manufacturing", "mechanical properties", and "fused deposition modelling". The following are some examples of fundamental principles. Others, such as "additive manufacturing" and "fused deposition modelling" are important and deserve more study.

Figure 15. Thematic evolution

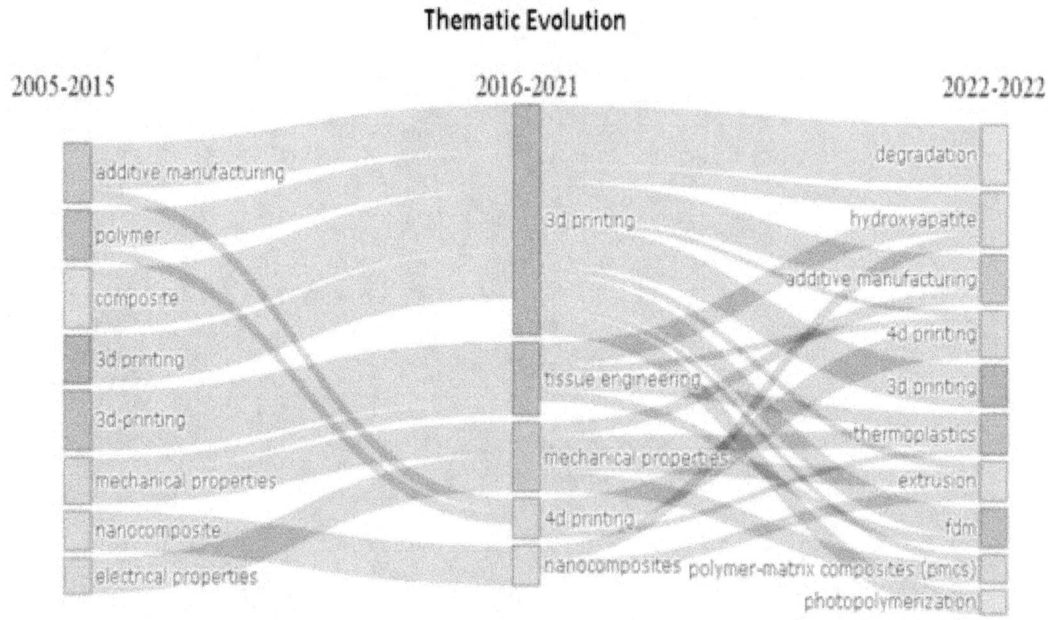

A sanky chart is used to demonstrate how multiple subjects are linked besides have advanced before. Each box on the chart symbolizes a subject, size and the theme of the boxes is related to the incidence with which the issue appears. The the waves connect each case, illustrating the theme's progress dashes, the better the relationship among the two themes, the thicker the linking. Academics can quickly identify the furthermost notable contributions meanwhile each flow has a different height based on its amount. Figure 15 shows a Sanky diagram that we developed to assess the theme progression.

Generally, it can be concluded that 3D printing topics become increasingly frequent throughout time, most probable due to the increased interest of researchers from many sectors. It's value mentioning that the phrase "3D printing" first appeared around 2005-2015, was perfected over the next 10 years, and has been popular ever since. Furthermore, this was the most popular topic between 2016 and 2021.

The United States published the furthermost articles (1559 papers), tracked by China (1284), the United Kingdom (303), and India (301), all of which produced far less papers than the United States and China. As a result, these two countries accounted for around 60% of total output. Other nations of origin were Germany, South Africa, France, and Italy as presented in Figure 16. Because the productivity of the other nations not depicted in figure 16 fluctuated from one to 100 articles alone, the productivity pattern implies an extremely skewed distribution of research productivity.

The United States (20159 citations), China (9697), Australia (4413), Singapore (3523), and the United Kingdom are the most often referenced countries (3505). Other countries represented included Italy (2091), Canada (1916), Austria (1768), France (1512), and Germany (1475). (Figure 17). Taking note of a problem, in furtherance of to be attributed to a country's scientific output, involves that a minimum one of the authors be affiliated with a research educational institution of that country, tells us that it is likely that the identical scientific paper is reported for several countries; additionally, it is possible to see

Figure 16. Number of articles classified by country of origin

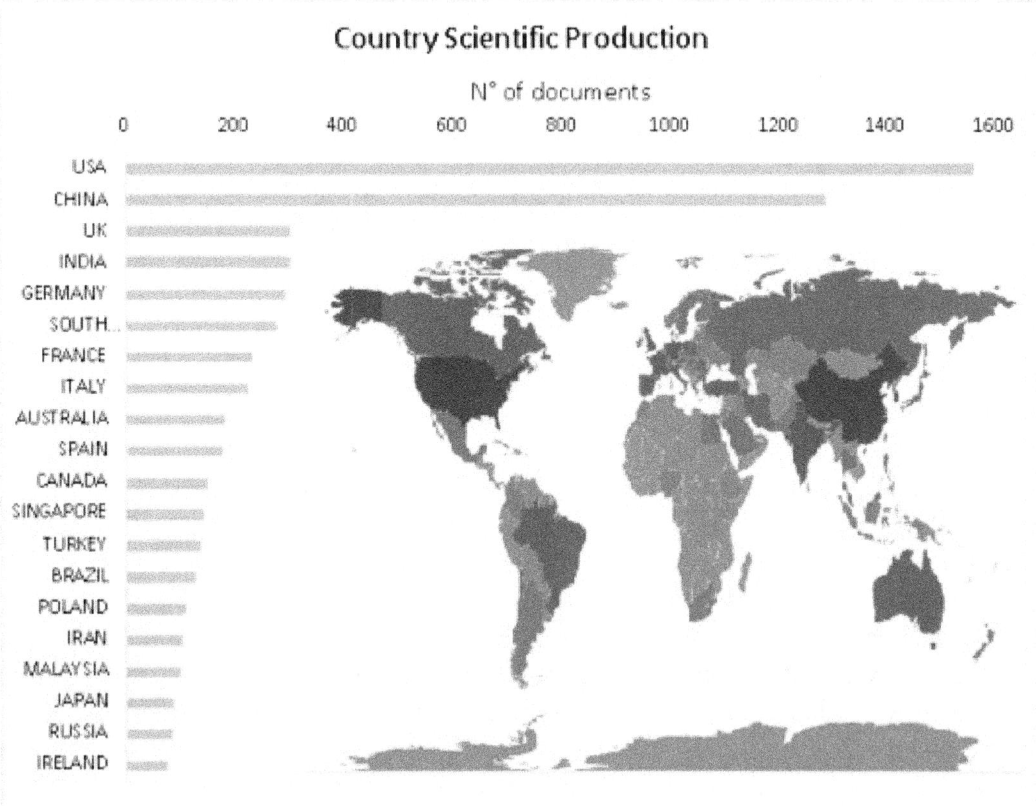

Figure 17. Number of citations according to country of origin

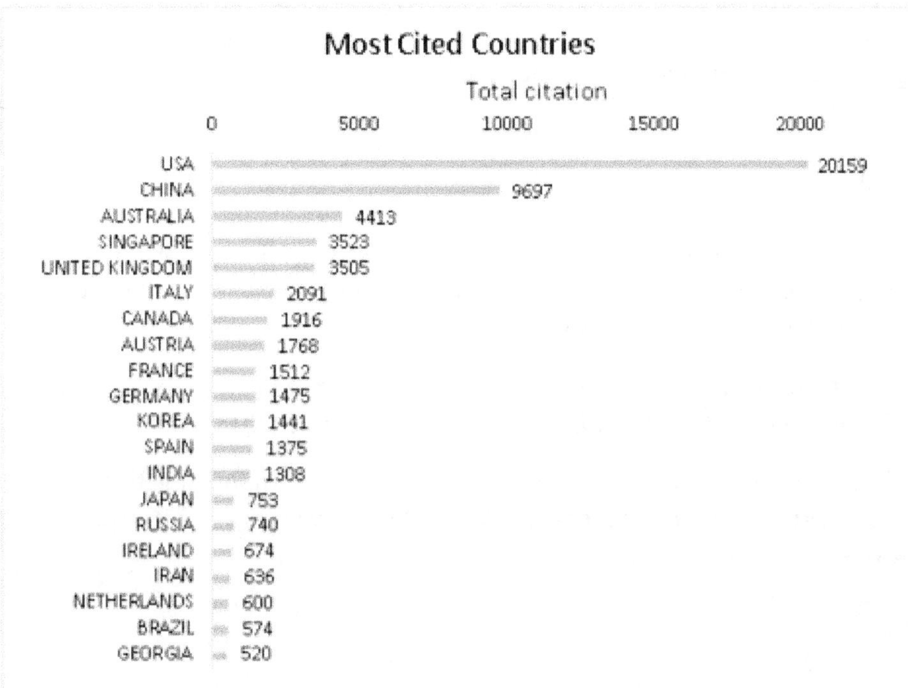

in this study that there are papers that are shared by multiple writers, implying that the similar scientific paper may be reported for more than one country.

Articles are the most common sort of publication used to convey this data, followed by review papers. The three most prominent journals in terms of published documents are Polymers, Materials and Additive manufacturing. However, despite the fact that it is not listed in top 5, Advanced Materials have the highest citation.

In terms of source local impact by TC index, the Composite part B: Engineering certainly takes supremacy over the other publications, followed by ACS Applied Materials and Interfaces, Advanced Materials. It is vital to note that, globally, Polymers is the journal where academics most want to publish and promote their work.

CONCLUSION

The top 10 Scientifics who have involved and written the furthermost on the issue of 3D printed polymer composite in all of its facets, also the accompanying citations and the major sources of publication were identified. Based on the data, this issue begins to gather traction in 2014, with the number of articles gradually growing over the preceding several years. Considering that the issue of 3D printing is becoming more important, articles available in best peer review journals are the primary power vehicles required by various writers to promote their output. The nations where these papers came from, also were similarly discovered, displaying remarkable contact between people who research besides the roots of the work created, and additional investigations has already been developed in 3D printed polymer composite. The current research is of particular theoretic and applied value since it adds to an improved knowledge not only of the major work that has been created in the domain of 3 d printed polymer composite, but either of the key sources of publishing in the field, also in relation to nations or of the main journals, also notably of the important writers who have been published. Assuming the significant interaction of academics of diverse nationalities with one another, this work is also an essential addition to thinking on collective progress to be developed, foremost to increased stages of growth besides competitiveness for these organizations, institutions and countries involved.

REFERENCES

An, Y., Myung, J. H., Yoon, J., & Yu, W. (2022). Three-dimensional printing of continuous carbon fiber-reinforced polymer composites via in-situ pin-assisted melt impregnation. *Additive Manufacturing, 55*, 102860. doi:10.1016/j.addma.2022.102860

Bandyopadhyay, A., & Heer, B. (2018). Additive manufacturing of multi-material structures. *Materials Science and Engineering R Reports, 129*, 1–16. doi:10.1016/j.mser.2018.04.001

Blok, L. G., Longana, M. L., Yu, H., & Woods, B. K. S. (2018). An investigation into 3D printing of fibre reinforced thermoplastic composites. *Additive Manufacturing, 22*, 176–186. doi:10.1016/j.addma.2018.04.039

Chacón, Caminero, Núñez, García-Plaza, García-Moreno, & Reverte. (2019). Additive manufacturing of continuous fibre reinforced thermoplastic composites using fused deposition modelling: Effect of process parameters on mechanical properties. *Composites Science and Technology, 181*(8).

Chen, Liu, Tian, Zhu, Yan, Shi, Kong, Qi, & Zhou. (2022). 3D-Printed Anisotropic Polymer Materials for Functional Applications. *Advanced Materials, 34*(5).

Diouf-Lewis, A., Farahani, R. D., Iervolino, F., Pierre, J., Abderrafai, Y., Lévesque, M., Piccirelli, N., & Therriault, D. (2022). Design and characterization of carbon fiber-reinforced PEEK/PEI blends for Fused Filament Fabrication additive manufacturing. *Materials Today. Communications, 31*, 103445. doi:10.1016/j.mtcomm.2022.103445

Feng, Y., Zhu, Sh., Mei, D., Li, J., Zhang, J., Yang, Sh., & Guan, Sh. (2020). Application of 3D Printing Technology in Bone Tissue Engineering: A Review. *Current Drug Delivery, 18*(7), 847–861. doi:10.2174/18755704MTExsNDcy2 PMID:33191886

Fu, Y., Kan, Y., Fan, X., Xuan, Sh., & Yao, X. (2022). Novel designable strategy and multi-scale analysis of 3D printed thermoplastic fabric composites. *Composites Science and Technology, 222*, 109388. doi:10.1016/j.compscitech.2022.109388

Gackowski, B. M., Phua, H., Sharma, M., & Idapalapati, S. (2022). Hybrid additive manufacturing of polymer composites reinforced with buckypapers and short carbon fibres. *Composites. Part A, Applied Science and Manufacturing, 154*, 106794. doi:10.1016/j.compositesa.2021.106794

Gackowski, B. M., Sharma, M., & Idapalapati, S. (2022). Processing and characterization of tailorable 3D-printed acrylonitrile butadiene styrene composites with hybrid buckypapers. *Composites Communications, 32*, 101162. doi:10.1016/j.coco.2022.101162

Heidari-Rarani, Rafiee-Afarani, & Zahedi. (2019). Mechanical characterization of FDM 3D printing of continuous carbon fiber reinforced PLA composites. *Composites Part B: Engineering, 175*(15).

Jammalamadaka & Tappa. (2018). Recent Advances in Biomaterials for 3D Printing and Tissue Engineering. *Journal of Functional Biomaterials, 9*(1), 22.

Kuang, X., Chen, K., Dunn, C. K., Wu, J., Li, V. C. F., & Qi, H. J. (2018). 3D Printing of Highly Stretchable, Shape-Memory, and Self-Healing Elastomer toward Novel 4D Printing. *ACS Applied Materials & Interfaces, 10*(8), 7381–7388. doi:10.1021/acsami.7b18265 PMID:29400445

Kumekawa, N., Mori, Y., Tanaka, H., & Matsuzaki, R. (2022). Experimental evaluation of variable thickness 3D printing of continuous carbon fiber-reinforced composites. *Composite Structures, 288*, 115391. doi:10.1016/j.compstruct.2022.115391

Ligon, S. C., Liska, R., Stampfl, J., Gurr, M., & Mülhaupt, R. (2017). Polymers for 3D Printing and Customized Additive Manufacturing. *Chemical Reviews, 117*(15), 10212–10290. doi:10.1021/acs.chemrev.7b00074 PMID:28756658

Ngo, T. D., Kashani, A., Imbalzano, G., Nguyen, K. T. Q., & Hui, D. (2018). Additive manufacturing (3D printing): A review of materials, methods, applications and challenges. *Composites. Part B, Engineering, 143*, 172–196. doi:10.1016/j.compositesb.2018.02.012

Parandoush, P., & Lin, D. (2017). A review on additive manufacturing of polymer-fiber composites. *Composite Structures*, *182*, 36–53. doi:10.1016/j.compstruct.2017.08.088

Parker, M., Inthavong, A., Law, E., Waddell, S., Ezeokeke, N., Matsuzaki, R., & Arola, D. (2022). 3D printing of continuous carbon fiber reinforced polyphenylene sulfide: Exploring printability and importance of fiber volume fraction. *Additive Manufacturing*, *54*, 102763. doi:10.1016/j.addma.2022.102763

Pritchard, A. (1969). Statistical bibliography or bibliometrics. *The Journal of Documentation*, *25*(4), 348–349.

Rafiee, Farahani, & Therriault. (2020). *Multi-Material 3D and 4D Printing: A Survey*. Academic Press.

Shahrubudin, N., Lee, T. C., & Ramlan, R. (2019). An overview on 3D printing technology: Technological, materials, and applications. *Procedia Manufacturing*, *35*, 1286–1296. doi:10.1016/j.promfg.2019.06.089

Sun, Yue, Yu, Shao, Peng, Zhou, Demoly, Zhao, & Qi. (2022). *Machine Learning-Evolutionary Algorithm Enabled Design for 4D-Printed Active Composite Structures*. Academic Press.

Valino, A. D., Dizon, J. R. C., Espera, A. H. Jr, Chen, Q., Messman, J., & Advincula, R. C. (2019). Advances in 3D printing of thermoplastic polymer composites and nanocomposites. *Progress in Polymer Science*, *98*, 101162. doi:10.1016/j.progpolymsci.2019.101162

Wang, P., Zou, B., Ding, Sh., Zhuang, Y., Liu, J., & Li, L. (2022). Functionally graded polyetheretherketone-based composites additively manufactured by material extrusion using a transition interface design method. *Composites. Part A, Applied Science and Manufacturing*, *158*, 106977. doi:10.1016/j.compositesa.2022.106977

Wright, W. J., Koerner, H., Rapking, D., Abbott, A., & Celik, E. (2022). Rapid fiber alignment quantification in direct write printing of short fiber reinforced composites. *Composites. Part B, Engineering*, *236*, 109814. doi:10.1016/j.compositesb.2022.109814

Wu, H., Fahy, W. P., Kim, S., Kim, H., Zhao, N., Pilato, L., Kafi, A., Bateman, S., & Koo, J. H. (2020). Recent developments in polymers/polymer nanocomposites for additive manufacturing. *Progress in Materials Science*, *111*, 100638. doi:10.1016/j.pmatsci.2020.100638

Xiao, Y., Zhang, J., Fang, B., Zhao, X., & Hao, N. (2022). Acoustics-Actuated Microrobots. *Micromachines*, *13*(3), 481. doi:10.3390/mi13030481 PMID:35334771

Zaheri, Fenner, Russell, Restrepo, Daly, Wang, Hayashi, Meyers, Zavattieri, & Espinosa. (2018). Revealing the Mechanics of Helicoidal Composites through Additive Manufacturing and Beetle Developmental Stage Analysis. *Advanced Functional Materials, 28*(33).

Chapter 3
Properties and Applications of Natural Fiber–Reinforced 3D–Printed Polymer Composites

Srinivasan V. R.
https://orcid.org/0000-0003-2325-8349
Dayananda Sagar College of Engineering, India

Prakash C. P. S.
Dayananda Sagar College of Engineering, India

Praveena B. A.
Nitte Meenakshi Institute of Technology, India

ABSTRACT

There is a huge scope for natural fiber-reinforced composite materials, which can be used in 3D printing technology. Non-degradable materials used in the additive manufacturing field can be replaced by the innovative application of natural fiber-reinforced composite materials. This chapter introduces the advantages of 3D-printed natural fiber composite materials compared to polymer materials. The homogeneity and thermal stability of fiber reinforcement are the critical characteristics of the composite material. Also, the standardization of material testing has considerable space for research. Continuous natural fiber-printed composite and short natural fiber-printed composite are the major focuses for researchers in the field of additive manufacturing. Various fiber materials yield diverse results in the final product manufactured. Various additive manufacturing techniques can be applied and compared for the same reinforcements and matrix material.

INTRODUCTION

Amid growing concerns about the use of non-renewable resources, some sectors have given significant recognition to ecologically responsible methods, goods and recycled or scrap utilization. Minimize, reuse,

DOI: 10.4018/978-1-6684-6009-2.ch003

recycle, recover, remodel and reprocess are essential components of sustainable production. A really excellent definition of external regulation that has been put into place in Europe from 2015 to resolve those environmental challenges is the need that end-of-life cars be reused or recycled 90 percent of the time, utilized for regenerative braking 10 percent of the time and balanced in landfill (Paul et al., 2015). Numerous studies are being conducted around the world to address environmental problems by utilizing materials and composite parts made from renewable resources (Faruk et al., 2012). Composites prepared from the polymeric matrix and natural fibers are of particular relevance in this situation because they can offer the required output and qualities at a reasonable price. Though natural fiber reinforced composites (NFRCs) have a bright future, there are still a number of material and manufacturing issues that must be resolved if durability and effectiveness are to be maintained over the long run. The fabrication of natural fiber reinforced composites, which have uses in the automobile, electronics, packing, building and medical industries, presently uses a variety of synthetic polymers as matrices. Organic materials in such composite have ecological and financial advantages over synthetic reinforcement agents like glass fibers because they are widely available, recycled, compostable and most importantly they are relatively affordable (Mohammed et al., 2015). For elevated engineering disciplines, thermoplastic-based natural fiber reinforced composites have lately engrossed increasing attention (Faruk et al., 2012; Mohammed et al., 2015). Comprehensive environmental compliance of natural fiber reinforced composites is still a serious difficulty because these non-biodegradable polymer composites provide considerable disposal issues. The optimum method in this situation is to create natural fiber reinforced composites utilizing recyclable polymer matrix made from renewable or non-renewable materials or existing non-biodegradable polymer materials from renewable natural resources (Mohanty et al., 2000; Shen et al., 2009). However, because of their poor thermal, mechanical and physical qualities, these bio-composite materials are not widely used. Using natural fibers in polymeric composite structures has the primary advantage of obtaining higher stiffness-to-weight ratios, recycled content, good biocompatibility, insulation and carbon dioxide neutralization in comparison to their respective traditional equivalents, which include glass fibers and carbon fibers (Madsen & Lilholt, 2003; Satyanarayana et al., 2009).

These can sometimes be weaker than composite materials built with synthesized reinforcing agents, dependent on the matrix and fiber blend, but they are acquiring a lot of relevance because of the advantages listed previously. Additionally, compared to manufactured fibers like glass fibers (54.70 MJ/kg), the manufacturing of natural fibers including flax (9.50 MJ/kg) always requires lesser energy (Joshi et al., 2004). In addition to the discussion of life cycles of natural and glass fiber reinforcing composite structures, the author of the report found that natural fiber reinforced composites allow the inclusion of significant amounts of fibers, which decreases the total utilization of synthesized polymeric matrix and promotes lightweight design and energy recovering following end-of-life combustion (Merlini et al., 2011; Jawaid et al., 2013; Cao et al., 2012; Majid et al., 2010; Jacob et al., 2006). Thermoplastic materials, as opposed to thermoset plastics and elastomers with complicated cross-linked structures, provide much more design freedom and straightforward methodological approaches. On the other hand, hardly few investigations have been conducted on the usage of agronomic wastes such as hulls of soybean. The majority of natural fibers were unusable around 473 K, hence natural fiber reinforced composites cannot be manufactured with matrices that require a higher processing temperature (Facca et al., 2006).

NATURAL FIBER REINFORCED POLYMER COMPOSITE MATERIALS, PROPERTIES AND THEIR APPLICATIONS

Types and Properties of NFRCs

According to the origin, natural fibers are typically divided into categories such as animal fibers, plant fibers and mineral fibers (Grand View Research; Stamboulis et al., 2001; Dicker et al., 2014). A fundamental distinction between plant and animal fiber is that plant fiber is predominantly contained of proteins whereas animal fiber is primarily contained of cellulose. The characteristics and ingredients of natural fibers are listed in Table 1.

Table 1. Characteristics and composition of natural fibers (Grand View Research; Stamboulis et al., 2001; Dicker et al., 2014)

Fiber	Density, g/cm³	Cellulose weight, %	Crystallinity of cellulose, %	Lignin, weight. %	Angle of Microfibril ,°
Bast					
Hemp	1.47	72	70	7	4
Flax	1.5	67.5	70	2.35	8
Jute	1.4	66.5	65	12.5	8
Kenaf	1.55	31.39	-	13	12
Ramie	1.55	72.4	-	0.7	7.75
Leaf					
Sisal	1.47	72	60	9	17.5
Pineapple	1.2	76	52	10.5	11
Abaca	1.5	59.5	-	9.5	22.5
Banana	1.35	54	50	7.5	17.5
Fruit/Seed					
Coir	1.2	37.5	30	42.5	39.5
Kapok	1.3	24	-	17	5
Cotton	1.6	86.85	-	0.75	25
Grass/Reed					
Bamboo	0.85	40	50	26	9.5
Switch grass	1.42	32	-	-	-
Miscanthus	1.41	37.5	-	-	-
Agriculture remains					
Wheat straw	-	41.5	-	16	-
Soy hull	-	49	-	13.5	-
Corn stover	-	39	-	14	-

Natural fibers offer reinforcement properties with low density, good strength and stiffness. Natural fibers normally have low densities and thus result in low weight (Duigou et al., 2020). Natural fibers are the appropriate ingredients for fillers in both engineering and non-engineering purposes since they are likely to be lighter with a density of about 1.10 g/cm^3 and 1.60 g/cm^3 and are inexpensive with prices ranging from \$0.20/kg to \$4.0/kg. Contradictory to pricing and densities, as that of the natural fiber type shifts between stems, leaves and fruits to agriculture waste, the cellulose content and crystalline nature also changes. For example, jute fiber produced from plants contains crystallinity of cellulose of fifty to seventy percent, but hemp and bast fibers contain crystallinity of cellulose of fifty to ninety percent. For goods like automobile parts or fiberboard, bast fibers are frequently utilized to attain excellent mechanical qualities (elastic constant, strength and rigidity) (Anandjiwala & Blouw, 2007). The major parts of plant fiber are generally cellulose (sixty to eighty percent), hemicellulose (five to ten percent), and lignin (five to ten percent). The other constituents are wax, tannin, humidity (up to twenty percent) and solvent organic substances, which have been shown to differ greatly based on their kind and provenance (Nagarajan et al., 2013). Reference (Mussig et al., 2010) contains additional information on the makeup of natural fibers.

The natural fiber is a material made of crystals, stiff celluloid microfibril suspended in lignin and hemicellulose-based, flexible, non-crystalline form. The content, microfibril orientation, crystalline nature and inner structures of fibers affect their characteristics, which in turn affect natural fiber reinforcement (Taj et al., 2007; Nabi Saheb & Jog, 1999; Eder & Burgert, 2010; Muthuraj et al., 2015; Ramesh et al., 2017; Bourmaud et al., 2018; Ahmad et al., 2015). The main ingredient of natural fiber, cellulose, does have stiffness and strength of 139 GPa and less than 2 GPa, correspondingly (Shah, 2013). Nevertheless, because the microfibril orientation significantly affects the rigidity of these natural fibers, fibers with such a higher cellulose quantity and a lower microfibril orientation are known to have a significant reinforcement impact in polymeric composite materials. From the different natural fiber sources, bast fibers typically contain high cellulose concentration (between 31 and 77 weight percent) and lesser microfibril orientations. Additional components of natural fibers, including gelatin and hemicellulose, control things like moisture absorption, moist strength, ballooning and fiber bundling incorporation (Bogoeva-Gaceva et al., 2007). Hence, thorough natural fiber characterization is crucial to attain the intended strength in natural fiber reinforced polymer composite.

Table 2 provides a summary of key features of natural fibers, including fiber diameters, bundle width and fiber length. As mentioned in table 2, natural fibers come in a broad variety of fiber diameters, fiber bundles width and length in relation to various shapes. This leads to significant changes in the characteristics of polymeric composite materials made with these fibers. Additionally, the inherent differences in natural fibers present substantial difficulties for production process optimization when the fibers are utilized as reinforcing agents. To accomplish optimal results in natural fiber reinforced polymers, proper feedstock materials evaluation and management are crucial.

The mechanical qualities are directly impacted by the wide differences of fiber structures including densities and microfibril angle (MFA). The innate mechanical properties of those natural fibers determine the mechanical characteristics of natural fiber reinforced polymer composites (Slootmaker & Müssig, 2010; Burgert, 2006). Highly dense fibers frequently display greater strength and rigidity than lesser dense fibers. Similar to synthetic fibers, young's modulus of natural fibers mainly relies on microfibril angle. Less microfibril angle leads to stiff fibers. This makes sense because a lower microfibril angle allows cellulose fibers to be practically aligned to the load axis and can support greater load, increasing rigidity. Owing to its higher rigidity, such low-microfibril angle fibers do, nevertheless, typically display

Table 2. Important characteristics of natural fibers (Taj et al., 2007; Nabi Saheb & Jog, 1999; Eder & Burgert, 2010; Muthuraj et al., 2015; Ramesh et al., 2017; Bourmaud et al., 2018; Ahmad et al., 2015)

Fiber	Diameter/Width, µm	Width of Bundle, µm	Fiber length, mm
Wood	27	-	2.4
Flax	40	330	77
Hemp	22	262	31.5
Jute	17	262	3
Kenaf	31	138	6
Abaca	26	500	7
Sisal	25	225	4
Coir	20	210	225

brittle behavior. But on the other side, natural fibers with higher microfibril angles often display higher levels of strength due to their substantial permanent deformations (Martinschitz et al., 2008; Keckes

Table 3. Mechanical properties of natural and synthetic fibers (Martinschitz et al., 2008; Keckes et al., 2003)

Fiber	Elongation %	Tensile strength MPa	Elastic modulus GPa
Natural fiber			
Jute	1.65	493.4	33.4
Sisal	7.95	467.5	23.5
Banana	5	672	24.85
Flax	2.6	1088.5	54
Kenaf	1.5	572.5	44
Pineapple	7.65	398.5	65
Cotton	6	530	8.55
Coir	30	205	4.65
Hemp	26.45	915	46.4
Ramie	2.9	675	66.5
Wool	30	147	2.85
Spider silk	17.5	923.5	12
Kapkok	2.6	69	2.85
Abaca	6	496	42
Bamboo	4.8	185	14
Synthetic fiber			
E glass	2.45	2750	70
Aramid	3.45	3075	65
Carbon	1.6	4000	235

et al., 2003). The diameter of the fiber and the level of cellulose polymerization are additional elements that affect the mechanical characteristics of natural fibers. Table 3 gives the quick description of the mechanical characteristics of various fibers. Microfibril angle has a significant effect on the mechanical properties such as elastic modulus, strain and ductility. As the microfibril angle increases, the tensile strength swiftly decreases. Some of the natural fibers posses lower hygromorph actuation property due to greater microfibril angle value. The hygroscopic expansion is subjective by the biochemical arrangement of the fibers and the angle of the microfibrils (Duigou et al., 2020).

Fiber orientation influences the tensile strength, impact strength and bending strength of the NFRCs. The pull out of fiber is influenced by the fiber orientation impacts the machinability of the NFRCs. Until 90° of fiber orientation angle, as the fiber angle increases, the modulus of transverse of the composite declines. This is due to the arrangement of the major axis of fiber and practical load established with the fiber arrangement of 0°. For the fiber position from 0° to 90°, crosswise load affects fiber and exerts low confrontation to the distortion for evenly applied stress (Lasikun et al., 2018; Kumar & Prakash, 2021).

The natural fiber properties clearly demonstrate the significant variance between the fibers, in contrast to the relatively uniform characteristics of manufactured fibers. A really wide range in the mechanical characteristics of natural fibers might raise serious questions about the resulting mechanical dependability of composite materials manufactured. Since the mechanical characteristics of fiber bundles might vary substantially from those of single fibers, it is unclear if the parameters (Table 3) were determined on fiber bundles or single fibers. The data discloses that while the density of natural fibers is generally similar, there are significant differences in their mechanical characteristics. When it comes to specific strength, spider silk has been found to be between 876 MPa and 973 MPa. As much as 1835 MPa in maximum strength, flaxseed fiber can potentially be as weak as 343 MPa. Flax has a similarly broad spectrum in terms of its young's modulus; (8-100 GPa). Natural fibers, in general, have a wide range of mechanical properties and are significantly weaker in tension than synthetic fibers, according to this argument. Flaxseed, pineapple and ramie are a few examples of natural fibers that can match the young's modulus of synthetic fabrics. The inherent mechanical qualities as well as other features of natural fiber are compiled in Table 4, influence both the manufacturing and the qualities of natural fiber reinforced polymer composites.

Table 4. Essential properties of natural fibers which influence treating of NFRC (Pervaiz & Sain, 2003; Eichhorn et al., 2001; Summerscales et al., 2010; Zini & Scandola, 2011).

Class	Essential properties
Geometry	Length, Thickness, Cross sectional area, form and arrangement.
Mechanical	Bending strength, Young's modulus, Yield, Fatigue and Elongation characteristics.
Physical	Density, Surface quality, crystallinity, angle of microfibril, absorption of water and Inflammation.
Chemical	Concentration of various ingredients and contaminations, cultivation and polymerization.
Thermal	Moisture content, Coefficient of thermal extension and Thermal conductivity.

The factors, including interaction, heat resistance, humidity susceptibility and fibers-matrices adhesion, directly affect the attributes of natural fiber reinforced polymer composites (Lau et al., 2018; Koronis et al., 2013; Wambua et al., 2003). For instance, the hydro-philicity and hydro-phobicity of natural fibers and matrices, correspondingly, limit fibers-matrices compatibility and can lower the mechanical characteristics of such composite materials (Mohanty et al., 2000; Bogoeva-Gaceva et al., 2007). As a result, numerous physicochemical techniques for altering fibers (Bogoeva-Gaceva et al., 2007) and adding binding reagents throughout manufacturing (Koronis et al., 2013) were devised.

Advantages, Disadvantages and Applications of NFRCs

Many diverse sectors use natural fiber as reinforcing agents in polymeric materials and most of the key benefits of these natural fiber reinforced polymer composites are listed in: (Faruk et al., 2012; Mohammed et al., 2015; Mohanty et al., 2000; Shen et al., 2009; Madsen & Lilholt, 2003; Satyanarayana et al., 2009; Joshi et al., 2004; Wambua et al., 2003; Eichhorn & Young, 2004; Al-Maadeed & Labidi, 2014; Mohanty et al., 2002; Pervaiz & Sain, 2003).

Natural fibers are most advantageous for the ecology because they are renewable, Carbon dioxide neutral and biodegradable. For instance, substituting natural hemp fibers for manufactured glass fiber can result in a reduction of 3 tons of Carbon dioxide/ton of substance (Pervaiz & Sain, 2003).
Leftovers which are easily accessible and safe following improved energy recovering by burning.
Natural fibers can decrease the average cost of polymeric materials because they are cheap.
Minimize the usage of petroleum-based materials by substituting down to 50 percent and more per natural fibers in compression moulding and injection moulding, respectively.
Improve workplace safety and make production methods somewhat safe (reduction dermal and respiratory irritations).
Ecologically responsible production and dumping methods are used by natural fiber reinforced.
Natural fibers are soft, which extends the lifespan and productivity of manufacturing plants and machinery.
E-glass fibers cause waste concerns, so natural fibers are the ideal substitute.
Natural fabrics are lighter, which significantly reduces the mass and increases the fuel economy for automobile applications.
Compared to E-glass, some natural fibers have greater effective strength, stiffness, resilience and elasticity.
Exceptional sound absorbing characteristics can be found in natural fiber reinforced polymers.
Utilizing crop wastes effectively creates revenue and work prospects.
Natural fiber reinforced polymer composites offer effective thermal insulation.

Natural fiber reinforced polymer composites have the following drawbacks: (Pervaiz & Sain, 2003; Eichhorn et al., 2001; Summerscales et al., 2010; Zini & Scandola, 2011; Promper, 2010)

Higher humidity absorption (depending on fiber type) leads to inflammation, which affects the dependability and effectiveness of items.
Natural fibers with poor permeability (hydro-philic fibers and hydro-phobic matrices) frequently have inadequate fiber-matrix interface adhesion, which lowers mechanical characteristics.
Certain biodegradable polymers and some fibers are incompatible.
Huge scattering in natural fiber reinforced polymer properties is primarily caused by large differences in fiber qualities and characteristics.
Easily vulnerable to pest and fungus infestations.
Easily combustible and has a slow decomposition temperature.

Natural fiber qualities and properties can differ considerably depending on the climate and farming circumstances (harvesting period, solid grade, fertilizer, geographical area, weather, etc.).
Natural fiber reinforced polymers have a low heat conductance over glass fiber reinforcing composite materials.
Geographic accessibility and complicated distribution network.
Climate and agricultural production differences cause price changes and demand.
Non-structural parts often use natural fiber reinforced polymers because of their weak thermal characteristics, and the operating temperature is 473 K.

After its first manufacturing in 1908s, natural fiber reinforced polymer composites nurtured significant popularity in the construction and automobile industries, largely because of its lightness, affordable, as well as other characteristics (Zini & Scandola, 2011). To create natural fiber reinforced polymer composites, a number of natural fibers including flax, jute, sisal, coco, hemp, hardwood, kenaf and bamboos were employed. These matrices comprise thermosets, thermoplastics, elastomers and protein. The manufacturing processes and attributes of natural fiber reinforced polymer composites are strongly influenced by the fiber or fiber bundle features. The fiber features can be seen in Tables 3 and 4. The distribution of fibers in polymer matrices and length are crucial for achieving homogeneous mechanical characteristics, particularly for transportation components. To obtain desirable mechanical characteristics without impeding the molding route, a blend of natural fibers with diverse sizes and distributions, including flax kenaf or hemp, sisal is generally employed (Promper, 2010). Hardwood fibers which have been processed with acrylates are now used for serial manufacturing in the automobile sector; the technique for other fibers is still being developed. With the ease of fabrication and significant enhancement in fuel economy and decrease in greenhouse gases, many top automakers, utilize natural fiber reinforced polymer composites for a wide range of materials, including seat backs, boot linings, door handles boards, sound insulating material panels, orchestral panel supports and engine insulations. For the floor protective trimming of Mercedes A class, for instance, banana fiber reinforcing composite materials were employed (DaimlerChrysler, 2019). Composite materials made from polypropylene and polyurethane are frequently used to make car parts because of their simple processing, lightweight and superior mechanical qualities. Nylon, polystyrene and polyethylene are among other common matrix materials. In order to produce items including tyre coverings and tailgate trim, the market is continually working to replace synthetic polymer with sustainable materials such polylactic acid (PLA) and polybutylene succinate, which are reinforcing by the fibers of flax, wool and kenaf (Koronis et al., 2013; Graupner et al., 2009). As an alternate, such recyclable natural fibers as reinforcement composite materials are also utilized to create kayaks, laptops and smartphone coverings and windmill rotors (Graupner et al., 2009). These recyclable composite materials were discovered to have characteristics similar to composite materials reinforced with polycarbonate. As a result, it seems that strong vehicle components can be produced using polymers that are entirely recyclable and sourced from renewable sources. As illustrated in Table 5, further uses for natural fiber reinforced polymer composites include packing, home furnishings, sporting goods, electronic devices and musical equipments (Shah, 2013).

Snowboard boots and cycle frames are made from hybrid composite materials reinforced by hemp, carbon and jute fiber because they have great vibration damping and mechanical qualities. Flax-carbon (1:3) composite materials for sports equipment, 4:1 flax-to-carbon composite materials for cycle frames and flaxseed composite materials for snowboards are a few instances of sports equipment (Dicker et al., 2014).

Table 5. Typical applications of NFRCs in different industrial sectors (Ramesh et al., 2017; Bourmaud et al., 2018; Ahmad et al., 2015; Shah, 2013)

Industrial Sector	Products
Automotive, aerospace and transport	Engine parts, Door panels, headlights, dash board, mat, decks, wheel pans, railway coach interior, boat body and architecture molds.
Constructions	Window panes, Bridge, roof tile, partition panels, false ceiling, walls, flap frames.
Electronic equipments	Computer, mobile phones and laptop coverings
Sports equipments	Bicycle parts, gloves, table tennis boards, baseball bats and hockey sticks.
Household items	Chairs, fencing items, doors, boxes, Bath tubs, floor mats and blankets
Material handling equipments	Storage boxes, bins, food containers and gas cylinders.

FABRICATION OF NFRC THROUGH 3D-PRINTING

Natural fibers are the raw materials that resemble hair and are extracted from plants, animals or mineral resources. The fundamental distinction between plants and animals fiber is that plant fibers are primarily formed of cellulose, while animal fiber is mainly built of protein. Rice husk, wheat, hemp, bamboo and cotton are a few examples of natural fibers of plant origin. According to Kunanopparat et al., (2008), increasing the fiber concentration of wheat composite made it lesser responsive to heat processing in terms of its mechanical properties. The physiological adherence was said to be greatly influenced by the operating temperatures, and this physiological adherence had a considerable influence on the mechanical characteristics of the composite materials. According to Chen et al., (2020), bamboo fibers exhibit higher percentages of fiber displacement reactions and matrices failures and drastically reduced proportions of fracturing characteristics as a result of their hierarchy fiber woven structures. As opposed to solely using PLA dependent thermoplastics, Oliveira et al., (2019) found that composite materials using natural cotton fiber as filler materials had smaller environmental impacts. The natural cotton composite materials also showed improvement all in all from the standpoint of eco-efficiency. According to Mazzanti et al., (2020), the inclusion even of tiny quantities of hemp fiber promotes PLA deterioration to a degree comparable to those of wastewaters in freshly made PLA that has not been dried. Sustainable organic fiber is a type of natural fiber which is widely available and replenishes over a predetermined length of time. Because of their biodegradability, affordability, acceptable mechanical qualities and lightweight, natural fibers are frequently using it as a minimal cost substitute for synthetic fiber (such as nylon) (Stoof, 2016). Natural fiber typically has a lower density of around 1.10 and 1.60 g/cm^3, which results in lightweight materials. Typically, plant fiber comprises sixty to eighty percent cellulose, five to twenty percent hemicellulose and the remaining materials lignin, wax and moisture (up to twenty percent) make up the remainder (Nagarajan et al., 2013). In addition, cellulose content or crystallinity should be taken into consideration while selecting polymer reinforcements. For instance, to improve the mechanical qualities of categories in order of elasticity, toughness and rigidity, bast fibers having 50–90 percentage cellulose crystallinity are used in different automobile parts (Bledzki & Gassan, 1999). Rice husk fiber has recently been used as reinforcement in polymer composites, creating natural fiber reinforced polymer composites. Lignocellulose fiber strength could be enhanced because of their high cellulose content, which could enhance the characteristics of composite material (Mazzanti et al., 2019). Rosa et al., (2009) asserts that rice husk cellulose and tannin constituents allow it to be treated at high temperatures than hardwood. Additionally, because of the qualities of ligno-cellulose, it offers a relatively important amount of fiber that may be used as fillers or a straight replacement for hardwood (Kim et al., 2009). It is strongly advised to use rice husk as reinforced fibers in composites because it typically goes to waste and has no other significant economic interest. Achievement in combining rice husk with reinforced composites or hardwood substitutes in an extensive range of applications may ultimately result positive impacts on the environment, such as sustainable forestry.

Natural fibers were subsequently frequently included as additives in filament for 3D printing (Wang et al., 2017). It is also becoming more and more common in a number of industrial fields, including constructions, automobile industry, heat insulation composites and sound-absorbing components (Koronis et al., 2013; Santoni et al., 2019). Manufacturers are constantly looking for innovative components and refining manufacturing procedures to create better goods that are as efficient, sustainable and waste-free as possible. When thermoplastic polymers are reinforced with natural fibers, odd problems can arise. First, when being continuously heated to a temperature exceeding 473 K, the ligno-cellulose in natural fiber degrades (Moran et al., 2007; Mazzanti et al, 2017). Consequently, whenever selecting an appropriate thermoplastic, the melting point is to be considered. To avoid the hydrolysis of polymer matrix and evaporation while blending, fiber dryness is crucial before any process (Elsawy et al., 2017; Van den Oever et al., 2010). Fiber reinforced composite materials are mainly prepared by the continual extruder method and are utilized in 3D printing. Nevertheless, these have produced a number of difficulties, including an uneven mixture of the fillers and the polymeric matrices (Mazzanti et al., 2019), temperature regulation (Filgueira et al., 2017) and the development of holes throughout production (Osman & Atia, 2018; Depuydt et al., 2019). They ultimately cause problems, such as nozzle blockage in 3D printers (Filgueira et al., 2017) and variable mechanical characteristics. In order to increase the interface adhesion among the polymeric matrix and fillers, chemicals including crosslinking agent (Mazzanti et al., 2019), chemical methods (Stoof et al., 2017; Liu et al., 2019) and bonding agent (Guo et al., 2018) are added into the solution of the matrix and fillers. In addition, the impact of manufacturing and printing settings hasn't been well studied until lately. Every polymeric material, kind of fibers, concentration level and standardized existence are described in depth along with important references. Additionally, the effects of additives on the composite materials, including chemical treatment (Liu et al., 2019; Guo et al., 2018; Milosevic et al., 2017), compatibilizers (Filgueira et al., 2017), plasticizers (Depuydt et al., 2019; Stoof et al., 2017) and hardening agents (Nguyen et al., 2018) are demonstrated. For example, Nguyen (2018), observed that the hardening reagent of nitrile rubber made a significant contribution to improved mechanical efficiency. According to Stoof et al., (2017), the use of plasticizers can enhance layer adherence and minimize pore diameter, surface quality and the mechanical characteristics of the composites. Additionally, it is evident that the fiber content never went above 30wt%. Increased fiber content ultimately leads to much more complex manufacturing in 3D printing, including non-homogenous natural fiber reinforced polymer mixtures which may clog the nozzles (Filgueira et al., 2017) and raise the melting temperature, which necessitates more energy for extruding. Furthermore, when the number of polymeric matrices which can wet the fibers diminishes, a smaller polymer matrix can lead to more brittle filaments (Nguyen et al., 2018).

PROCESSING OF NATURAL FIBERS AND FABRICATION OF FILAMENTS FOR 3D PRINTING

Pre-Fabrication of Fibers

The majority of plant dependent natural fibers employed for natural fiber reinforced polymers composites are ligno-cellulosic, comprising cellulose and lignin making up the bulk of the fibers and other elements like tannin, waxes and humidity making up smaller percentages based on the composition of the fiber. When coupled with polymeric matrix, such fibers are prone to instability (Panthapulakkal et al., 2005). Natural fiber surfaces are hydrophilic, which means they attract water due to their high hydroxyl content. Fiber and polymeric matrices stability is raised by this, including the detail that majority polymer matrices in compounded were hydrophilic or resisted liquids in their composition. This contradiction has important consequences. Firstly, the mechanical characteristics of the materials can suffer from poor adhesion at the fiber/polymer contact.

Panthapulakkal et al., (2005) recently commented on fiber composite material that displays debonding in the absence of any fiber treatments or coupling agents. Poor debonding performance resulted in poor mechanical characteristics of composite material. Moreover, it is more difficult to create a homogeneous mix due to the weak compliance of the fiber and polymer matrices. In the polymer matrices, the fibers are frequently clumped together and formed lumps (George et al., 2001). Agglomeration is a problem that a not only impacts the mechanical characteristics of composites but it also creates difficulties while creating feedstock for 3D printing since blockage occurs repeatedly. Natural fibers have the affinity to agglomerate in the polymer matrix due to formation of hydrogen bondage. To improve the fiber distribution in the polymer matrix, alteration and functionalization is very essential. The hydroxyl groups can be altered with more hydrophobic molecules to strengthen the attraction amongst the polymer matrix and the altered fibers. It decreases the chances of hydrogen bonding and improve the distribution in the matrix. The approach of functionalization of the fibers depends on the application, which can avoid agglomeration to create a homogeneous blend of the fiber and the polymer matrices (Shesan et al., 2019). As a result, there is been a huge amount of research on the coexistence of fibers and polymers. Numerous pre-fabrication techniques for fiber were evaluated and found to be successful (Herrera-Franco & Drzal, 1992; Qaiss et al., 2014). The majority of techniques fell into one of the two groups: (1) changing the fiber surface or (2) changing the polymer matrices.

The goal of fiber surface treatment is to reduce the fiber hydrophilicity and so as to enhance the interface interaction among the fibers and polymer matrices. To increase adherence, this procedure could also turn the fiber surface rough. There are numerous techniques used to modify the fiber surfaces, and these techniques are further divided into physical, chemical and mechanical surface modifications. Mechanical surface treatment is the simplest type among the 3 kinds (Fuqua et al., 2012). To produce a rough surface and increase adherence, the fiber is coiled or swaged. Due to their minimal price and favorable effects on fibers, these techniques are typically widely used. Consequently, the majority of pre-fabrication for fibers is chemical.

Reduction of number of hydroxyl groups on the fiber exterior and its hydrophilicity is the primary goal of any chemical treatment. Other procedures can make the fiber less hydrophilic at one end while creating chemical connections with the polymeric matrices at the others. Presently, a variety of chemical procedures are utilized to prepare fiber for compounding (Valadez-Gonzalez et al., 1999). Chemical methods can differ depending on the characteristics of fibers and polymers employed. The most well-liked surface modification among the aforementioned techniques is alkaline treatment (Fuqua et al., 2012). This seems to be more utilized because it is economical. The major mode underlying alkaline treatment is the elimination of fiber components such as tannin, oil, lignin, waxes and hemi cellulose, which is relatively weak than cellulose and inhibits interaction with the polymer matrices. According to Royan et al. (2018), it has been shown that the surfaces of the fiber are rough and also the characteristics of the composite have enhanced after elimination of these elements. The variables determine how effectively this process works, including the alkaline concentrations, treatment duration, type of alkaline and treatment temperatures. Studies on composites have revealed a variety of characteristics and consequences due to the wide range of alkaline treatment settings (Mahjoub et al., 2014). Regarding the physical processing of the fibers, this includes techniques like heating, gamma radiations, electron beam processing and UV ozone processing (Trindade et al., 2008; Faulstich de Paiva & Frollini, 2006). Additionally, such procedures can lessen hydroxyl groups while boosting adherence. With UV/O_3 treatment, there may even be a rise in established sets on the treated fiber exterior, which would help the treated fiber chemical interaction with the matrices. Although surface treatment of the fibers is beneficial, these techniques are typically combined for enhancing adherence of the fiber. To improve adhesion, this method alters the construction of the polymer matrix. Maleic anhydride is the most extensively utilized compatibilizing compound in research. A particular kind of polymer, such as PP or HDPE, is frequently grated with MA and combined with the composite. This technique is becoming increasingly popular as the performance of composite improves. Researchers have found that compatibilizing agents can greatly improve mechanical properties (Duy et al., 2013).

Manufacturing of Filament

Natural fiber reinforced polymer filaments for the 3D printing technique is currently made using procedure, the same as how traditional filaments are made. Nevertheless, the mechanical characteristics of composite materials are significantly influenced by composite fabrication. The characteristics and suitability of the finished filaments are directly influenced by procedures including sifting the fibers, drying and blending techniques. These procedures, like drying and surface preparation, make sure that the fiber and polymer matrices are well-bonded. Pre-mixing also guarantees a uniform mixture, producing a harder substance. As a result, the existing techniques for making composite materials are covered in this chapter well before the processes for making filament. Sieving the fiber powders are the initial part in building the natural fiber reinforced polymer composites (Duy et al., 2013). The size of the particles of the fiber must be as fine as feasible for natural fiber reinforced polymer composites to be employed in 3D printing, particularly in the fused deposition modeling method, to prevent contamination at the printing nozzles. In order to guarantee zero larger particles are blended in with the composites, the powders may occasionally be filtrated based on the origin of the fiber. This is a crucial stage since what happens next directly impacts the filament quality. According to Petchwattana et al., (2019), just those composite materials manufactured with 70-microns particle diameter are capable of printing with no problems. The components must then be dried. The raw ingredients are dried in a hot air oven for a predetermined amount of time or until their humidity is well below the predetermined level (Chen et al., 2015; Chen et al., 2016). Getting rid of any humidity in the fibers or polymer granules is the major motivation behind this. The relative humidity of the materials may be deleterious to subsequent operations, such as extruders. If the mixture has a higher moisture concentration, the rate of evaporation during extrusion or hot mixing could produce vapour bubbles and cavities. This will have an impact on the porosity of the composite in addition to its mechanical characteristics. It is common knowledge that natural fiber reinforced polymer composites are sensitive to moisture and can bloat and even fracture if there is an adequate quantity of moisture intake. According to Chen et al., (2016) the permeability of composite materials significantly rose as the amount of fiber content did. In order to reduce permeability, it is necessary to guarantee that there is a minimal amount of moisture present throughout filament processing.

After the polymer and fibers have been completely treated, the polymer matrices and fibers are mixed. Various combination strategies, such as first dryness blending, then melt blending and ultimately actively extruding the filaments, were used to determine the results (Zhang et al., 2018; Le Guen et al., 2019). The majority of blending processes, according to the literature analysis, take a similar approach. Prior to being combined in dual extruders, the polymeric grains, fiber powders and compatibilizing reagent are first left to dry (Zhang et al., 2018; Badouard et al., 2019). After that, they are blended with another drying blender or through hands. Whenever compounding is necessary, a twin-screw extruder performs noticeably better work than a single-screw extruder. Twin-screw extruder is frequently used in industries where an extruder is used to blend and combine polymers with colors or fillers because of having superior compound capabilities (Giles et al., 2004). Due to this, several investigators melt and blend the composites using a twin-screw extruder. The filaments are easily made by feeding the composites grains through an extruder and pressing the molten composites through a die with the necessary diameter once the mixture has been homogenized. The feeder for the fused deposition modeling process is often a filament with a diameter of 1.750 mm or 2.850 mm. As fused deposition modeling printers depend on programming that presume a consistent diameter of filaments, the size of these filaments must stay unchanged with no significant fluctuations. Too-smaller or too larger filaments result in under or over extrusions, correspondingly, which lowers the efficiency of the printed samples. Considering this, there is no much information in the research right now about how scientists make sure the filaments are extruded with uniform dimensions. Only a few recent studies were conducted on the topic of using natural fiber reinforced polymer composites for 3D printing procedures, notably the fused deposition modeling method. This demonstrates there is still a lot of potential for investigation in this field of study.

MOTIVATION AND CONCLUDING STATEMENTS

The industry for natural fiber-reinforced plastics is now expanding quickly because of its affordable pricing, as well as its physical, mechanical and ecologically favorable qualities. The ecological and longevity problems are solved by the development of natural fiber-reinforced polymers. With recent advancements in Europe, it is anticipated that the automobile industry, government organizations, and small eco-friendly businesses will remain the primary markets for fiber reinforced composites. By integrating fiber reinforced polymers with 3D printing, this latest innovation increases productivity even more. Compared to subtractive fabrication, 3D printing is less expensive and allows more complex design. Consequently, as materials perform better, more composite materials will be used in new applications.

It might be advantageous to use polymeric reinforced with natural fibers to decrease the price of filaments while maintaining its mechanical qualities. Additionally, adding recovered polymers and agricultural residue fibers to the composites improved their sustainable development. Moreover, it appears that 3D printing and natural fiber-reinforced polymeric materials have greatly enhanced the overall mechanical properties. However, in addition to the appropriate elements, the natural fiber filler must have a substantial quantity. Surprisingly, semi-crystalline materials, such as semi-crystalline polyethylene, proved to be favorable when coupled with natural fibers.

Additives, coupling agents and compatibilizers, are known to enhance the mechanical properties. This is primarily due to the fact that coupling agents tend to improve the contact between fibers and polymer matrices.

Additionally, the use of natural fiber results in poor manufacturing quality due to small limits and issues. To prevent moisture from producing voids throughout the manufacturing process, the component must be adequately dried well before the compound stage and before printing. Second, the natural fiber fillers must be sieved into tiny particle diameters (such as 70 microns) to prevent agglomeration from clogging the nozzles of the 3D printing machine and causing a blockage.

Additionally important for consistent and efficient mechanical properties of the finished goods is the uniform blending of the ingredients. Since there is little information available on printing with NFRC, several parameters using standard filament were explored. When printing using natural fiber reinforced polymer composites, researchers have mentioned employing printer temperature in the range from 453 K to 483 K. Additionally, data indicate that as layer height and infill % rose, mechanical strength rose as well. Additionally, it was discovered that raster angles of 45 enabled the component to balance out the force and create mechanical qualities that were more constant.

As was already said, the quality and characteristics of the components are significantly influenced by the parameter settings used in the additive manufacturing process. Because of this, creating high-quality goods with enhanced attributes depends on the natural fiber reinforced composite product optimization procedure. Because there's no lot of optimization work being done on the natural fiber reinforced polymer composites, there could be a significant need to bridge this gap in the future.

Future exploration in the production of natural fiber-matrix composite filaments must focus on enhancing the workability or flow characteristics. Manufacturing problems may significantly harm the mechanical qualities of the finished items. By altering the fiber's surface in various ways, mechanical, chemical, and physical surface treatments (such as spinning, acidic, alkali and ultraviolet) can strengthen composite materials by improving the fiber's strong adherence to the polymer matrices. Additional research will also be necessary to decrease the negative effects of shrinkage and warpage during the three-dimensional printing of HDPE composites.

REFERENCES

Ahmad, F., Choi, H. S., & Park, M. K. (2015). A review: Natural fiber composites selection in view of mechanical, light weight, and economic properties. *Macromolecular Materials and Engineering*, *300*(1), 10–24. doi:10.1002/mame.201400089

Al-Maadeed, M. A., & Labidi, S. (2014). Recycled polymers in natural fibre-reinforced polymer composites. In A. Hodzic & R. Shanks (Eds.), *Natural fibre composites Materials, Processes and properties* (pp. 103–114). Woodhead Publishing Limited.

Anandjiwala, R. D., & Blouw, S. (2007). Composites from bast fibres-prospects and potential in the changing market environment. *Journal of Natural Fibers*, *4*(2), 91–109. doi:10.1300/J395v04n02_07

Badouard, C., Traon, F., Denoual, C., Mayer-Laigle, C., Paës, G., & Bourmaud, A. (2019). Exploring mechanical properties of fully compostable flax reinforced composite filaments for 3D printing applications. *Industrial Crops and Products*, *135*, 246–250.

Bledzki, A. K., & Gassan, J. (1999). Composites reinforced with cellulose based fibres. *Progress in Polymer Science*, *24*(2), 221–274.

Bogoeva-Gaceva, G., Avella, M., Malinconico, M., Buzarovska, A., Grozdanov, A., Gentile, G., & Errico, M. E. (2007). Natural fiber eco-composites. *Polymer Composites*, *28*(1), 98–107. doi:10.1002/pc.20270

Bourmaud, A., Beaugrand, J., Shah, D. U., Placet, V., & Baley, C. (2018). Towards the design of high-performance plant fibre composites. *Progress in Materials Science*, *97*, 347–408. doi:10.1016/j.pmatsci.2018.05.005

Burgert, I. (2006). Exploring the micromechanical design of plant cell walls. *American Journal of Botany*, *93*(10), 1391–1401. doi:10.3732/ajb.93.10.1391 PMID:21642086

Cao, X. V., Ismail, H., Rashid, A. A., Takeichi, T., & Vo-Huu, T. (2012). Maleated natural rubber as a coupling agent for recycled high density polyethylene/natural rubber/kenaf powder biocomposites. *Polymer-Plastics Technology and Engineering*, *51*(9), 904–910. doi:10.1080/03602559.2012.671425

Chen, G,. & Luo, H. (2020). Effects of node with discontinuous hierarchical fibers on the tensile fracture behaviors of natural bamboo. *Sustainable Materials and Technologies, 26.*

Chen, R. S., Ab Ghani, M. H., Salleh, M. N., Ahmad, S., & Tarawneh, M. A. (2015). Mechanical, water absorption, and morphology of recycled polymer blend rice husk flour biocomposites. *Journal of Applied Polymer Science*, *132*(8), 41494.

Chen, R. S., Ahmad, S., & Gan, S. (2016). Characterization of Rice Husk-Incorporated Recycled Thermoplastic Blend Composites. *BioResources*, *11*(4), 8470–8482.

Chen, R. S., Ahmad, S., & Gan, S. (2016). Rice husk bio-filler reinforced polymer blends of recycled HDPE/PET: Three-dimensional stability under water immersion and mechanical performance. *Polymer Composites*, *39*, 2695–2704.

Chen, R. S., Salleh, M. N., Ab Ghani, M. H., Ahmad, S., & Gan, S. (2015). Biocomposites Based on Rice Husk Flour and Recycled Polymer Blend: Effects of Interfacial Modification and High Fibre Loading. *BioResources*, *10*(4), 6872–6885.

Daimler Chrysler. (2019). *Awarded for banana fibre use in Mercedes a class - NetComposites*. https://netcomposites.com/news/2005/august/19/daimlerchrysler-awarded-for-banana-fibre-use-in-mercedes-a-class/

De Oliveira, A. S., de Macedo, N. J. R., & Rosa, D. S. (2019). Eco-efficiency of Poly (lactic acid)-Starch-Cotton composite with high natural cotton fiber content: Environmental and functional value. *Journal of Cleaner Production*, (217), 32–41.

Depuydt, D., Balthazar, M., Hendrickx, K., Six, W., Ferraris, E., Desplentere, F., & Ivens, J., Van Vuure, & A.W. (2019). Production and characterization of bamboo and flax fiber reinforced polylactic acid filaments for fused deposition modeling (FDM). *Polymer Composites*, *40*, 1951–1963.

Dicker, M. P. M., Duckworth, P. F., Baker, A. B., Francois, G., Hazzard, M. K., & Weaver, P. M. (2014). Green composites: A review of material attributes and complementary applications. *Composites. Part A, Applied Science and Manufacturing*, *56*, 280–289. doi:10.1016/j.compositesa.2013.10.014

Duigou, A. L., Correa, D., Ueda, M., Matsuzaki, R., & Castro, M. (2020). A review of 3D and 4D printing of natural fibre biocomposites. *Materials & Design, 194*(108911), 1–26.

Duy, T. T., Dang, N. M., Thuc, C., Thuc, H., & Dang, T. T. (2013). Study of Mechanical Properties of Composite Material Based on Polypropylene and Vietnamese Rice Husk Filler. *Journal of Chemistry, 2013*, 1–6.

Eder, M., & Burgert, I. (2010). Natural fibres - function in nature. In J. Mussig (Ed.), *Industrial Applications of Natural Fibres: Structure, Properties and Technical Applications* (pp. 23–40). John Wiley & Sons, Ltd.

Eichhorn, S. J., Baillie, C. A., Zafeiropoulos, N., Maikambo, L. Y., Ansell, M. P., Dufresne, A., & … . (2001). Current international research into cellulosic fibres and composites. *Journal of Materials Science, 36*(9), 2107–2131.

Eichhorn, S. J., & Young, R. J. (2004). Composite micromechanics of hemp fibres and epoxy resin microdroplets. *Composites Science and Technology, 64*(5), 767–772. doi:10.1016/j.compscitech.2003.08.002

Elsawy, M. A., Kim, K. H., Park, J. W., & Deep, A. (2017). Hydrolytic degradation of polylactic acid (PLA) and its composites. *Renewable & Sustainable Energy Reviews, 79*, 1346–1352.

Facca, A. G., Kortschot, M. T., & Yan, N. (2006). Predicting the tensile strength of natural fibre reinforced thermoplastics. *Composites. Part A, Applied Science and Manufacturing, 37*(10), 1660–1671. doi:10.1016/j.compositesa.2005.10.006

Faruk, O., Bledzki, A. K., Fink, H. P., & Sain, M. (2012). Biocomposites reinforced with natural fibers: 2000-2010. *Progress in Polymer Science, 37*(11), 1552–1596. doi:10.1016/j.progpolymsci.2012.04.003

Faulstich de Paiva, J. M., & Frollini, E. (2006). Unmodified and modified surface sisal fibers as reinforcement of phenolic and lignophenolic matrices composites: Thermal analyses of fibers and composites. *Macromolecular Materials and Engineering, 291*(4), 405–417.

Filgueira, D., Holmen, S., Melbø, J. K., Moldes, D., Echtermeyer, A. T., & Chinga-Carrasco, G. (2017). Enzymatic-assisted modification of thermomechanical pulp fibers to improve the interfacial adhesion with poly (lactic acid) for 3D printing. *ACS Sustainable Chemistry & Engineering, 5*(10), 9338–9346.

Fuqua, M. A., Huo, S., & Ulven, C. A. (2012). Natural fiber reinforced composites. *Polymer Reviews (Philadelphia, Pa.), 52*, 259–320.

George, J., Sreekala, M. S., & Thomas, S. (2001). A review on interface modification and characterization of natural fiber reinforced plastic composites. *Polymer Engineering and Science, 41*(9), 1471–1485.

Giles, H. F. Jr, Mount, E. M. III, & Wagner, J. R. Jr. (2004). *Extrusion: The Definitive Processing Guide and Handbook.* William Andrew Publishing.

Grand View Research. (n.d.). *Natural fiber composites (NFC) market size, share & trends analysis report by raw material, by matrix, by technology (injection molding, compression molding, pultrusion), by application, and segment forecasts.* Academic Press.

Graupner, N., Herrmann, A. S., & Müssig, J. (2009). Natural and man-made cellulose fibre-reinforced poly(lactic acid) (PLA) composites: An overview about mechanical characteristics and application areas. *Composites. Part A, Applied Science and Manufacturing*, *40*(6), 810–821.

Guen, L. M., Hill, S., Smith, D., Theobald, B., Gaugler, E., Barakat, A., & Mayer-Laigle, C. (2019). Influence of Rice Husk and Wood Biomass Properties on the Manufacture of Filaments for Fused Deposition Modeling. *Frontiers in Chemistry*, *7*(735), 1–10.

Guo, R., Ren, Z., Bi, H., Song, Y., & Xu, M. (2018). Effect of toughening agents on the properties of poplar wood flour/poly (lactic acid) composites fabricated with Fused Deposition Modeling. *European Polymer Journal*, *107*, 34–45.

Herrera-Franco, P. J., & Drzal, L. T. (1992). Comparison of methods for the measurement of fibre/matrix adhesion in composites. *Composites*, *23*(1), 2–27.

Jacob, M., Francis, B., Thomas, S., & Varughese, K. T. (2006). Dynamical mechanical analysis of sisal/oil palm hybrid fiber-reinforced natural rubber composites. *Polymer Composites*, *27*(6), 671–680. doi:10.1002/pc.20250

Jawaid, M., Khalil, H. P. S. A., Hassan, A., Dungani, R., & Hadiyane, A. (2013). Effect of jute fibre loading on the mechanical and thermal properties of oil palm-epoxy composites. *Composites. Part B, Engineering*, *45*(1), 619–624. doi:10.1016/j.compositesb.2012.04.068

Joshi, S. V., Drzal, L. T., Mohanty, A. K., & Arora, S. (2004). Are natural fiber composites environmentally superior to glass fiber reinforced composites? *Composites. Part A, Applied Science and Manufacturing*, *35*(3), 371–376. doi:10.1016/j.compositesa.2003.09.016

Keckes, J., Burgert, I., Fruhmann, K., Müller, M., Kollin, K., & Hamilton, M. (2003). Cell-wall recovery after irreversible deformation of wood. *Nature Materials*, *2*(12), 810–814. doi:10.1038/nmat1019 PMID:14625541

Kim, S., Kim, H. J., & Park, J. C. (2009). Application of recycled paper sludge and biomass materials in manufacture of green composite pallet. *Resources, Conservation and Recycling*, *12*(53), 674–679.

Koronis, G., Silva, A., & Fontul, M. (2013). Green composites: A review of adequate materials for automotive applications. *Composites. Part B, Engineering*, *44*(1), 120–127. doi:10.1016/j.compositesb.2012.07.004

Koronis, G., Silva, A., & Fontul, M. (2013). Green composites: A review of adequate materials for automotive applications. *Composites. Part B, Engineering*, *44*(1), 120–127.

Kumar, A. L., & Prakash, M. (2021). The effect of fiber orientation on mechanical properties and machinability of GFRP composites by end milling using cutting force analysis. *Polymers & Polymer Composites*, *29*(9), 178–187.

Kunanopparat, T., Menut, P., Morel, M. H., & Guilbert, S. (2008). Plasticized wheat gluten reinforcement with natural fibers: Effect of thermal treatment on the fiber/matrix adhesion. *Composites. Part A, Applied Science and Manufacturing*, *39*(12), 1787–1792.

Lasikun, Ariawan, D., Surojo, E., & Triono, J. (2018). Effect of fiber orientation on tensile and impact properties of Zalacca Midrib fiber-HDPE composites by compression molding. *AIP Conference Proceedings*, (1), 1–5.

Lau, K., Hung, P., Zhu, M.-H., & Hui, D. (2018). Properties of natural fibre composites for structural engineering applications. *Composites. Part B, Engineering*, *136*(1), 222–233. doi:10.1016/j.compositesb.2017.10.038

Liu, H., He, H., Peng, X., Huang, B., & Li, J. (2019). Three dimensional printing of poly (lactic acid) bio based composites with sugarcane bagasse fiber: Effect of printing orientation on tensile performance. *Polymers for Advanced Technologies*, *30*(4), 910–922.

Madsen, B., & Lilholt, H. (2003). Physical and mechanical properties of unidirectional plant fibre composites-an evaluation of the influence of porosity. *Composites Science and Technology*, *63*(9), 1265–1272. doi:10.1016/S0266-3538(03)00097-6

Mahjoub, R., Yatim, J. M., Sam, A. R. M., & Hashemi, S. H. (2014). Tensile properties of kenaf fiber due to various conditions of chemical fiber surface modifications. *Construction & Building Materials*, *55*, 103–113.

Majid, R. A., Ismail, H., & Taib, R. M. (2010). Effects of polyethylene-g-maleic anhydride on properties of low density polyethylene/thermoplastic sago starch reinforced kenaf fibre composites. *Iranian Polymer Journal*, *19*, 501–510.

Martinschitz, K. J., Boesecke, P., Garvey, C. J., Gindl, W., & Keckes, J. (2008). Changes in microfibril angle in cyclically deformed dry coir fibers studied by in-situ synchrotron X-ray diffraction. *Journal of Materials Science*, *43*(1), 350–356. doi:10.100710853-006-1237-7

Mazzanti, V., de Luna, S. M., Pariante, R., Mollica, F., & Filippone, G. (2020). Natural fiber-induced degradation in PLA-hemp biocomposites in the molten state. *Composites. Part A, Applied Science and Manufacturing*, *137*, 105990.

Mazzanti, V., Malagutti, L., & Mollica, F. (2019). FDM 3D printing of polymers containing natural fillers: A review of their mechanical properties. *Polymers*, *11*(7), 1094.

Mazzanti, V., & Mollica, F. (2017). In-process measurements of flow characteristics of wood plastic composites. *Journal of Polymers and the Environment*, *25*, 1044–1050.

Merlini, C., Soldi, V., & Barra, G. M. O. (2011). Influence of fiber surface treatment and length on physico-chemical properties of short random banana fiber-reinforced castor oil polyurethane composites. *Polymer Testing*, *30*(8), 833–840. doi:10.1016/j.polymertesting.2011.08.008

Milosevic, M., Stoof, D., & Pickering, K. L. (2017). Characterizing the mechanical properties of fused deposition modelling natural fiber recycled polypropylene composites. *Journal of Composites Science*, *1*(7), 1–14.

Mohammed, L., Ansari, M. N. M., Pua, G., Jawaid, M., & Islam, M. S. (2015). A review on natural fiber reinforced polymer composite and its applications. *International Journal of Polymer Science*, *2015*, 1–15. doi:10.1155/2015/243947

Mohanty, A. K., Misra, M., & Drzal, L. T. (2002). Sustainable Bio-Composites from renewable resources: Opportunities and challenges in the green materials world. *Journal of Polymers and the Environment*, *10*, 19–26.

Mohanty, A. K., Misra, M., & Hinrichsen, G. (2000). Biofibres, biodegradable polymers and biocomposites: An overview. *Macromolecular Materials and Engineering*, *276/277*(1), 1–24. doi:10.1002/(SICI)1439-2054(20000301)276:1<1::AID-MAME1>3.0.CO;2-W

Moran, J., Alvarez, V., Petrucci, R., Kenny, J., & Vazquez, A. (2007). Mechanical properties of polypropylene composites based on natural fibers subjected to multiple extrusion cycles. *Journal of Applied Polymer Science*, *103*, 228–237.

Mussig, J., Fischer, H., Graupner, N., & Drieling, A. (2010). Testing methods for measuring physical and mechanical fibre properties (plant and animal fibres). In J. Mussig (Ed.), *Industrial Applications of Natural Fibres: Structure, Properties and Technical Applications (pp. 269–310)*. John Wiley & Sons, Ltd. doi:10.1002/9780470660324.ch13

Muthuraj, R., Misra, M., & Mohanty, A. K. (2015). Studies on mechanical, thermal, and morphological characteristics of biocomposites from biodegradable polymer blends and natural fibers. In M. Misra & J. Pandey (Eds.), *Biocomposites design and mechanical Performance*. doi:10.1016/B978-1-78242-373-7.00014-7

Nabi Saheb, D., & Jog, J. (1999). Natural fiber polymer composites: A review. *Advances in Polymer Technology*, *18*(4), 351–363. doi:10.1002/(SICI)1098-2329(199924)18:4<351::AID-ADV6>3.0.CO;2-X

Nagarajan, V., Mohanty, A. K., & Misra, M. (2013). Sustainable green composites: Value addition to agricultural residues and perennial grasses. *ACS Sustainable Chemistry & Engineering*, *1*(3), 325–333. doi:10.1021c300084z

Nagarajan, V., Mohanty, A. K., & Misra, M. (2013). Sustainable green composites: Value addition to agricultural residues and perennial grasses. *ACS Sustainable Chemistry & Engineering*, *1*, 325–333.

Nguyen, N. A., Bowland, C. C., & Naskar, A. K. (2018). A general method to improve 3D-printability and inter-layer adhesion in lignin-based composites. *Applied Materials Today*, *12*, 138–152.

Osman, M. A., & Atia, M. R. (2018). Investigation of ABS-rice straw composite feedstock filament for FDM. *Rapid Prototyping Journal*, *24*(6), 1067–1075.

Panthapulakkal, S., Law, S., & Sain, M. (2005). Enhancement of processability of rice husk filled high-density polyethylene composite profiles. *Journal of Thermoplastic Composite Materials.*, *18*, 445–458.

Paul, V., Kanny, K., & Redhi, G. G. (2015). Mechanical, thermal and morphological properties of a bio-based composite derived from banana plant source. *Composites. Part A, Applied Science and Manufacturing*, *68*, 90–100. doi:10.1016/j.compositesa.2014.08.032

Pervaiz, M., & Sain, M. (2003). Carbon storage potential in natural fiber composites. *Resources, Conservation and Recycling*, *39*(4), 325–340.

Petchwattana, N., Channuan, W., Naknaen, P., & Narupai, B. (2019). 3D printing filaments prepared from modified poly(lactic acid)/teak wood flour composites: An investigation on the particle size effects and silane coupling agent compatibilisation. *The Journal of Physiological Sciences; JPS*, *30*(2), 169–188.

Promper, E. (2010). Natural fibre-reinforced polymers in automotive interior applications. In J. Mussig (Ed.), *Industrial Applications of Natural Fibres: Structure, Properties and Technical Applications* (pp. 423–436). John Wiley & Sons, Ltd.

Qaiss, A. E. K., Bouhfi, D. R., & Essabir, H. (2014). Natural fibers reinforced polymeric matrix: Thermal, mechanical and interfacial properties. *Biomass and Bioenergy Processing and Properties*, 225–245.

Ramesh, M., Palanikumar, K., & Reddy, K. H. (2017). Plant fibre based bio-composites: Sustainable and renewable green materials. *Renewable and Sustainable Energy Reviews, Elsevier*, *79*, 558–584. doi:10.1016/j.rser.2017.05.094

Rosa, S. M. L., Santos, E. F., Ferreira, C. A., & Nachtigall, S. M. B. (2009). Studies on the properties of rice-husk-filled-PP composites - Effect of maleated PP. *Materials Research*, *12*(3), 333–338.

Royan, N. R. R., Sulong, A. B., Yuhana, N. Y., Chen, R. S., Ab Ghani, M. H., & Ahmad, S. (2018). UV/O_3 treatment as a surface modification of rice husk towards preparation of novel biocomposites. *PLoS One*, *13*(5).

Santoni, A., Bonfiglio, P., Fausti, P., Marescotti, C., Mazzanti, V., Mollica, F., & Pompoli, F. (2019). Improving the sound absorption performance of sustainable thermal insulation materials: Natural hemp fibres. *Applied Acoustics*, *150*, 279–289.

Satyanarayana, K. G., Arizaga, G. G. C., & Wypych, F. (2009). Biodegradable composites based on lignocellulosic fibers-An overview. *Progress in Polymer Science*, *34*(9), 982–1021. doi:10.1016/j.progpolymsci.2008.12.002

Shah, D. U. (2013). Developing plant fibre composites for structural applications by optimising composite parameters: A critical review. *Journal of Materials Science*, *48*(18), 6083–6107. doi:10.100710853-013-7458-7

Shen, L., Haufe, J., & Patel, M.K. (2009). *Product overview and market projection of emerging bio-based plastics.* PRO-BIP 2009, Final Report.

Shesan, O. J., Stephen, A. C., Chioma, A. G., Neerish, R., & Rotimi, S. E. (2019). Fiber- Matrix Relationship for Composites Preparation. In Renewable and Sustainable Composites (pp. 1-30). INTECH Open Access Publisher.

Slootmaker, T., & Müssig, J. (2010). SEM catalogue for animal and plant fibres. In *Industrial Applications of Natural Fibres: Structure, Properties and Technical Applications* (pp. 311–336). John Wiley & Sons, Ltd. doi:10.1002/9780470660324.ch14

Stamboulis, A., Baillie, C. A., & Peijs, T. (2001). Effects of environmental conditions on mechanical and physical properties of flax fibers. *Composites. Part A, Applied Science and Manufacturing*, *32*(8), 1105–1114. doi:10.1016/S1359-835X(01)00032-X

Stoof, D. (2016). *Development of Recycled Polypropylene Composite Materials for Applications in 3D Printing* [Ph.D. Thesis]. University of Waikato, Hamilton, New Zealand.

Stoof, D., Pickering, K., & Zhang, Y. (2017). Fused deposition modelling of natural fibre/polylactic acid composites. *Journal of Composites Science*, *1*(8).

Summerscales, J., Dissanayake, N. P. J., Virk, A. S., & Hall, W. (2010). A review of bast fibers and their composites.Part 1-Fibers as reinforcements. *Composites. Part A, Applied Science and Manufacturing*, *41*(10), 1329–1335.

Taj, S., Munawar, M. A., & Khan, S. (2007). Natural fiber-reinforced polymer composites. *Proceedings of the Pakistan Academy of Sciences*, *44*(2), 129–144.

Trindade, W. G., de Paiva, J. M. F., Leao, A. L., & Frollini, E. (2008). Ionized-Air-Treated Curaua Fibers as Reinforcement for Phenolic Matrices. *Macromolecular Materials and Engineering*, *293*(6), 521–528.

Valadez-Gonzalez, A., Cervantes-Uc, J. M., Olayo, R., & Herrera-Franco, P. J. (1999). Effect of fiber surface treatment on the fiber–matrix bond strength of natural fiber reinforced composites. *Composites. Part B, Engineering*, *30*(3), 309–320.

Van den Oever, M. J. A., Beck, B., & Müssig, J. (2010). Agrofibre reinforced poly (lactic acid) composites: Effect of moisture on degradation and mechanical properties. *Composites. Part A, Applied Science and Manufacturing*, *41*(11), 1628–1635.

Wambua, P., Ivens, J., & Verpoest, I. (2003). Natural fibres: Can they replace glass in fibre reinforced plastics? *Composites Science and Technology*, *63*(9), 1259–1264. doi:10.1016/S0266-3538(03)00096-4

Wang, X., Jiang, M., Zhou, Z., Gou, J., & Hui, D. (2017). 3D printing of polymer matrix composites: A review and prospective. *Composites. Part B, Engineering*, *110*, 442–458.

Zhang, Q., Cai, H., Zhang, A., Lin, X., Yi, W., & Zhang, J. (2018). Effects of lubricant and toughening agent on the fluidity and toughness of poplar powder-reinforced polylactic acid 3D printing materials. *Polymers*, *10*(932), 1–11.

Zhao, D. X., Cai, X., Shou, G. Z., Gu, Y. Q., & Wang, P. X. (2016). Study on the preparation of bamboo plastic composite intend for additive manufacturing. *Key Engineering Materials*, *667*, 250–258.

Zini, E., & Scandola, M. (2011). Green composites: An overview. *Polymer Composites*, *32*(12), 1905–1915.

KEY TERMS AND DEFINITIONS

Additive Manufacturing: Material is added layer by layer to produce the workpiece.
Biodegradable: Can be disintegrated by bacteria or other living beings and hence avoiding toxic waste.
Cellulose: Organic composite of the primary cell wall of the plant.
Crystallinity: Level of structural order of solid.
Hydrolysis: Chemical response in which a water molecule disintegrates single or many chemical bonds.
Matrix: The component of a composite material.
Microfibril: A minute fibril, containing of cellulose and glycoproteins.
Polylactic Acid: An aliphatic polyester manufactured through polymerization of lactide.

APPENDIX

Table 6. Table of abbreviations

Abbreviations	Meaning
3D	Three Dimensional
NFRC	Natural Fiber Reinforced Composite
PLA	Polylactic Acid
UV/O_3	Ultraviolet-Ozone
MA	Maleic Anhydride
HDPE	High-Density Polyethylene
PP	Polypropylene

Chapter 4
Mechanical and Tribological Properties of Polymer Composites Developed by FDM

Pradeep Kumar G. S.
CHRIST University (Deemed), India

Sachit T. S.
Symbiosis Institute of Technology, Symbiosis International University (Deemed), India

Mruthunjaya M.
JSS Academy of Technical Education, India

Harish Kumar M.
CHRIST University (Deemed), India

Raghu Yogaraju
BMS College of Engineering, India

Sasidhar Jangam
CHRIST University (Deemed), India

ABSTRACT

In the coming years, researchers and manufacturers will be more concerned with the demand for products that are easy to construct and can be rendered seamlessly, even at complicated geometries, with the touch of a button. These requirements will be met by the advent of additive manufacturing. This will serve as the catalyst for a revolution. Additive layer manufacturing (ALM), also known as additive manufacturing (AM) or more commonly referred to as 3D printing, is a modern fabrication technology that uses a variety of raw materials to produce items such as medical implants and aircraft wing components by printing layers of material based on 3D digital models. In this chapter, the authors provide a comprehensive overview of the mechanical and tribological behaviour of ceramic, metallic, and fiber-reinforced polymer composites that are made by a range of additive manufacturing (AM) techniques. These composites can be used in a variety of applications, including aerospace, automotive, and medical.

DOI: 10.4018/978-1-6684-6009-2.ch004

INTRODUCTION

Nature provides every new thought if perceived carefully. Composites are one such form of materials that exist in nature. Composite material is generally amalgamation of two or more materials, having different properties, which do not blend into each other and provide better results in comparison to original individual material. The natural substance like wood and parts of human body are best examples of composite materials. Utilization of composite materials for human need have been evolving from 1500 BC and is still in continuation (Johnson, 2018). In the recent past the advancement in composite material has been evolving very rapidly. Ease of compatibility with various forms of utilization has seen major boost and has been applied to newer domains of application. Initially the usage of composite materials was restricted in aerospace, defense and high value goods industry; however, with its usage in various forms it has been evolved in every domain and we can see its applications in and around us. The advantages of using composite materials are low cost, better performance, light weight, environment compliance and safe for usage (M. P. Todor & Kiss, 2016; P. Todor et al., 2017)

Composite materials consists of three basic components viz; matrix, reinforcement and interphase region (Ghosh, 1991). The improvement in desired properties can be established with these parameters by choosing the right form of matrix, reinforcement and the manufacturing process (Jim, 2015). Figure 1 and 2 shows classification of composites based on matrix and reinforcement respectively. The categorization of matrix materials is in three forms, metal, ceramics and polymer. Polymer composites are widely used in current applications in various fields. There are two forms of polymer matrix available which are thermoplastic and thermoset plastic. Thermosets are kind of plastics that transform from liquid to solid due to certain reactions or curing. This change from liquid to solid state is irreversible and permanent. Once the plastic is completely set, heating it will lead in degrading the material. Melting of material won't occur in thermoset as the degradation temperature is near to melting temperature and hence degradation occurs. On the other hand, thermoplastics are melt process able plastics. The processing of these plastics takes place by heating. Addition of quantifiable amount of heat leads in increase in temperature above the melting point where melting occurs and it liquefies and softens for the processing. Withdrawal of heat from this plastic leads in glasslike solidification. The process of heating and solidifying can be carried out for these plastics, however, repeated cycles leads in deterioration of properties of plastic. Reinforcements used in composites are categorized into fibers, flakes or particles. Addition of each of these reinforcements help in enhancing the properties of the composites and can be used for certain applications. Fibers as reinforcement is most commonly used reinforcement in polymers matrix like ABS, PLA, PC, PEEK etc. Fibers provide great strength to matrix because of its aspect ratio in terms of L/D, enhanced shear stress transfer due to its alignment in matrix, forms a great bonding between matrixes providing good interface. There are numerous forms of fibers used as reinforcement; among them most commonly used are carbon fibers, glass fibers, aramid fibers and boron fibers. There are numerous methods for development of composites. The method of manufacturing is chosen depending on the part to be developed, material chosen, design and application.

Open contact moulding is commonly used form of manufacturing process which includes hand layup and spray up process. These both processes are used in tandem to reduce the labor cost. The process is basically used in making large components which are simple in their design. Another form of manufacturing processes are resin infusion processes which are more sophisticated than open moulding processes and more automated. Resin transfer moulding, reaction injection moulding, vacuum-assisted resin transfer moulding and resin film infusion are some of the processes under this category. Compression moulding

Figure 1. Matrix categorization of composites

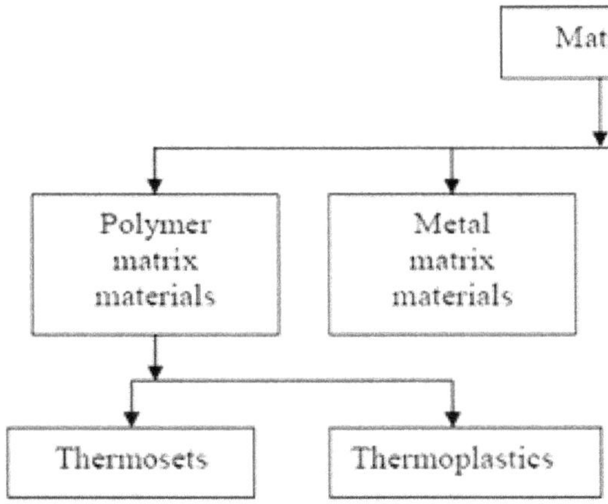

Figure 2. Reinforcement categorization of composites

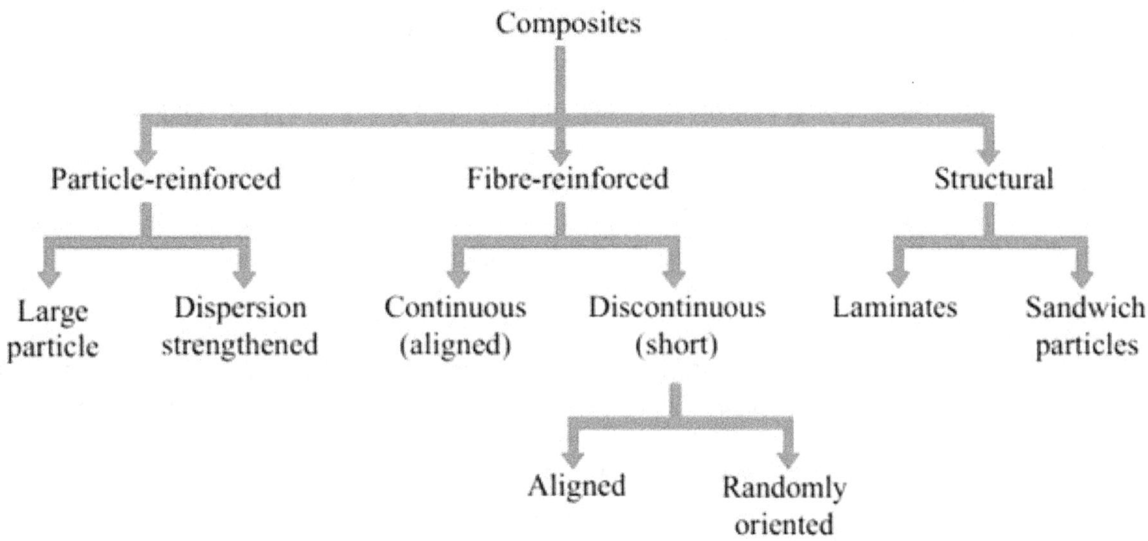

is another form of manufacturing process that is being used in development of high quality composite parts. The placement of material into the mould is done either through an automated process or manual process. These moulds are pressed against each other by applying hydraulic pressure. The cycle time depends on the size of the part to be developed and its thickness. Injection moulding is another commonly used form of manufacturing process which is widely used in development of thermoplastic composites. Filament winding, pultrusion process and automated fiber placement are some other processes that are used commonly. The new form of manufacturing process that has emerged in recent times is known as Additive Manufacturing (AM) process. This process of manufacturing, fabricates the parts layer by layer in bottom up approach. The ease of user interface and its adaptability to develop any kind of complex shape has made the industry and researchers to approach to this method of manufacturing from conventional systems. The process of AM enables in approaching diversified field of materials and gives greater advantages in terms of developing complex geometries, precision parts, reduced lead time and easy to approach in terms of development of any complex part (Gao et al., 2015). This chapter in the further sections discusses about AM, materials and characteristics.

ADDITIVE MANUFACTURING

A 3D CAD file is used in the additive manufacturing (AM), commonly referred to as 3D printing, process to build objects layer by layer. Layers are cut into the created 3D model so that each layer can be deposited one at a time. Stereolithography (SLA), created by 3D Systems in 1987. Wohlers & Gornet, 2015, was the first AM technique that was commercially available. Over the past ten years, the AM process has developed a number of different technologies. Laminated object manufacturing (LOM) was created by Cubital and Helisys, Fused Deposition Modeling (FDM) by Stratasys, Digital Light Processing (DLP) by Envision TEC, and Selective Laser Sintering (SLS) by DTM Corp., a subsidiary of 3D Systems. The classification of AM technologies has been done either by considering the deposition technique or materials used by Taormina et al., 2018. The advent of AM technology began 2 decades ago; however, its standardization has been done very recently in ASTM52900-2015. The 7 main categories of classification of AM are binder jetting, material extrusion, direct energy deposition, material jetting, powder bed fusion, sheet lamination and vat photo polymerization.

The process of additive manufacturing in any of the above mentioned categories has similar steps to be followed from modeling until print. The four major steps involved in development of product are; CAD software for designing the model, pre processing, printing and post processing.

i. CAD software for designing the model: Computer Aided Design Software has a great impact in the process of Additive Manufacturing. This software is used for the purpose of designing and testing of models which would be used for real world applications. Some of the most commonly used software are:

AutoCAD: this is one of the most widely used software for 3D design and has been extensively used by industrial experts.

Creo: this is one of the market leader in product designs software and has wide range of functional options.

Solid Works: used predominantly for engineering object design. This has very wide variety of features.

ii. Pre processing involves two major steps which must be completed in the period between design and manufacturing. Simulation modeling and preparing files for 3D printing are the two steps that make a model ready to be printed.
iii. The process of printing seems to be very different from one process of printing to another depending on the type additive technology. The general from of 3D printing process involves laying of powder material one layer over the other using binding liquid. The process of additive manufacturing is binder jetting.

The other forms of AM technology looks different from the one mentioned above. Stereolithography (SLA) cures the layers of photopolymer resin by using laser instead of binders. In the process of development, the build platform is lowered in the pool of resin into which the product id traced with laser path. Once the process starts curing of first layer the build platform is fractionally lowered to cure the next layer in process. Another form of AM process which is widely used in Fused Deposition Modeling (FDM), wherein a thermoplastic is used to heat the material and lay it on the build platform one over the other. Once the layer laid of build platform completely dries next layer is laid upon this layer. Other forms of AM technology includes SLS< DMLS and EDM.

iv. Post processing is the final step of manufacturing that involves various methods and is time consuming. There are numerous steps involved in this, however the general category of this process falls under three categories:

   a) Removal of build: Disengagement of surplus material from the fabricated part and platform.
   b) Part separation: Removing the object from the build platform, separating parts, and removing any support structures used to aid the build process.
   c) Debinding: The developed parts are soaked in certain solutions to get the required texture and to remove any excess material which would not be removed physically.

POLYMER COMPOSITES BY FDM

The desirable properties of polymers such as process ability and lower density have made this material to be used as matrix material for greater applications. The advantages with respect to mechanical and thermal properties are amicable.

Usage of additive manufacturing process in industrial scale leads to reduction in production time and cost. Additionally, it supports the idea of polymer recycling, which is the main objective to decrease landfilling. However, the amount of polymer recycling is limited by the necessary mechanical and thermal qualities of 3D printed items. 199 However, it is significantly necessary to employ plastics like PLA, poly(butylenessuccinate), and poly(hydroxyalkanoates) (PHAs) in place of typical petro-based plastics in additive manufacturing without compromising the necessary qualities. The table below lists the most popular polymers used in the AM process.

High-speed aeroplanes, rockets, and other relevant space and electronics are currently made using high-temperature resins as composite materials. When a composite (such a carbon/epoxy composite) has fibre reinforcements spread throughout a weak matrix, the reinforcements share the bulk of the load. Therefore, the strength and stiffness of the composites are greatly dependent on these parameters of fibers (Lee & Jain, 2012). The main benefit of fibre reinforced composites is their high strength and modulus, which are on par with or superior to many conventional metallic materials. These composite materials outperform metallic materials significantly due to their lower specific gravity, higher strength-to-weight and modulus-to-weight ratios. Additionally, many composite laminates have good fatigue strength to weight ratios and fatigue damage tolerance. (Govaert, 2005; Menachem, 2007) Utilization of fibers has been widely used in many forms. Development of fibers through conventional methods offers certain resistance in comparison to additive manufacturing processes. Achieving complex shapes and maintaining the cost to minimal can be achieved using AM. Combining the technology of fiber reinforcement

Table 1. Properties of polymers

Sl No	Polymers	Properties and functions
1	Acrylonitrile butadiene styrene (ABS)	Creation of ABS takes place by polymerization process of styrene and acrylonitrile in the presence of polybutadiene. This is a classic example of amorphous thermoplastic. Due to its excellent printability, it is one of the polymers that is frequently used in 3D printing. The printability of ABS has been investigated using a variety of printing conditions, including processing temperature, bed temperature, and printing speed.
2	Polypropylene (PP)	Chain-growth polymerization is used to create polypropylene, a durable thermoplastic polymer that is a member of the polyolefin family. In addition to being slightly harder and more heat resistant than PE, it also exhibits certain properties like non-polarity, semi-crystallinity, and resistance to chemical reaction in polymer. The application of this polymer can be widely seen in day-to-day products and its application can be found in diverse industries such as packaging, labelling, textiles and automotive.
3	Polyethylene (PE)	The polyolefin family also includes polyethylene (PE), which shares many of the same characteristics as polypropylene (PP). They have a wide range of applications due to their excellent ductility, impact strength, outstanding water barrier, and chemical resistant properties. In packaging applications, thermoplastic petroleum-based polymers are commonly utilised to create plastic sheets, bags, membranes, and containers. Low density polyethylene (LDPE), linear low density polyethylene (LLDPE), high density polyethylene (HDPE), high molecular weight polyethylene (HMWPE), and ultra high molecular weight polyethylene are some of the several types of PE based on chemical structure, density, and branching (UHMWPE). Among the different forms of PE, UHMWPE is frequently utilised as a feedstock in additive manufacturing to create a variety of complicated shaped things, including antifriction goods for mechanical engineering and pharmaceuticals.
4	Polyethylene Terephthalate (PET)	PET is a thermoplastic, colourless, semi-crystalline polymer made from petroleum that is a member of the polyester family. PET is utilised in a variety of items, including substrate for solar cells, engineering plastic in electronics and fibres for clothes in the textile industry. The method of condensation polymerization is used for the generation of PET. Recycling is common in these materials as it exhibits non biodegradability. Due to its mechanical qualities, including increased ductility, lower resistance to friction, good resistance to abrasion, excellent dimensional stability, and high tracking resistance, PET is frequently utilised as a feedstock in additive manufacturing (AM) procedures to create complexly shaped objects.
5	Polycarbonate (PC)	A class of thermoplastic polymers having connected carbonate groups are referred to as polycarbonates. A reaction between bisphenol A and phosgene results in the creation of PCs, which have the qualities of being robust, strong, temperature- and impact-resistant, and optically transparent. PCs are utilised in a variety of industrial and commercial applications, including electronics, building supplies, data storage, and complex shaped objects. Polycarbonates are a type of thermoplastic polymers with linked carbonate groups. PC is capable of massive plastic deformations without cracking, in contrast to the majority of thermoplastics on the market. For the aforementioned applications, 3D printed items are created on PCs.
6	Thermoplastic elastomers (TPE)	Thermoplastic elastomers are a type of copolymer, where one component is typically rubber and the other is plastic. The thermoplastic and elastomeric characteristics of TPE are therefore combined. While the majority of elastomers fall within the thermosets category. The stretch ability of thermoplastic elastomers up to moderate elongations and shape memory, which offer a longer service life and superior physical qualities, are their advantages. Only the type of crosslinking connection in their chemical structures differentiates thermoplastic and thermoset elastomers. The kind of crosslinking has a significant impact on the materials' elastic characteristics. Styrene block copolymers (SBC), thermoplastic polyolefin elastomers (TPPOE), thermoplastic vulcanizates (TPV), thermoplastic polyurethanes (TPPU), and thermoplastic vulcanizates (TPV) are some examples of the commercial TPEs that are readily available.
7	Polyamide (PA)	A macromolecule known as polyamide has repeating units joined by amide bonds. Application of polyamide can be found in various industries such as automotive and textile. The material is commonly used for development of products like kitchen utensils, carpets and sportswear. Polyamides are created by step growth polymerization. The 2 most common forms of polyamides used in AM are aramids and nylon.
8	Poly(lactic acid) (PLA)	Lactides and lactic acid monomer (2-hydroxypropanoic acid) can be converted into poly lactic acid utilising a variety of polymerization processes, including condensation polymerization and ring opening polymerization. Poly lactic acid is a member of the thermoplastic aliphatic polyester family. Using a conventional fermentation process, lactic acid is removed from natural resources such sugarcane, corn, sweet potatoes, cassava, and sugar beet pulp. By altering polymer chain orientation, PLA characteristics can change. Depending on polymer chain arrangements the division of PLA can be done as crystalline polymers, amorphous and semicrystalline polymers. PLA has found its application as 3D printing feedstock material due to all the above advantages.

and AM paves a greater opportunity for development of complex and strong parts. The mixture of fiber and matrix can take place at various proportions depending on the affinity of fibers towards the matrix and can lead in development of strong parts with improved mechanical properties. There are certain parameters that has be studied to optimally developed reinforced polymers. Some of the crucial parameters to be considered in AM are thickness of layer, speed with which infill takes place, temperature of nozzle and raster angle. Fiber reinforced composites are most used polymer composites which have been widely used in industries. Prominent industries are aerospace, motorsports, biomedicine, automobile, construction and many more. The usage of fiber reinforced composites have been increasing slowly in AM process and studies are focused on development on parts in consideration with its orientation, process parameters, nozzle temperature and other subjects.

MECHANICAL PROPERTIES OF POLYMER COMPOSITES DEVELOPED BY FDM

Various research studies were carried out to study the effects of fiber reinforcement in polymer matrix developed by 3D printing technology. One such study was carried out by placing carbon fiber between the PLA polymers layers with an intention of enhancing the strength and fatigue life. It was evident from the experiments that increasing the amount of glass fiber in polymer led in generation of greater voids through 3D printing process and negatively impacted the tensile results (Der Klift et al., 2016). Formation of poor bond between PLA and fiber has significant impact on mechanical characteristic. Utilization of certain methods such as surface treatment of fibers using methylene dichloride helps in enhancement of tensile and flexural strength (Li et al., 2016). In yet another similar study, tensile properties were studied by reenforcing glass fiber into PLA. A policy of rule of mixture was introduced to estimate the percent of glass fiber to be added and it was noted that addition of glass fiber to a content of 4 wt% gave a tensile strength value of 83 MPa. The actual findings gave a result of 71 MPa of tensile strength. Comparative study was done by addition of glass fiber to polypropylene and later was mixed with maleic anhydride polyolefin (POEg-MA) (Sodeifian et al., 2019). Addition of glass fiber to polypropylene increased the tensile strength and modulus, however the flexibility was reduced. Addition of maleic anhydride polyolefin resulted in decreased strength and modulus and increase in flexibility. Studies related to addition of carbon fiber to acrylonitrile butadiene styrene in different composition was carried out by Ning et al., 2015. The components were fabricated using Fused Deposition Technology. The results showed an increment in tensile strength and modulus of reinforced polymer. However, in comparison to pure polymer samples, the reinforced polymer showed reduction in ductility and yield strength. Different volume contents of glass fiber was added to ABS matrix to analyze the process ability and engineering properties (Zhong et al., 2001). Noted that with increased volume content of glass fiber the tensile strength and surface rigidity increased. In another similar study addition of glass fiber to polypropylene, similar results were obtained with parts fabricated by FDM (Carneiro et al., 2015).

F. O. et al., 2013 analyzed mechanical properties of unsaturated polyester by addition of woven mat GF by varying the fiber loads from 5 to 30% in terms of 5%. Increment in various mechanical properties were enhanced with incremental addition of GF. Reduction in impact strength was seen when the GF Vf was raised by 25%. At 25% Vf, the maximum strain was discovered. Erden et al., 2010 using a novel technique developed woven E-GF reinforced polyester resin to study the mechanical properties. To polyester resin, oligomeric siloxane was added at various concentrations of 1, 2, and 3 wt%, respectively. Enhancement in mechanical properties such as shear, flexural and tensile was noted with addition of oligomeric siloxane. The optimal combination of composite developed was found to be with addition of 3 wt% oligomeric siloxane to polyester, which gave better results in comparison to other combinations. An increment in tensile strength, interlaminar shear strength, natural frequency and flexural strength was noted with addition of reinforcement. Al-alkawi et al., 2012 studied the wear behavior of E-GF-reinforced woven strand mats at a range of temperatures, including 40C, 50C, and 60C. S-N curve provided a result of decreased tensile and fatigue strength by increasing temperature to 60^{0}C having 33% fiber volume fraction. Tensile strength % decrease factors were lower than fatigue strength percentage reduction factors for various temperature regions. Awan et al., 2009 characterized tensile behavior for GF reinforced polyester composite have varied fiber cross section of 1,2 and 3 ply nature. Composites with a higher GF weight percentage have stronger reinforcement. Each specimen had a distinct cross-sectional area, thickness, and number of layers. Increased addition of reinforcement led to enhanced stiffness and strength of the composites. Three-ply reinforcement showed the highest tensile strength. Chen et al., 2006 tried to study the mechanical behavior of newly developed composite by mixing polyphenylene sulphide and polyamide as matrix material. Glass fiber of different volume was added to this matrix from 5 to 30 wt% respectively. Maximum tensile strength was found for the composite with 30 wt% of GF addition and maximum flexural strength was found for 25 wt% of GF. The maximal impact strength was found to be at 0% of fibre reinforcement in comparison to composites that contained fibre. Yuanjian & Isaac, 2008 developed GF reinforced polyester composites with two fibre geometries. Orientation of 45 degree with 42 wt% of fiber was one and the other was orientation of 0/90 by adding 47wt% of fiber. The developed composites were used for studies related to low velocity impact and tension. The outcome showed that as impact energy increased from 0 to 25 J in 5 J increments, the residual tensile strength and stiffness decreased. The maximum properties were maintained up to a test energy of 10 J, and as the test energy was increased to 20 J, the above qualities underwent a significant reduction. The two geometries had comparable impact damage. Damage to the matrix is caused by the low impact energy of GF-reinforced composites. At different impact damage energies of 1.4 J, 5 J, and 10 J, respectively, the tension-tension fatigue failure test was conducted. A S-N curve is used.

TRIBOLOGICAL PROPERTIES OF POLYMER COMPOSITES DEVELOPED BY FDM

El-Tayeb et al., 2008 evaluated glass fiber reinforced composite by considering different sliding velocities. At NTP conditions load of 30, 60 and 90N were applied with equivalent sliding velocities of 2.8, 3.52, and 3.9 m/s. The evaluation studies by addition of chopped GF to parallel/anti-parallel polyesters was carried out. The results showed that P-orientation sliding showed lesser friction coefficient for light weights and greater speeds in comparison to anti parallel orientation. A reduced (16%) rate of mass loss was seen in AP-orientation in comparison to P-orientation. El-Tayeb & Yousif, 2007 examined multi-pass two-body abrasive wear properties of CGRP composites. Sliding velocities were chosen between 0.157 to 0.314 m/s with load applicability from 5N to 25N. Against water-resistant SiC abrasive paper, the test subject slides in a dry contact environment. With increase in speed and load, reduction in wear rate was observed. AP-orientation increased the CGRP composite's resilience to abrasion. In comparison to various orientations, SEM analysis showed that anti parallel arrangement of fibers provided minimum wear rate. There were also no fiber damages. Quintelier et al., 2006 GF-reinforced polyester composites were tested for friction and wear under dry conditions with loading ranging from 60 to 300 N and a constant speed of 10 mm/s. Transverse orientation provided greater friction in comparison to parallel orientation as per SEM data. Mathew et al., 2007 made a comparative study between various resins and fibers by changing the orientation. The numerous forms of resins used were polyester, vinyl ester, and epoxy resin. Various wrap knit fibers were directly used along with E- glass fiber. Various forms of designs used were biaxial, triaxial and quadratic. Better results were obtained with biaxial composite. Kishore et al., 2000 studied the effects of sliding velocity and load of bi-directionally weaved epoxy composite reinforced by e-glass fiber. Various filler parameters were chosen for the experiment. Loads of 42, 140 and 190N were considered. In the block-on-roller test, it was discovered that the composite with oxide particles outperformed rubber particles in terms of wear resistance when there was little load. On the other hand, under more stressful circumstances, rubber particles outperformed oxide particles in terms of wear resistance. Chand et al., 2000 in a similar study as previous one, the authors carried out experiments to understand the effect of addition of short E-GF-reinforced polyester composites. The parameters considered for the study were sliding speeds, abrasive particle sizes, and applied stresses. It was noted that, with increased addition of reinforcement the wear rate reduced with respect to decrease in total volume content. They discovered lesser wear loss with higher GF content. Suresha et al., 2006 assessed friction and wear behavior of epoxy reinforced with GF woven mats. The tests were carried out for various speeds and loads. Five weight percent of each of the inorganic fillers, graphite and SiC particles, were added. Findings from SEM show that SiC-filled composites have the highest wear resistance and the lowest coefficient of friction of any type of composite, followed by graphite-filled composites. Yousif et al., 2008 experimented two methods viz; pin-on-disc and block-on-ring to study the wear and friction behavior of fiber reinforced polymer composite. Load variation was done from 30N to 100N for both the test methods. Orientation of fiber was done in both parallel and anti parallel ways. It was observed that, greater the water content made to develop greater surface roughness values. Suresha et al., 2006 examined friction and wear studies of glass fiber reinforced epoxy composite. The specimens considered were with and without SiC particles. The results revealed that at sliding distances between 2000 m and 4000 m, composites filled with (5 wt%) SiC particles exhibited a greater coefficient of friction and a higher resistance to wear than composites without SiC-filled particles. Mohan et al., 2011 tried to study wear behavior by considering sliding character for jatropa oil cake reinforced with E-GF. Composites were made by woven fabric for loads of 10 and 20N. The formal conclusion with respect to increased sliding distance was increased wear loss of the material. The sliding distance was 2000m for 10N load. Patnaik et al., 2010 studied wear properties of glass fiber reinforced epoxy composites along with reinforcing certain particles like SiC, Al_2O_3 and bark dust of pine. The various combination tried were epoxy 50wt% and glass fiber 50 wt%, epoxy 40 wt% along with glass fiber 50 wt% and alumina 10 wt%, glass fiber 50 wt% + epoxy 40 wt% + dust of pine bark 10 wt% and finally a combination of 50 wt% glass fiber + 40 wt% epoxy and 10 wt% SiC. Tests were conducted by variation of load from 50 to 75N with sliding distances between 200 and 600 m. They came to the conclusion that the composite of glass fiber 50 wt% + epoxy 40 wt% + dust of pine bark 10 wt% had better wear resistance for all sliding distances.

CONCLUSION

With the extensive research carried out in the field of additive manufacturing there still remains no proper evidence in applying certain procedures for optimal product to be developed. FDM one of the foremost manufacturing process for polymers and its composites is easily available and applicable to vast area of research in numerous fields like automobile, aerospace, marine, household applications, electronic packaging and many more. Due to varied process parameters the availability of similar results with same material developed by different researchers are varied. Research studies has been largely carried out in analyzing the various factors that are to be considered in development of parts by AM, understanding the mechanical behavior and also development of certain functional parts. However, there remains a huge void in thermal studies of polymer nanocomposites. The attainment of required strength for the developed polymers through FDM is not in-line with conventionally manufactured parts. Certain studies have pointed out in difficulty towards extraction of certain polymer composites through FDM process due to its reactivity with the nozzle. Considering all these factors the possibility for development of novel polymer nanocomposite has a huge opportunity.

Addition of reinforcement has seen certain defects in its manufacturing. To overcome these defects and to enhance the desired properties of the polymer, a novel polymer nanocomposite development can be taken as research study. Dimensional stability, surface roughness, tensile and impact strength, flexural strength, microstructure studies can be extensively carried out.

REFERENCES

Al-alkawi, H. J., Al-fattal, D. S., & Ali, A. H. (2012). Fatigue behavior of woven glass fiber reinforced polyester under variable temperature. *Exilir Mech. Engg., 53*, 12045–12050.

Awan, G.H., Ali, L., & Ghauri, K.M. (2009). Effect of various forms of glass fiber reinforcements on tensile properties of polyester matrix composite. *J Faculty Eng Techno, 16*, 33–39.

Carneiro, O. S., Silva, A. F., & Gomes, R. (2015). Fused deposition modeling with polypropylene. *Materials & Design, 83*, 768–776. doi:10.1016/j.matdes.2015.06.053

Chand, N., Naik, A., & Neogi, S. (2000). Three-body abrasive wear of short glass fibre polyester composite. *Wear, 242*(1–2), 38–46. doi:10.1016/S0043-1648(00)00398-7

Chen, Z., Liu, X., Lu, R., & Li, T. (2006). Mechanical and tribological properties of PA66/PPS blend. III. Reinforced with GF. *Journal of Applied Polymer Science, 102*(1), 523–529. doi:10.1002/app.24253

El-Tayeb, N. S. M., & Yousif, B. F. (2007). Evaluation of glass fiber reinforced polyester composite for multi-pass abrasive wear applications. *Wear, 262*(9–10), 1140–1151. doi:10.1016/j.wear.2006.11.015

El-Tayeb, N. S. M., Yousif, B. F., & Yap, T. C. (2008). An investigation on worn surfaces of chopped glass fibre reinforced polyester through SEM observations. *Tribology International, 41*(5), 331–340. doi:10.1016/j.triboint.2007.07.007

Erden, S., Sever, K., Seki, Y., & Sarikanat, M. (2010). Enhancement of the mechanical properties of glass/polyester composites via matrix modification glass/polyester composite siloxane matrix modification. *Fibers and Polymers, 11*(5), 732–737. doi:10.100712221-010-0732-2

F. O., A., P. O., A., & O. O., O. (2013). Mechanical Properties of a Polyester Fibre Glass Composite. *International Journal of Composite Materials, 2*(6), 147–151. doi:10.5923/j.cmaterials.20120206.06

Gao, W., Zhang, Y., Ramanujan, D., Ramani, K., Chen, Y., Williams, C. B., Wang, C. C. L., Shin, Y. C., Zhang, S., & Zavattieri, P. D. (2015). The status, challenges, and future of additive manufacturing in engineering. *CAD Computer Aided Design, 69*, 65–89. doi:10.1016/j.cad.2015.04.001

Ghosh, S. K. (1991). Manufacturing engineering and technology. *Journal of Materials Processing Technology, 25*(1), 112–113. doi:10.1016/0924-0136(91)90107-P

Jim, W. (2015). *The science and technology of composite materials - Curious.* Australian Academy of Science. https://www.science.org.au/curious/technology-future/composite-materials

Johnson, T. (2018). *History of Composites: The Evolution of Lightweight Composite Materials.* ThoughtCo. https://www.thoughtco.com/history-of-composites-820404#:~:text=The

Kishore, S., Sampathkumaran, P., Seetharamu, S., Vynatheya, S., Murali, A., & Kumar, R. K. (2000). SEM observations of the effects of velocity and load on the sliding wear characteristics of glass fabric-epoxy composites with different fillers. *Wear, 237*(1), 20–27. doi:10.1016/S0043-1648(99)00300-2

Lee, L., & Jain, R. (2012). *Fiber Reinforced Polymer (FRP) Composites for Infrastructure Applications.* Springer Science+Business Media. www.springer.com

Li, N., Li, Y., & Liu, S. (2016). Rapid prototyping of continuous carbon fiber reinforced polylactic acid composites by 3D printing. *Journal of Materials Processing Technology, 238*(2000), 218–225. doi:10.1016/j.jmatprotec.2016.07.025

Mathew, M. T., Padaki, N. V., Rocha, L. A., Gomes, J. R., Alagirusamy, R., Deopura, B. L., & Fangueiro, R. (2007). Tribological properties of the directionally oriented warp knit GFRP composites. *Wear, 263*(7-12), 930–938. doi:10.1016/j.wear.2006.12.001

Meijer, H. E. H., & Govaert, L. E. (2005). Mechanical performance of polymer systems: The relation between structure and properties. *Progress in Polymer Science (Oxford), 30*(8–9), 915–938. doi:10.1016/j.progpolymsci.2005.06.009

Menachem, L. (2007). Handbook of Fiber Chemistry. In Encyclopedic Dictionary of Polymers (3rd ed.). CRC Press. doi:10.1007/978-0-387-30160-0_2928

Mohan, N., Natarajan, S., Kumaresh Babu, S. P., & Siddaramaiah. (2011). Investigation on sliding wear behaviour and mechanical properties of jatropha oil cake-filled glass-epoxy composites. *Journal of the American Oil Chemists' Society, 88*(1), 111–117. doi:10.100711746-010-1657-z

Ning, F., Cong, W., Qiu, J., Wei, J., & Wang, S. (2015). Additive manufacturing of carbon fiber reinforced thermoplastic composites using fused deposition modeling. *Composites. Part B, Engineering, 80*(1), 369–378. doi:10.1016/j.compositesb.2015.06.013

Patnaik, A., & Satapathy, A. (2010). Investigations on Three-Body Abrasive Wear and Mechanical Properties of Particulate Filled Glass Epoxy Composites Composites for wind turbine blade View project Development and selection of biomaterial for hip joint replacement View project. *Researchgate. Net, 5*(2), 37–48.

Quintelier, J., De Baets, P., Samyn, P., & Van Hemelrijck, D. (2006). On the SEM features of glass-polyester composite system subjected to dry sliding wear. *Wear, 261*(7–8), 703–714. doi:10.1016/j.wear.2006.01.006

Sodeifian, G., Ghaseminejad, S., & Yousefi, A. A. (2019). Preparation of polypropylene/short glass fiber composite as Fused Deposition Modeling (FDM) filament. *Results in Physics, 12*(August), 205–222. doi:10.1016/j.rinp.2018.11.065

Standard Terminology for Additive Manufacturing – General Principles – Terminology (ASTM52900), International Organization for Standardization: Geneva, Switzerland. (2015). https://www.iso.org/obp/ui/#iso:std:iso-astm:52900:ed-1:v1:en

Suresha, B., Chandramohan, G., Prakash, J. N., Balusamy, V., & Sankaranarayanasamy, K. (2006). The Role of Fillers on Friction and Slide Wear Characteristics in Glass-Epoxy Composite Systems. *Journal of Minerals & Materials Characterization & Engineering, 05*(01), 87–101. doi:10.4236/jmmce.2006.51006

Suresha, B., Chandramohan, G., Rao, P. R. S., Sampathkumaran, P., Seetharamu, S., & Venkateswarlu, V. (2006). Friction and slide wear characteristics of glass-epoxy and glass-epoxy filled with SiCp composites. *Indian Journal of Engineering and Materials Sciences, 13*(6), 535–541.

Taormina, G., Sciancalepore, C., Messori, M., & Bondioli, F. (2018). 3D printing processes for photocurable polymeric materials: Technologies, materials, and future trends. *Journal of Applied Biomaterials & Functional Materials, 16*(3), 151–160. doi:10.1177/2280800018764770 PMID:29609487

Todor, M. P., & Kiss, I. (2016). Systematic approach on materials selection in the automotive industry for making vehicles lighter, safer and more fuel–efficient. *Applied Engineering Letters, 1*(4), 91–97.

Todor, P., Bulei, C., & Kiss, I. (2017). An Overview on Fiber-Reinforced Composites Used in the automotive industry. *International Journal of Engineering, XV*(2), 181–184.

Van Der Klift, F., Koga, Y., Todoroki, A., Ueda, M., Hirano, Y., & Matsuzaki, R. (2016). 3D Printing of Continuous Carbon Fibre Reinforced Thermo-Plastic (CFRTP) Tensile Test Specimens. *Open Journal of Composite Materials, 06*(01), 18–27. doi:10.4236/ojcm.2016.61003

Wohlers, T., & Gornet, T. (2015). History of additive manufacturing. Wohlers Report 2015.

Yousif, B. F., & El-Tayeb, N. S. M. (2008). Wear and friction characteristics of CGRP composite under wet contact condition using two different test techniques. *Wear, 265*(5–6), 856–864. doi:10.1016/j.wear.2008.01.016

Yuanjian, T., & Isaac, D. H. (2008). Combined impact and fatigue of glass fiber reinforced composites. *Composites. Part B, Engineering*, *39*(3), 505–512. doi:10.1016/j.compositesb.2007.03.005

Zhong, W., Li, F., Zhang, Z., Song, L., & Li, Z. (2001). Short fiber reinforced composites for fused deposition modeling. *Materials Science and Engineering A*, *301*(2), 125–130. doi:10.1016/S0921-5093(00)01810-4

Chapter 5
Fused Deposition Modelling of Polylactic Acid (PLA)-Based Polymer Composites: A Case Study

Murugesan Palaniappan
College of Engineering, Imam Mohammad Ibn Saud Islamic University, Saudi Arabia

Satyanarayana Tirlangi
https://orcid.org/0000-0001-8265-8743
Visakha Institute of Engineering and Technology, India

M. Jinnah Sheik Mohamed
Nellai College of Engineering, India

R. M. Sathiya Moorthy
PERI Institute of Technology, India

Srinivas Viswanth Valeti
Aditya Engineering College, India

Sampath Boopathi
https://orcid.org/0000-0002-2065-6539
Muthayammal Engineering College, India

ABSTRACT

In this chapter, a case study on polymer composites made using the fused deposition modelling (FDM) process is exemplified. The fundamentals of the additive manufacturing process, such as stages, applications, classifications, disposal methods, material selection, general principles of FDM, and selection process parameters, have been explained. The case study focuses on creating practical four-dimensional feedstock filament prototypes out of polylactic acid (PLA), polyvinyl chloride (PVC), wood powder,

DOI: 10.4018/978-1-6684-6009-2.ch005

and Fe3O4 powder for making quality FDM printed polymer composites. The procedure for making composites using the best combination of FDM process parameters has been illustrated using real-time experimental approaches.

INTRODUCTION

Additive manufacturing, also known as 3D printing, is a new technique that will revolutionise the way that products are designed and made. Products are manufactured in three dimensions using a method of mass production known as "3D printing." Due to the setup's minimal cost, additional adjustments are simple to implement. Numerous industries, including agriculture, the automotive industry, aerospace, and healthcare, have used 3D printing. In the manufacturing sector, this technology has developed into a more adaptable and potent method (Khan & Shaikh, 2014). The effectiveness and potential of additive manufacturing technologies demonstrate superior outcomes over traditional production methods.

Figure 1. Various Stages of Additive Manufacturing Process

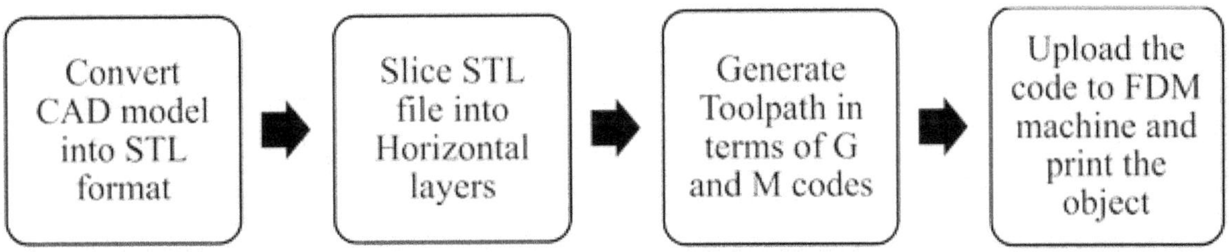

The process of creating objects via 3D printing is known as additive manufacturing (AM). The various stages of additive manufacturing process are shown in Figure 1. The fusing of layers and the materials used in pattern creation are used to categorise AM procedures. Binder jetting, material extrusion, sheet lamination, powder bed fusion, and vat photopolymerization are a few of the important procedures. AM is a process of digital mass-production techniques that has revolutionised the industrial sector. It uses 3D CAD models for manufacturing, which eliminates the need for any additional tools. AM has changed the way that items are made so that they are both sturdy and light. One of the low-cost AM processes for a variety of thermoplastics and thermoplastic composites has received a lot of attention is fused deposition modelling (FDM). For FDM applications, a wide variety of polymeric materials, including polylactic acid (PLA), acrylonitrile butadiene styrene (ABS), polyamide (PA6), and polyvinyl chloride, are accessible. One of the common materials for FDM applications is PLA. Numerous research organisations have thoroughly investigated biocompatibility problems. For the purpose of creating multi/hybrid composites that support 4D applications, PLA was reinforced with PVC, wood powder, and magnetite powder (Fe_3O_4) (Mazurchevici et al., 2020; Testad et al., 2013). When activated by an external magnetic field, the material matrix is intended to develop the ability to self-assemble (as stimulus). The newly created composite matrix's 4D capacity has been added using Fe_3O_4 powder. Twin-screw extrusion was used to generate the material matrix as feedstock filament, which was then used in an FDM setup for 3D printing. The classification of additive manufacturing processes is illustrated in Figure 2.

Figure 2. Classification of additive manufacturing processes

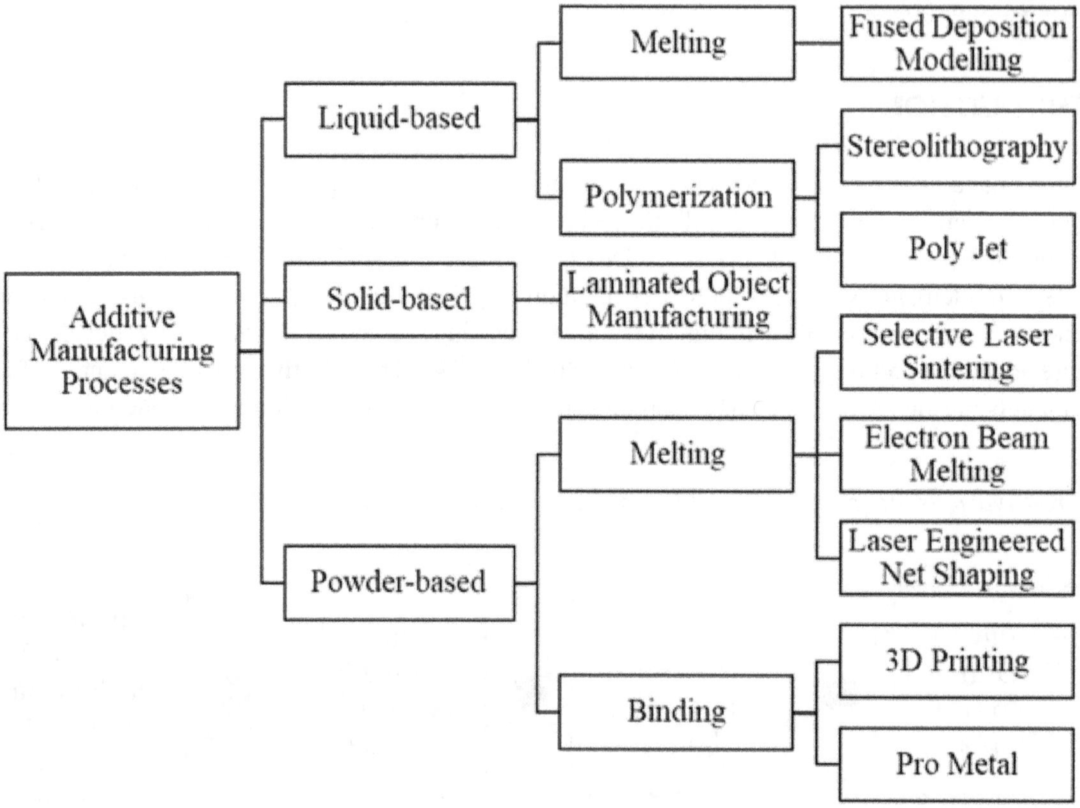

Time-related characteristics in 3D-printed items lead to 4D behaviour. It is possible to design FDM, an additive manufacturing process for printing things, to behave like meta-materials. These are materials that, because of their shape, geometry, orientation angle, or joint angles, have unique properties. External magnetic fields, such as those produced by magnetic fields, can have an impact on these composites' properties. Depending on the complexity of the structure and the strain rate, 3D printed shape memory polymer (SMP) reacts to diverse environmental stimuli. Due to its biocompatibility, it opens up a new area of application where a stable composite is produced by combining inorganic and organic materials.

Net-shaped objects created by 3D printing utilising multiple printing techniques and various metals or powders have the potential to exhibit 4D behaviour. Magnetic Ni-Mn-Ga alloy that can be educated to exhibit SMP behaviour. A significant factor in the rate of expansion and contraction overstrain is 3D shape produced on FDM. The metal composite part created using additive manufacturing technology and cured by certain post-treatment receives the SME from the martensitic transition. The alloy's composition is essential for the desired qualities. A suitable outcome is provided to the object by post-processing composite produced in three dimensions. SMP can be programmed to react to a variety of environmental cues, including pressure, temperature, heat, and pH. A 4D property is also provided by hydrogel in response to osmotic pressure (Mazzanti et al., 2019).

There are two varieties of SMP, one of which contains some innate characteristics that react to outside stimuli. The other has mechanical training for varied applications to external situations. A prime example of time-responsive materials is scaffolds manufactured for biological applications. When applying osmotic pressure, SMP can also be employed to obtain shape-changing behaviour. To achieve the desired configuration, many piezoelectric smart materials have been supplemented with a polymer matrix. The L/R ratio of the nanotubes is crucial to the behaviour of the composite. Additionally, offering self-healing qualities, shape memory polymers are essential for soft actuators. Smart materials used in 4D printing are well suited for use in the medical industry as scaffolds, drug delivery systems, autonomous actuators, heat-responsive tissues, etc. A variety of materials with 4D properties are printed.

The Miura origami structure is well-known for having 4D features based on its geometry. Typically, a fold or edge pattern determines how they behave. Meta-materials' structural characteristics, such as joint angles and edge lengths, are what give them their intelligence. With some magnetic material, such as Fe_3O_4, vulcanised rubber with elastomeric properties can be employed. It is possible to print the matrix produced by composites utilising additive manufacturing methods. According to experts at the University of Bristol, it will be useful in static applications as well as electronics. Numerous applications have made substantial use of shape memory metals and polymers. With these materials, different stimuli are used depending on how correctly they respond.

ADDITIVE MANUFACTURING

Different additive manufacturing techniques exist, and they can be divided into groups according on how they function, what materials they use, and how they are put together. Laminated object manufacturing (LOM), selective laser sintering, selective laser melting, stereolithography, and ultrasonic additive manufacturing (UAM) are some of the procedures.

Digitally controlled and automated AM techniques require little human intervention. The first method to show 3D printing using CAD data was the stereolithography apparatus (SLA). AM is a crucial technique that is used to create usable prototypes in a short amount of time. The pieces are produced digitally without using any tools, which reduces the time requirements. A quick production method that lessens supply chain complexity is 3D printing.

The automotive, medical, aerospace, and construction sectors have changed how they view additive manufacturing. In more recent years, AM has also been employed to create durable, affordable biomedical implants. AM can be used to create a variety of pharmaceutical and medical products. It has been employed in the pharmaceutical sector to produce oral medications, implants with hard or soft structures, transdermal medications, therapeutic instruments, prosthetics, diagnostic tools, etc. AM is also used in the marine industry.

FUSED DEPOSITION MODELING

Schematic setup of FDM process is illustrated in Figure 3. In this style of printing, the chosen goods are consistently dismissed. ABS and/or polylactic acid (PLA) are used in the FDM method to fabricate the parts. The industry considers this technique appropriate since it allows for the production of parts with complex critical forms without the use of specialised tooling. The fabrication tool path is generated by the

FDM machine software after analysis of the geometry. According to the tool path, the heated nozzle head moves in the X and Y directions while semi-molten plastic beads are ejected and incredibly thin layers are essentially established on the base. The Extrusion Manufacturing Device is employed as a primary method of producing goods. Through the use of this tooling technology, the process has advanced and become more effective in terms of time savings. The investigation is being conducted on a variety of materials, including metal powders, abrasives, polymers, and ceramics. Real-time manufacturing (RT) allows for up to a 50%-time reduction over traditional manufacturing methods. Numerous post-processing methods have also been developed to increase efficiency. The items' quality and dimensional accuracy can both be enhanced via RT (Wickramasinghe et al., 2020).

Figure 3. Schematic setup of FDM Process

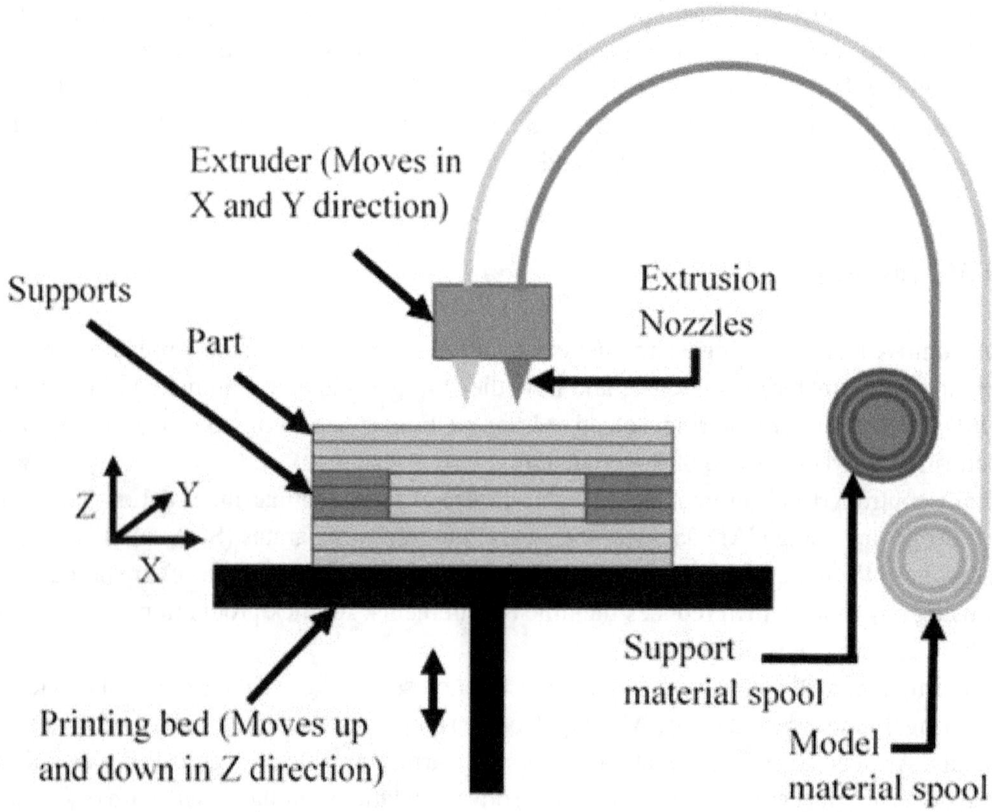

The aerospace sector, automobile and interior design components, and implant and medical applications have been the most widespread uses. The aerospace and automobile industries have long employed fused deposition modelling (FDM). FDM was also used in the production of the intake manifold systems for SAE Formula vehicles. Additionally, the electrical sector uses FDM to create circuit boards. The FDM material has been employed in the biomedical industry due to its biocompatible characteristics. FDM is also used in complex constructions, as is evident. Utilizing stereolithography, they use ABS as well as thermoplastics.

Fused deposition modelling (FDM) is now frequently used to create 3D scaffolds and biomaterials used in tissue engineering. On a biodegradable substance with possible use in tissue engineering, many studies are being done. FDM and electrospinning were coupled to create a coronary artery bypass graft. Additionally, the 3D PCL-TCP mesh for the drug delivery system was made using the FDM technique. The aerospace sector also makes use of RP in some way. With the use of 3D printing, aeroplane prototype models are also produced. Additionally, a change in composite metal and polymer filaments has been seen. When creating models and prototypes for the automotive industry, fused deposition modelling (FDM) is a popular technique (Syrlybayev et al., 2021).

These are used to create wires and wire channels. Using ink-jet and micro-nozzle technology, conductors and insulators can both be produced. Transistors and polymer-based semiconductors are two examples of such products in the electronic industry. FDM is often used during the manufacturing and formulation of foods. A reliable supply of conventional toys is now available. For the toy industry, materials with the ability to eradicate infections are highly desired. They employ a variety of techniques, including SLA, SLS, and LOM, but mostly extrusion-based techniques like FFF and FDM. The advantages and disadvantages of FDM process is displayed in Figure 4.

Figure 4. Advantages and disadvantages of FDM Process

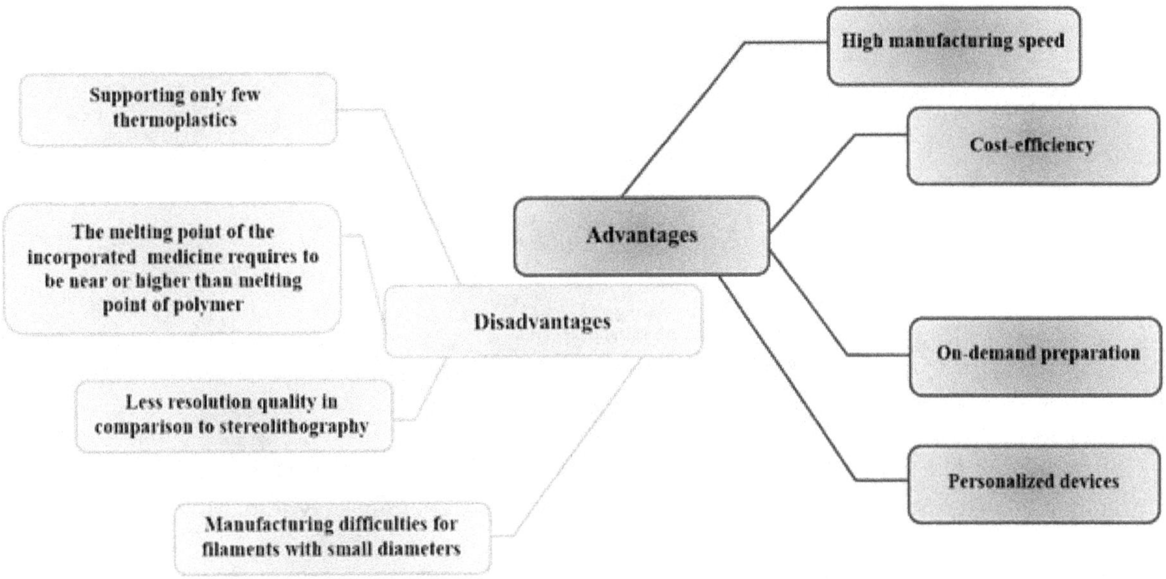

PARAMETERS OF FDM

The various process parameters for the FDM process is demonstrated in the Figure 5. On various FDM levels, three different types of parameters are taken into account to obtain effective and ideal results. The terms "layer thickness," "air gap," and "part orientation" refer to these variables. Raster angle (in degrees) and width (in mm) and their relationships. The FDM is a method based on topic matter elimination. This approach combines the banishing process with computer-aided design (CAD), material science, and computer numerical control. The initial step in creating a product is solid modelling, which must be done using 3D CAD software. The outcomes of FDM processing are continually being improved and refined by researchers. Model tip size, slice height, raster width, model build temperature, model raster angle, part fill style, and part interior style are some of the parameters that must be changed during the pre-processing of the STL (Comb et al., 1994). These parameters have been optimized by taguchi, response surface method, and random search optimization methods (Boopathi, 2019, 2021; Sampath et al., 2021).

Figure 5. Parameters used in FDM process

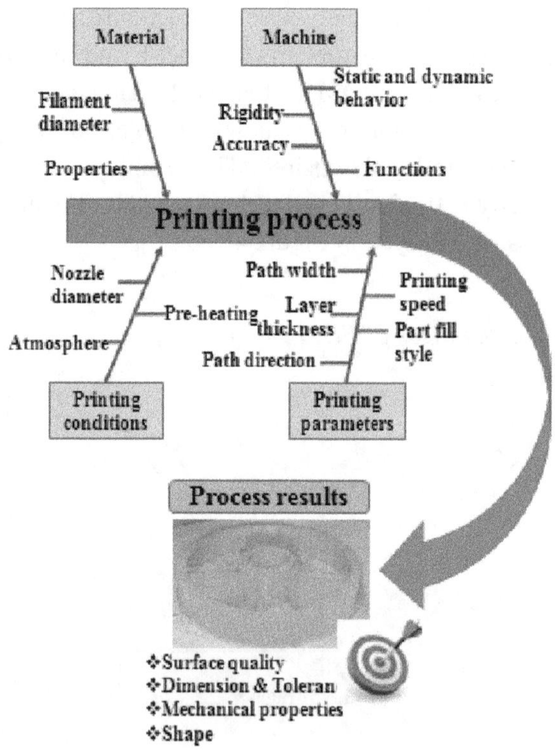

Slice height: The thickness of the layers during part production is referred to as this. This can also be thought of as the head moving in z-directions relative to the coordinate system. More time is needed for thinner materials.

Model tip size: The size of the extrusion tip of the tip. Through this tip, the material is extruded out. The diameter of that tip is known as the Model tip size. Model build temperature: The liquefier's internal temperature is governed by the heating element's temperature. (Raster width: This refers to the size of the model tip to the layer. Modelling on the CAD software, pre-processing, fabrication of components on an FDM machine, and removal of aid from manufactured materials are the four steps of FDM. Infill pattern, infill density, printing speed, extrusion temperature, and air gap are additional factors. Numerous studies have been conducted recently in the area of optimising parameters for increased effectiveness. To increase effectiveness, several studies on variables like nozzle diameter, extrusion temperature, velocity, layer thickness, etc. were carried out. The overall change in the tensile strength, construction time, and surface roughness was the result of these factors.

SELECTION OF MATERIALS FOR FUSED DEPOSITION MODELING

AM, also known as layered manufacturing, rapid prototyping, and solid freeform manufacturing, is a method of plastic forming. Depending on the material, size, and needs of the manufactured parts, several machining processes, such as drilling, milling, grinding, and turning, are used.

Materials are added layer by layer to create products in additive manufacturing. For AM, various types of materials are utilised. On the basis of their complexity and shape, 3D CAD designs are divided into smaller chunks. The machine then employs these segments in an additive manufacturing process to create the part. In addition to other alloys, polymers, and metals are frequently utilised in additive manufacturing. Combinations of different alloys with bioactive glasses are frequently used in the AM process and techniques. According to experts at the Massachusetts Institute of Technology (MIT) in the US, chemical vapours can effectively dissolve the top layers of parts, transforming the solid layers into a semi-molten condition that allows them to be machined into ABS 3D printed parts. ABS material can withstand temperatures between -40 °C and 100 °C, which offers a suitable range for a variety of procedures and uses. In general, the ABS material is chosen for its impact resistance and toughness. The majority of industries use ABS materials, including those in the automobile, aerospace, toy, and electrical housing sectors. Preservative assembly, another name for 3D printing, was invented in the 1980s. Fused deposition modelling is one method that can be used to create 3D objects. An FDM printer uses computer-aided design (CAD) files to produce goods. The FDM technique is used as a manufacturing procedure in almost all industrial industries. FDM is made of a variety of thermoplastics, resins, polymers, alloys, and metals. Various industries also use polymer composites for the same purposes. The bulk work has been finished. using military aircraft engines from the FDM Aerospace Industry as UAV propellers UGV Rover Chassis Prototype Electronics for the Automotive Industry Sensors PCBM Manufacturing of Unrelated Food Manufacturing of toys, interior design, and medical implants and tissue engineering in Biomedicine Medical Devices Propellers for the maritime industry used in surgery ABS-based models for prototypes. ABS material is more usable and ideal due to its quality. They are mostly used in the aerospace sector. When choosing FDM, factors such as corrosion resistance, high-grade strength, high machinability, resistance to heat and chemical dampness, and a variety of colour possibilities are crucial.

The description of the other FDM based materials is as: The most popular FDM material, ABS-plus Thermoplastics, is 40% stronger than regular ABS polymers and is environmentally stable. Thermoplastics made of ABS are translucent and have excellent mechanical qualities. Wherever the need for light transmission flow exists, they are frequently used. (ABS-M30 Thermoplastic materials recover from impacts and have a high tensile strength. The layers have a strong relationship. They outperform other ABS materials in terms of strength by 25% to 75%. (ULTEM-9085: They are used in the marine and aerospace industries. They possess flame-retardant qualities. (ABS-M30i Thermoplastics: Due of their great strength, they are used in the medical, pharmaceutical, and food packaging industries. Additionally, this substance is sterilisable. (PPSF (poly-phenyl sulfone) Thermoplastic: Because of its high thermal conductivity, the biomedical, automotive, and aerospace sectors. With the aid of fibre-optic manganese disodium diolamine, conductive Materials and capacitive sensors that are used in biomedical devices, human interface devices, etc. are manufactured. Electrical components can also be made out of polycarbonate. These components are combined with copper mesh and nickel alloy. Ultem 1010 and polyamide are other materials that are utilised. Fused deposition modelling makes heavy use of polylactic acid (PLA). Applications for PDLA, high impact polystyrene, and high-density polyethylene vary. Polymer-based FDM materials include acrylonitrile styrene acrylate (ASA), polyetherimide, and nylon 12(Chavan et al., 2017; Dey & Yodo, 2019; Lin et al., 2018; Norani et al., 2021). The materials of FDM processed parts are used in the food business so that they can be sanitised. The autoclave, ethylene oxide, hydrogen peroxide, and the other four methods were used for sterilization.

Other materials include: A thermoplastic polyurethane with excellent toughness and abrasion resistance is called a TPU 92A elastomer. They are applied to flexible hose, tube, and air duct applications, among others. Antero 840CN03 (PEKK) has electrostatic dissipative and mechanical qualities that make it ideal for use in industrial and space applications. Antero 800NA is chemical resistant and has a great strength to weight ratio. They are used in metal forming tools, aerospace, and aircraft. A nylon polymer bonded with carbon fibre is called Nylon 12CF. They are the stiffest and have the highest strength-to-weight ratio. used for replacing metal and applying tooling. High tensile and flexural strength, rigidity, and biocompatibility characterise PC-ISO, a material. used in the production of medical equipment, food, and medicine packaging.

CONVENTIONAL PROCESSING OF FDM PARTS

Mechanical post processing is the use of items through machining processes such as turning, milling, grinding, drilling, and so on. The materials melt and break as a result of the extreme heat produced by conventional machining. If the surrounding circumstances and environment are maintained properly, traditional machining can be done on the materials. Chips show that the material has undergone plastic deformation. In machining, the tool's edge is heated, which transfers heat and melts the material. The machine's geometry and machining processes also contribute to the product improvement brought on by post processing. Because thermoplastic polymers are recyclable, they must first be heated and melted to return to their original state before being utilised.

The use of polymer-based materials has greatly increased as a result of 3D printing. FDM combines a 3DPB variant with traditional machining techniques for the following step. Some of the more important procedures, such as machining and injection moulding, are included in the post-processing techniques. The characteristics of conventional machining depend on a number of variables, including an understanding of the machine and the method used to create it. Recent years have seen a rise in inspection of many technologies, including vibration assisted machining, micromachining, turning, milling, boring, and drilling.

NON-CONVENTIONAL PROCESSING OF FDM PARTS

Traditional machining caused a number of problems for ABS polymers. Innovations have been developed to use a number of unconventional methods. Electro discharge machining (EDM), vapour finishing, and abrasive water jet (AWJ) are among the techniques used. Chemical procedures, such as chemical treatment and coating, are used in post-processing techniques to improve surfaces. Since laser cutting is a no-contact method, there is no need to physically link the tool to the project. Because there is no contact, the heat and friction produced by cutting the pieces result in no tool wear. The way we manufacture goods in India is changing as a result of laser cutting and other unconventional manufacturing methods. It speeds up manufacturing and increases machine productivity (Camposeco-Negrete, 2020; Lagoda & Gabor, 2018). In order to increase performance and lower costs, manual painting was the first type of chemical processing carried out in the UK. This method was specifically created for parts whose surface briefly reflows after being exposed to vapours. Sustainable development seeks to produce effective and improved outcomes, and unconventional machining techniques are achieving these goals. Non-traditional machining can always be combined with the use of renewable resources, recyclable materials, non-hazardous materials, etc. There are various procedures.

Processes like the applied voltage, electrolyte flow rate, feed rate, etc. are examples of non-traditional parameters. Thermal, chemical, and electrical energy are the main energy types used in unconventional processes. A variety of chemical machining procedures are currently used to increase the quality of FDM-produced products.

It was discovered after doing a literature review that ABS thermoplastic material is widely employed in 3D printing applications. Additionally, the higher layers of parts can be effectively disintegrated by chemical vapours, converting the solid layers into a semi-molten condition. After cooling, this caused the surface to reflow and the layers to resettle, creating a smooth surface. However, there isn't much research available that reports using chemical vapours to machine ABS parts. Additionally, there is less information available about the effects of unconventional machining methods on the surface properties and dimensional accuracy of polymers. Additionally, because the material used to make FDM parts is brittle, using traditional machining techniques results in the loss of complex details in the pieces.

RECYCLING OF ADDITIVE MANUFACTURING PARTS

The use of plastic components made via additive manufacturing (AM) is harmful to the environment in numerous ways. AM makes use of a highly machinable thermoplastic polymer material. Due to the production and disposal of AM components, our soil and groundwater are becoming poisoned. The European Commission was established with the goal of using only recyclable plastic by the year 2030. Different types of machining and treatment are used in the mechanical recycling of plastic parts. Due to their compatibility with attributes including corrosion resistance, low density, and high strength lathe machining, thermoplastics have been employed. Recovery is the gathering of waste products from various sources that can be recycled. For effective recycling, the materials must be collected and separated properly. The second stage is preparation, during which the recyclable materials are handled according to certain rules and procedures. There may be a number of ways to recycle materials, as well as a number of other ways to process them. One of the final processes in the recycling process is compounding, which involves disassembling materials into smaller parts like monomers and composite materials (Naveed, 2021; Yan & Gu, 1996).

Waste has increased along with the use of RP and FDM-based production. Plastic serves as the primary component of the majority of the materials used in FDM. A lot of plastic is released during post-processing, which is done after production. AM 3D printing techniques are harmful to both the environment and the user's health. Plastic waste from AM 3D printers is dumped into the environment and turns into a food source. The circular economy follows the principles of resource conservation, reuse, and reduction. The management of such trash is an urgent necessity for businesses and manufacturers. Most plastic wastes (PET, PP, and PS) are blends of polymers.

Recycling of FDM-produced parts necessitates a thorough understanding of the effects on the material's quality and usefulness. Today, FDM and other AM manufacturing techniques take into account the Plan-Do-Check-Act (PDCA) methodology from life cycle assessment (LCA). With the aid of recycling, additive manufacturing (AM) and FDM hold the future for sustainable production and a circular economy. Less than 10% of the items produced using FDM are recyclable. Numerous studies have demonstrated the superiority and biodegradability of materials like PLA plastic over the more widely used ABS.

The accumulation of ABS plastic and its effects on the climate and environment are shown by sustainability assessments and life cycle assessments. Although ABS material has been claimed to be biocompatible, other materials perform better. Since ABS plastic cannot decompose, recycling is required. One of the key elements to consider is biocompatibility. The majority of recycled ABS plastic is not biodegradable or biocompatible. Thermal recycling, materials recycling, and chemical recycling are a few of the more popular recycling methods.

CASE STUDY

To mechanically combine with TSE FDM feedstock filament made of PLA, PVC, Fe_3O_4, and wood powder for a commercial open-source 3D printer. Rheological and thermal analysis of TSE feedstock using the melt flow index (MFI) and differential scanning calorimetry (DSC), respectively. For the purpose of choosing the best feedstock filament for 3D printing, feedstock filament wire must undergo parametric optimization for its mechanical, morphological, and magnetic properties. Utilizing prepared feedstock filaments to print functional prototypes on FDM while doing parametric optimization for mechanical, morphological, thermal, and magnetic qualities. Various Stages in the case study of the case study is shown in Figure 6.

Different Stages of Work

The study focuses on creating practical 4D feedstock filament prototypes out of PLA, PVC, wood powder, and Fe_3O_4 powder. There are five sections to this study:

Stage 1: For feedstock filament based on hybrid blend (PLA/PVC/wood powder/Fe3O4 powder in a single composite), TSE method was parametrically optimised for mechanical, morphological, rheological, thermal, and magnetic properties.

Stage 2 Four distinct feedstock filaments—PLA, PLA/PVC, PLA/wood powder, and PLA/Fe_3O_4 powder—were individually investigated in this multi-material matrix for parametric optimization of the TSE process for mechanical, morphological, rheological, and thermal feedstock filament.

Stage 3 Creating functional prototypes for tensile and flexural 3D printing based on ASTM D638 Type IV and ATM D790, respectively, and fine-tuning the 3D printing process parameters for mechanical, thermal, and morphological properties for a hybrid functional prototype made of stage 1 feedstock filament based on a hybrid blend.

Stage 4 A multi-material functional prototype constructed of the multi-material feedstock from stage 2 is developed using 3D printing in accordance with ASTM D638 Type IV and ATM D790, respectively, and the 3D printing process parameter is optimised for mechanical, thermal, and morphological qualities.

Stage 5 In order to properly examine the material matrix life up to a 50% drop in mechanical qualities, the material matrix chosen on stages 1 and 2 was once again processed for the recycling issue.

Figure 6. Various Stages in the case study

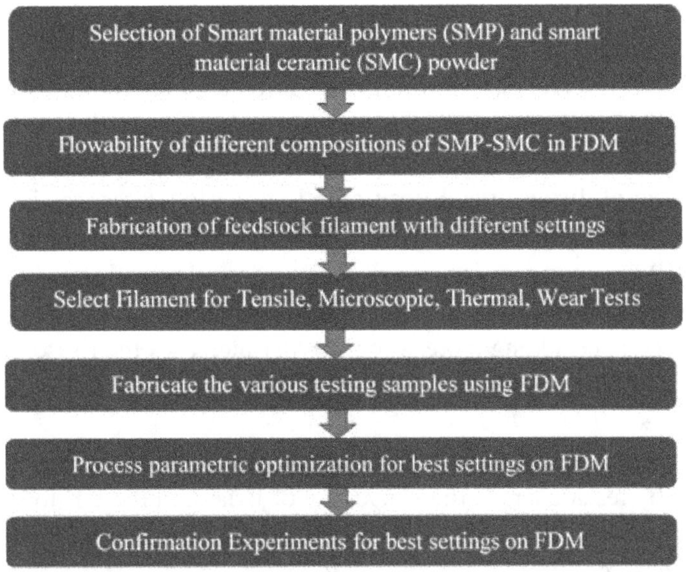

Selection of Smart Material Polymers (SMP) And Smart Material Ceramic (SMC) Powder

Sieve shaking of wood powder: Waste wood dust was gathered from a nearby factory and was made up of a variety of different-sized pieces of wood. Therefore, it was necessary to separate wood dust that was between 44 and 60 microns in size in order to reduce compatibility concerns during extrusion and 3D printing using FDM. To lessen the likelihood of nozzle choking, alternative sizes of wood particle reinforcement in the material matrix for the TSE and FDM printers have been created. The microscopic structure of fine wood dust (50 microns) recovered after sieving is substantially different from the microscopic structure of wood dust produced by burning wood (Choi & Samavedam, 2002; Sunny et al., 2019).

Mechanical blending of PLA-PVC-wood dust-Fe_3O_4 powder: When making the blend, coconut oil was utilised as a binder to help wood dust and Fe_3O_4 powder adhere to the surface of the PLA-PVC material. Based on commercial field practises, coconut oil has been employed as a substitute binding agent.

Flowability Testing for SMP-SMC in FDM

Vibration sample magnetometer testing (VSM): By applying a 1 Tesla external magnetic field to a VSM setup (Make: Lakeshore 7404 VSM), all samples of feedstock filaments of wire produced by the TSE machine were examined for their magnetic characteristics. It was possible to acquire optimal extrusion circumstances by observing the various magnetic properties, such as magnetization, coercivity, and retentivity.

DSC analysis: Using a DSC setup (Model: Mettler, Toledo) for 3 heating cycles and 3 cooling cycles, thermal analysis for the best and worst samples according to mechanical attributes was carried out. Temperatures ranged from 30 °C to 250 °C during the heating cycle and from 250 °C to 30 °C during the cooling cycle.

Morphological properties testing: The cracked surface and porosity of the feedstock samples were analysed using metallurgical image analysis software while the samples were being examined under a microscope.

Process capability analysis for magnetic, surface hardness, and porosity of feedstock: The filaments were tested for VSM, process capabilities, and surface characteristics (such surface hardness and porosity).

Mechanical blending of different compositions: In order to address the issue of several layers of various filaments sticking to one another with high adhesion, PLA has been chosen as the basic polymer. Separate additions of PVC, wood powder, and Fe_3O_4 powder were made to PLA. The PVC content of PLA has varied from 10% to 25% by weight. A 2.5–10 weight percentage of wood powder has been added to the PLA matrix since adding more wood powder increases the risk of clogging the extrusion nozzle. Additionally, a multi-blend (PLA 50 wt., PVC 25 wt., magnetite 20 wt., and wood powder 5 wt.%) has been developed.

FDM printing of tensile, flexural, and circular disc: For the first time, tensile and flexural specimens were 3D printed by researchers at the University of Bristol in the UK to study their mechanical, morphological, pullout, and other characteristics. The samples were printed under controlled circumstances in accordance with ASTM D638 Type IV and D790 standard circumstances.

Conducting Various Tests

Mechanical testing on UTM: A universal tensile tester (UTM) with a 5000N capability was used to tensile test all 3D printed functioning prototypes. A shore D tester was used to examine the materials' Shore D hardness.

Thermal conductivity test: A steam chamber, a lee disc, heating apparatus, a pot for producing steam, a conduit to connect the pot with the stem chamber, and temperature sensors made up a lee disc apparatus.

Circular compression testing on UTM setup: To evaluate the disc's performance under high compression and high stress circumstances, a specifically designed circular fixture was 3D printed. The circular disc underwent compressive testing while being processed using the UTM setup (Shanta Engineering, India) with a 5000N capacity. The modulus of toughness of the printed discs was determined using the stress vs. strain plot.

Morphological testing: Porosity measurements were performed on samples following the ASTM B276 standard, and microstructure graphs were created using a toolmaker microscope.

Multi-material printing using FDM: A 3D printer that can manufacture tensile specimens in accordance with ASTM D638 Type IV has been created by researchers at the University of Bristol in the UK. Three steps of 3D printing are used in the procedure, along with the pre-fabrication of materials for four distinct materials in the needed diameter.

Six layers make up the standard tensile specimen: one of virgin PLA, two of each of wood powder-, PVC-, and Fe_3O_4-reinforced PLA.

Mechanical testing and Morphological testing: Using a UTM machine (Make: Shanta Engg., India) with a 5000 N capacity, composite materials were manufactured and tested for their flexural, tensile, and pullout properties. Diagrams of stress vs strain were drawn so that the material's mechanics could also be discussed. Hardness and cracked surface characterisation have been carried out using the Shore D tester and tool maker's microscope. Additionally, analysis of variance (ANOVA) has been performed using the statistical software programme MiniTab 17 to optimise mechanical properties.

Composite Preparation

The composites were created utilising the mechanical mixing technique, in which a binder was used to bind polymeric PLA grains to wood, Fe_3O_4 powder, and PVC (Figure 7). In this experiment, a mixture of polyethylene (PLA) sheets and wood powder (50 m) was baked for 30 minutes at a preheated temperature of 45 °C. The weight percentage approach was utilised to combine all of the reinforcements (PVC, wood powder, and Fe_3O_4 powder) that were used in the matrix. The desired length can be produced using about 50 g of polyethylene terephthalate (PLA), which can also be recycled for additional re-extrusion cycles. Prior trial experiments showed that 50 g or so was sufficient to produce recyclable materials and feedstock with the necessary length (Camposeco-Negrete, 2020; Equbal et al., 2011; Raffic et al., 2019).

Figure 7. Polymeric matrix (a) PLA, (b) PLA/wood, (c) PLA/PVC, (d) PLA/Fe_3O_4, (e) hybrid blend of PLA/PVC/wood/Fe_3O_4 powder

Porosity testing and fractured surface analysis

The metallurgical image analysis software (MIAS) package tool was used to analyse images of fractured samples obtained under the microscope for porosity holes using testing standards ASTM B276.1 and ISO 9001:2008, as published by the US National Institute of Standards and Technology (NIST). Porosity holes, which are pores in the inner workings of composite materials, are crucial to the materials' mechanical strength.

RESULTS INTERPRETATION

The only important characteristic for TSE processing is screw temperature, while all the other factors together only contributed 2%. For the BMW i3, optimised values for a variety of mechanical characteristics, such as peak load, breaking load, break elongation, strength at peak, and strength at break, have been provided. Shore D hardness is represented by the following values, in order: 41.97N, 4.43mm, 40.36N, 4.81mm, 18.03 MPa, 16.43 MPa, 5.89, 7.32, and 66.60 at shore D. The high porosity (20.41%) and low surface hardness (58.5 Shore D) of the samples from experiment number 17 suggested a poor composite matrix, which eventually led to poor mechanical characteristics (Boopathi, 2022a; Boopathi, Haribalaji, et al., 2022; Sampath et al., 2021). In contrast, trial number 10 revealed a low porosity value and high hardness (65.8 Shore D), which led to a better surface texture as observed.

After it was discovered that all samples of this metal had a super-paramagnetic nature, which is evident from their hysteresis loop, it was discovered that Fe_3O_4 can be profitably employed for self-assembly applications, according to a study published in the Journal of Organic and Inorganic Chemistry. SEM and EDAX investigations have shown that there is no correlation between the mechanical and magnetic properties (Boopathi, Jeyakumar, et al., 2022; Gunasekaran et al., 2022; Haribalaji et al., 2022; Myilsamy & Sampath, 2021; Sampath & Myilsamy, 2021). Since there aren't many Fe_3O_4 particles on the filament's surface, magnetite particles must have combined with the surface's inner matrix. because the presence of magnetite powder, which is contained within the composite matrix, is the only factor affecting the material's magnetic characteristics. Thermoplastic Extrusion Sleeve, a procedure, has been demonstrated by researchers at the University of Bristol in the United Kingdom to be effective.

According to the study, in order for adequate fusion to occur between succeeding layers, MFI must be kept at its lowest possible value for multi-material printing to be successful. Fe_3O_4 reinforcement in PLA has enhanced break elongation, percentage elongation, and peak elongation while reducing peak load, break load, strength at peak, and strength at break. The matrix's increased Fe_3O_4 composition has led to an improvement in its mechanical properties. PVC has been discovered to be the optimal material for multi-material 3D printing, and successive layers of the 3D printed items can have varying degrees of reinforcement depending on the necessary mechanical qualities. The rheological and mechanical findings have been supported by Shore D hardness and porosity measurements. As demonstrated by the DSC, virgin PLA

With a 100% infill density, a 45° infill angle, and a 70 mm/s infill speed, engineers at the University of Bristol in the UK have issued a paper on how to improve the strength and stability of glass for use in solar photovoltaics. A model based on Taguchi L9 OA may be missing several variables, according to an ANOVA model, which indicates poor predictability because the model was not significant due to error more than 10%. A single variable has less of an impact on the outcome than do input variable interactions, according to the historical data technique. A study published in the British Journal of Crystallography indicated that samples printed with low infill angles and minimal porosity performed mechanically better than samples with maximum porosity. The optimal printing conditions for a 3-D object have been found to be infill density of 100, infill angle of 45°, and infill speed of 90 mm/s. Pull-out samples have a strain capacity of 0.4094 while pull-out samples have a break elongation of 40.12 mm and a strain capacity of 0.8024 for flexural data. Hardness, porosity, and surface roughness are connected to mechanical qualities as determined by morphological testing of flexural and pull-out specimens.

Failure starts at the bottom row in flexural testing because strain rises with thickness and is greatest at the bottom layer. Consequently, it is advised to print at a slight angle. Maximum strength is attained when the number of load-sharing layers rises. Maximum heat conductivity was demonstrated by a sample 9 produced at an infill density of 100%, an infill angle of 90°, and an infill speed of 70 mm/sec. As infill densities rose, it was discovered that the sample's thermal conductivities improved. The results are based on an FDM setup with a 100% infill density, a 90° infill angle, and an infill speed of 70 mm/sec, and they have a 95% degree of confidence. According to a study, 3D printed samples differ in their hardness, fractured surface analysis, and surface roughness depending on the type of surface they are printed on. The substance has demonstrated extremely low electrical conductivity, which is advantageous for both these applications and structural engineering work. It has been noted that the ideal/recommended printing conditions for a circular disc 3D printer include an infill density of 60%, an infill angle of 45°, and an infill speed of 70%. However, the DOE suggests that sample number 1 be considered the ideal situation.

It has never been demonstrated that a printed multi-material specimen can have a peak strength of 46.28 MPa and a break strength of 41.65 MPa. The sample was printed at the greatest density of 100%, with a 45° low infill angle, and at a speed of 90 mm/s for the infill. For applications requiring high toughness, experimental condition 7 may be utilised to print prototypes of the suggested material. The multi-material tensile specimen for sample 7 had the highest modulus of toughness of any sample (4.762 MPa), whereas the first sample had the lowest modulus of toughness (Boopathi, 2022b; Kavitha et al., 2022; Yupapin et al., 2022). A multi-material tensile specimen's measurement of hardness on two surfaces showed that the virgin PLA layer had a higher hardness value.

According to a study, samples with low density had more porosity holes, which caused them to perform less well than samples with high density. According to research from the University of Illinois at Urbana-Champagne and the Swiss Federal Laboratories for Materials Science and Engineering that was published in the journal Applied Crystallography, high roughness of samples printed with low density was also one of the reasons for their poor performance. According to a study, the ideal circumstances for a 3D printed flexible plastic sheet were high infill density, a 45° infill angle, and a 50 mm/s infill speed.

According to the most recent data available from the British Sport Performance Agency, the British para-athlete is ranked as the best in the world for pulling and flexion strength, with a maximum pull-out strength of 18.11 MPa and a maximum flexural strength of 26.92 MPa (BSPA). According to a recent study, 3D printed samples of the multi-material components utilised in the iRobot Robotic Driving System have the highest modulus of toughness (RWDS). RWDS are excellent choices for crash applications because of its high ductility and ductile nature. According to experts at the University of Texas at A&M, high density and low infill angles promote better layer-by-layer material diffusion, which results in reduced void formation and better diffusion between various layers, which improves flexural and pull-out performance (Boopathi et al., 2021, 2023).

According to a study that was published in the Journal of Crystallography, the mechanical qualities of multi-material blended PLA matrix-based 3D printed prototypes are superior to those of hybrid blended PLA and functional prototypes. A multi-material blended PLA matrix-based 3D printer prototype that was 3D printed has demonstrated improved mechanical properties at interfaces where the PLA base polymer has demonstrated a better affinity for various layers or interfaces, leading to the observation of the fusion of one layer into another composite layer. A hybrid blended PLA matrix-based prototype has a worse Ra (about 60% less) than a 3D printed prototype. According to scientists at the University of Texas at A&M in the United States, a better Ra led to improved mechanical performance.

For the third cycle of recycling, most PLA composites lost between 50 and 60 percent of their initial peak strength, with the exception of the PLA/Fe_3O_4 and virgin PLA matrix (which lost only 22 percent and 36 percent of mechanical strength, respectively). The PLA composite matrix won't work for 3D printing after the third recycling cycle. Additionally, it was noted that PLA's ductility decreased and brittle failure occurred as the number of recycling stages increased. With the addition of Fe_3O_4, the PLA matrix had improved ductility, increased recycling life, and maintained a considerable amount of mechanical strength through numerous cycles. In the majority of the cases examined by researchers from the University of Bristol, there were additionally favourable effects on the modulus of toughness for PLA composite matrixes. According to a study by researchers at the University of Bristol in the UK, PLA composites retain between 85 and 90 percent of their mechanical strength for the first recycling (with the exception of PLA/PVC composites) and between 65 and 75 percent of their strength up to the second stage of recycling. The material matrix's composition did not alter as a result of recycling, according to DSC analysis.

SUMMARY

In this chapter, a case study on polymer composites made using the fused deposition modelling (FDM) process is exemplified. The fundamentals of the additive manufacturing process, such as stages, applications, classifications and material selection, are explained. The procedure for making composites using the best combination of FDM process parameters has been illustrated.

REFERENCES

Boopathi, S. (2019). Experimental investigation and parameter analysis of LPG refrigeration system using Taguchi method. In SN Applied Sciences (Vol. 1, Issue 8). doi:10.100742452-019-0925-2

Boopathi, S. (2021). Improving of Green Sand-Mould Quality using Taguchi Technique. *Journal of Engineering Research.* doi:10.36909/jer.14079

Boopathi, S. (2022a). Cryogenically treated and untreated stainless steel grade 317 in sustainable wire electrical discharge machining process: A comparative study. *Environmental Science and Pollution Research International*, 1–10. doi:10.100711356-022-22843-x PMID:36057706

Boopathi, S. (2022b). Performance Improvement of Eco-Friendly Near-Dry Wire-Cut Electrical Discharge Machining Process Using Coconut Oil-Mist Dielectric Fluid. *Journal of Advanced Manufacturing Systems*, 1–20. Advance online publication. doi:10.1142/S0219686723500178

Boopathi, S., Balasubramani, V., Kumar, R. S., & Singh, G. R. (2021). The influence of human hair on kenaf and Grewia fiber-based hybrid natural composite material: An experimental study. *Functional Composites and Structures*, *3*(4), 045011. doi:10.1088/2631-6331/ac3afc

Boopathi, S., Haribalaji, V., Mageswari, M., & Asif, M. M. (2022). Influences of Boron Carbide Particles on the Wear Rate and Tensile Strength of Aa2014 Surface Composite Fabricated By Friction-Stir Processing. *Materiali in Tehnologije*, *56*(3), 263–270. doi:10.17222/mit.2022.409

Boopathi, S., Jeyakumar, M., Singh, G. R., King, F. L., Pandian, M., Subbiah, R., & Haribalaji, V. (2022). An experimental study on friction stir processing of aluminium alloy (AA-2024) and boron nitride (BNp) surface composite. *Materials Today: Proceedings, 59*, 1094–1099. doi:10.1016/j.matpr.2022.02.435

Boopathi, S., Venkatesan, G., & Anton Savio Lewise, K. (2023). Mechanical Properties Analysis of Kenaf–Grewia–Hair Fiber-Reinforced Composite. In *Lecture Notes in Mechanical Engineering* (pp. 101–110). Springer. doi:10.1007/978-981-16-9057-0_11

Camposeco-Negrete, C. (2020). Optimization of printing parameters in fused deposition modeling for improving part quality and process sustainability. *International Journal of Advanced Manufacturing Technology*, *108*(7–8), 2131–2147. doi:10.100700170-020-05555-9

Chavan, S. A., & Anwar, P. M. A., & R, P. S. B. (2017). Review of Effect of Process Parameters on Mechanical Properties of 3D Printed Components. *Internation Journal of Advance Research in Science and Engineerng*, *6*(5), 461–467.

Choi, S. H., & Samavedam, S. (2002). Modelling and optimisation of rapid prototyping. *Computers in Industry*, *47*(1), 39–53. doi:10.1016/S0166-3615(01)00140-3

Comb, J. W., Priedeman, W. R., & Turley, P. W. (1994). FDM Technology Process Improvements. *Proceedings of Solid Freeform Fabrication Symposium*, 42–49.

Dey, A., & Yodo, N. (2019). A systematic survey of FDM process parameter optimization and their influence on part characteristics. *Journal of Manufacturing and Materials Processing*, *3*(3), 64. Advance online publication. doi:10.3390/jmmp3030064

Equbal, A., Sood, A. K., Ohdar, R. K., & Mahapatra, S. S. (2011). Prediction of dimensional accuracy in fused deposition modelling: A fuzzy logic approach. *International Journal of Productivity and Quality Management*, *7*(1), 22–43. doi:10.1504/IJPQM.2011.037730

Gunasekaran, K., Boopathi, S., & Sureshkumar, M. (2022). Analysis of a Cryogenically Cooled Near-Dry Wedm Process Using Different Dielectrics. *Materiali in Tehnologije*, *56*(2), 179–186. doi:10.17222/mit.2022.397

Haribalaji, V., Venkatesan, G., Asif, M. M., Pandian, M., Subbiah, R., & Boopathi, S. (2022). Investigation on corrosion and tensile Characteristics: Friction stir welding of AA7075 and AA2014. *Materials Today: Proceedings*, *66*, 743–748. doi:10.1016/j.matpr.2022.04.037

Kavitha, C., Geetha Malini, P. S., Charan Kantumuchu, V., Manoj Kumar, N., Verma, A., & Boopathi, S. (2022). An experimental study on the hardness and wear rate of carbonitride coated stainless steel. *Materials Today: Proceedings*. Advance online publication. doi:10.1016/j.matpr.2022.09.524

Khan, I., & Shaikh, A. A. (2014). A Review of FDM Based Parts to Act as Rapid Tooling. *International Journal of Modern Engineering Research*, *4*, 59–65.

Lagoda, A., & Gabor, R. (2018, October). Strength parameters of the ABS materials used in 3D printing. *AIP Conference Proceedings*, *2029*, 020036. Advance online publication. doi:10.1063/1.5066498

Lin, P. P. A., Willis, K., & Wilhelm, E. J. (2018). Intelligent 3D printing through optimization of 3D print parameters. *US Patent 10,073,424, 1*(19). https://patents.google.com/patent/US10073424B2/en%0Ahttps://patentimages.storage.googleapis.com/28/1d/07/a0fdacf1cf3db7/US10073424.pdf

Mazurchevici, A. D., Nedelcu, D., & Popa, R. (2020). Additive manufacturing of composite materials by FDM technology: A review. *Indian Journal of Engineering and Materials Sciences*, *27*(2), 179–192. doi:10.56042/ijems.v27i2.45920

Mazzanti, V., Malagutti, L., & Mollica, F. (2019). FDM 3D printing of polymers containing natural fillers: A review of their mechanical properties. *Polymers*, *11*(7), 1094. Advance online publication. doi:10.3390/polym11071094 PMID:31261607

Myilsamy, S., & Sampath, B. (2021). Experimental comparison of near-dry and cryogenically cooled near-dry machining in wire-cut electrical discharge machining processes. *Surface Topography: Metrology and Properties*, *9*(3), 035015. doi:10.1088/2051-672X/ac15e0

Naveed, N. (2021). Investigate the effects of process parameters on material properties and microstructural changes of 3D-printed specimens using fused deposition modelling (FDM). *Materials Technology*, *36*(5), 317–330. doi:10.1080/10667857.2020.1758475

Norani, M. N. M., Abdollah, M. F. Bin, Abdullah, M. I. H. C., Amiruddin, H., Ramli, F. R., & Tamaldin, N. (2021). 3D printing parameters of acrylonitrile butadiene styrene polymer for friction and wear analysis using response surface methodology. *Proceedings of the Institution of Mechanical Engineers, Part J: Journal of Engineering Tribology, 235*(2), 468–477. 10.1177/1350650120925601

Raffic, N. M., Babu, K. G., & Madhan, P. (2019). Application of taguchi's experimental design and range analysis in optimization of FDM printing parameters for PET-G, PLA and HIPS. *International Journal of Scientific and Technology Research*, *8*(9), 891–902.

Sampath, B., & Myilsamy, S. (2021). Experimental investigation of a cryogenically cooled oxygen-mist near-dry wire-cut electrical discharge machining process. *Strojniski Vestnik. Jixie Gongcheng Xuebao*, *67*(6), 322–330. doi:10.5545v-jme.2021.7161

Sampath, B., Naveenkumar, N., Sampathkumar, P., Silambarasan, P., Venkadesh, A., & Sakthivel, M. (2021). Experimental comparative study of banana fiber composite with glass fiber composite material using Taguchi method. *Materials Today: Proceedings*, *49*(5), 1475–1480. doi:10.1016/j.matpr.2021.07.232

Sunny, B. C., Benedict, S., Rajan, M. P., & Srinivas, M. (2019). Impact of printing parameters on energy consumption of 3D printers using IoT cloud architecture. *2019 IEEE 16th India Council International Conference, INDICON 2019 - Symposium Proceedings*, 1–4. 10.1109/INDICON47234.2019.9029069

Syrlybayev, D., Zharylkassyn, B., Seisekulova, A., Akhmetov, M., Perveen, A., & Talamona, D. (2021). Optimisation of strength properties of FDM printed parts—A critical review. *Polymers*, *13*(10), 1587. Advance online publication. doi:10.3390/polym13101587 PMID:34069144

Testad, I., Corbett, A., & Aarsland, D. (2013). ORE Open Research Exeter. *Journal of Cleaner Production*.

Wickramasinghe, S., Do, T., & Tran, P. (2020). FDM-Based 3D printing of polymer and associated composite: A review on mechanical properties, defects and treatments. *Polymers*, *12*(7), 1–42. doi:10.3390/polym12071529 PMID:32664374

Yan, X., & Gu, P. (1996). A review of rapid prototyping technologies and systems. *CAD Computer Aided Design*, *28*(4), 307–318. doi:10.1016/0010-4485(95)00035-6

Yupapin, P., Trabelsi, Y., Nattappan, A., & Boopathi, S. (2022). Performance Improvement of Wire-Cut Electrical Discharge Machining Process Using Cryogenically Treated Super-Conductive State of Monel-K500 Alloy. *Iranian Journal of Science and Technology. Transaction of Mechanical Engineering*. Advance online publication. doi:10.100740997-022-00513-0

Chapter 6
Additive Manufacturing Developments in the Medical Engineering Field

Sampath Boopathi
https://orcid.org/0000-0002-2065-6539
Muthayammal Engineering College, India

Richa Khare
Amity University, India

Jaya Christiyan K. G.
M.S. Ramaiah Institute of Technology, India

T. Vijay Muni
https://orcid.org/0000-0003-1363-7249
Koneru Lakshmaiah Education Foundation, India

Smriti Khare
https://orcid.org/0000-0002-0278-5302
Amity University, India

ABSTRACT

The additive manufacturing technology has been applied in various sectors: the manufacturing of industrial components, toys, medicine, medical-surgical instruments, and tissue engineering sectors. In the tissue engineering field, it has been intensively applied to make biomaterials, organs, and drugs. The fundamental procedures of the additive manufacturing process, the various additive manufacturing techniques, and advanced methods that have been applied in the making and synthesis of organs in the tissue engineering fields have been described. In this chapter, the computer-aided tissue modelling process, different fundamental and advanced biomaterials, and advanced scaffold manufacturing applications in emerging tissue engineering fields have been illustrated.

DOI: 10.4018/978-1-6684-6009-2.ch006

INTRODUCTION

The additive manufacturing method is an emerging technique in various commercial, industrial, and medical applications due to their technical flexibility, use of various materials, ability to make complex profiles, and low manufacturing cost. In contrast, the materials were separated using conventional and unconventional manufacturing methods such as milling, lathe, drilling, and grinding operations, as well as electrical discharge and electrochemical machining processes (Boopathi, 2022b). The pollutants and harmful contaminants have also been emitted during the material removal process as opposed to material addition processes (Boopathi, 2021). In additive manufacturing techniques, the product is made by adding materials layer by layer based on the model developed by the users. The four basic and important procedures of additive manufacturing techniques are: (1) developing three-dimensional (3D) models using computer-aided design (CAD) software packages according to user requirements, (2) slicing the objects; (3), selection of the materials based on the applications; and (4) model printing using a 3D printing machine. The basic components of 3D printing methods are (1) hardware, (2) software, and (3) materials. Materials selection is a very important process for selecting suitable components, slicing methods, 3D printing machines, and location of usage. Recently, additive manufacturing (AM) or three-dimensional printing (3DP) techniques are now used in many biomedical applications, such as the printing of patient-specific parts, devices, specific models, and implantations or organs. The 3D functional tissues and organs are made using biocompatible materials by 3DP techniques, which is called the bioprinting technique. The successful methods of bioprinting organs have been useful to human society in organ transplantation and cancer tissues. Some terminology is needed to understand this chapter by the readers. The materials that have been utilised in biological systems are called biomaterials. Any materials that are fully or partially inserted into the body to replace the existing or to support the organs are called implantation materials. Any material used to replace a limb, organ, or tissue of the body is called a prosthesis. The functional organs of the human body have been replaced by a medical device called an artificial organ (Roopavath & Kalaskar, 2017).

In this chapter, the additive manufacturing/3D printing techniques in the Tissue Engineering field, various materials for making different tissues, possible and advanced additive manufacturing methods in various medical fields, and a variety of medical applications have been elaborated.

3D PRINTING PROCESSES IN TISSUE ENGINEERING

In the tissue engineering process, the live cells, biomolecules, and polymers are properly mixed and form the engineering structure of the tissue in vivo (Figure 1). (Badekila et al., 2021).

Live cells are used to make the 'bio-ink' in which all cells are living in vivo. The bio-molecular building blocks are built by adding bio-ink with polymers for easy flow. It has been used to make tissues and organs. Then organs are built by tissue engineering polymers with the reference of computer-added tissue or organ models. Various additive manufacturing techniques have been used to fabricate bio-molecules in vivo format (Bongiovanni Abel et al., 2020). The computer-aided tissue engineering principles are also illustrated below.

Figure 1. Important Stages in the 3D-Printing Process in Tissue Engineering

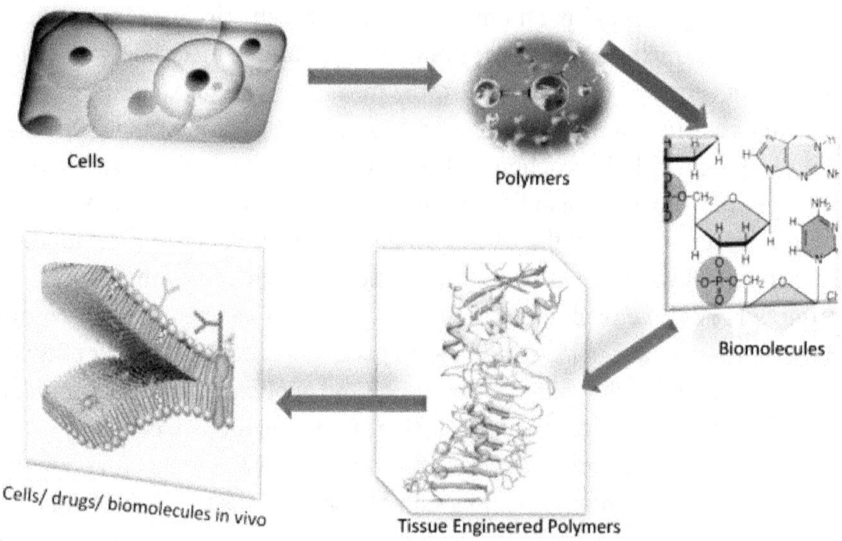

Computer Aided Tissue Engineering (CATE)

The various computer aided modelling, techniques, selection of bio-materials, bio-fabrication constraints, in-vivo validation, and manufacturing processes involved in developing tissues in medical fields are called computer-aided tissue engineering (CATE). There are two important processes in the 3D tissue development process: (1) tissue scaffold design and (2) tissue fabrication and optimization. Computer designing software is employed to analyse the structure and possibilities, and the manufacturing machine, the 3D printer, is involved in fabrication processes. The schematic arrangements of CATE are shown in Figure (2) (Sun et al., 2004).

Figure 2. Schematic Diagram: CATE

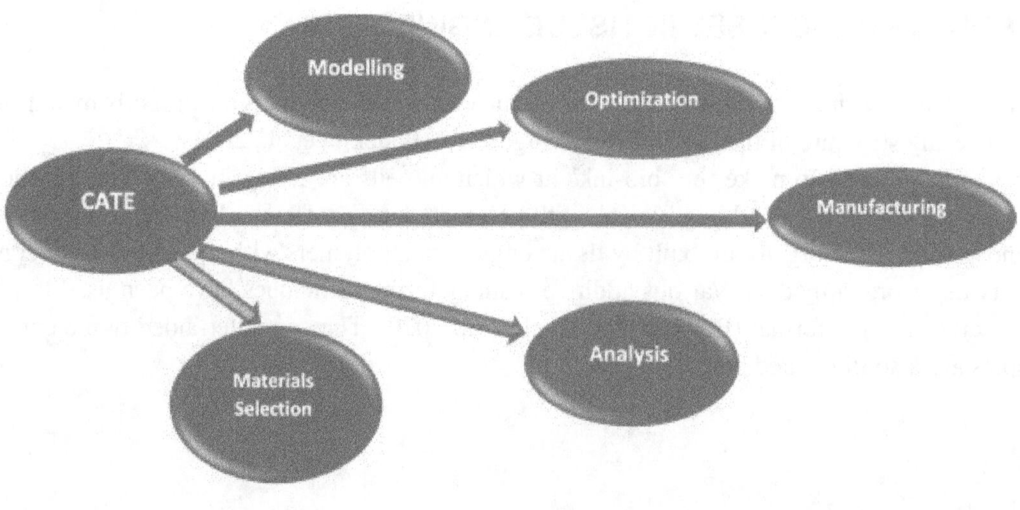

1. **Materials selection and Model Development**

CAD modelling of tissue and organs is defined by shape and skin surfaces by revised engineering principles. The defective patient parts are modelled based on medical imaging techniques. Computed tomography (CT) or magnetic resonance imaging (MRI) techniques are employed in reengineering processes. In the tomograms, cross-sectional images have been developed. In the binarization of grayscale tomograms, the tissue scaffold is designed by existing CAD software (3DSlicer, Materialise MIMICS, and InVesalius). The most popular and well-defined CAD 3D modelling formats are STL (Standard Tessellation Language), HGPL (HP Graphical Language), and CLI (Common Layer Interface). In STL format, the objects are meshes by vertex, edges, and lines on the triangular planes (Giannitelli et al., 2014).

2. **Computer Aided Analysis and Optimization**

In the tissue engineering field, various contradictory features have been characterised using the latest and most advanced modelling software. The lengthy and expensive processes have been performed while designing in-vitro and in vivo tissues using various design variables and constraints. Mechanical properties, such as porosity, stiffness, and biological and chemical activities, are called scaffold characteristics and are affected by different design variables, such as the selection of materials, dimensions, shape, and structure of the scaffold. Recently, finite element methods (FEM) have been applied to select the design variables for the optimum scaffold characteristics (Heljak et al., 2017). New optimization algorithms and artificial intelligence methods have been used to optimize the essential properties in tissue 3D printing processes (Mahmood et al., 2021).

3. **Computer-aided Additive Manufacturing**

After developing the tissue geometries, the 3D models are sliced into small 2D layers for easy fabrication. Many additive manufacturing methods are used to develop the tissue in vivo.

MATERIALS FOR 3D PRINTING IN MEDICINE

The following factors have been considered to select the materials for the medicine.

- Printable,
- Meet the properties after the fabrication process,
- Biocompatible,
- Exhibit tissue biomimicry,
- Form safe degradation by-products,
- Good degradation kinetics

Generally, materials for 3D printing in medicine are classified as metals, polymers, ceramics, and hydrogels. The classification of various materials used in the medical field is explained as follows.

1. **Biomaterials**

The various materials used in three-dimensional techniques in the medical field are illustrated in Figure 3(Williams, 2008). Examples of biomaterials used in medical fields are given below.

2. **Acrylonitrile Butadiene Styrene (ABS) for limb assistive devices using fused deposition modelling (FDM).**

    1. Titanium Striker Triathlon for the knee by selective laser melting (SLM).
    2. MUGEOT Titanium alloy for hip implants by selective laser melting (SLM).
    3. Joimax EndoLIF Titanium fusion cage by selective laser melting (SLM).
    4. OsteoFab PEEK cranial by selective laser melting (SLM).

Figure 3. Various Organs Versus Bio-materials

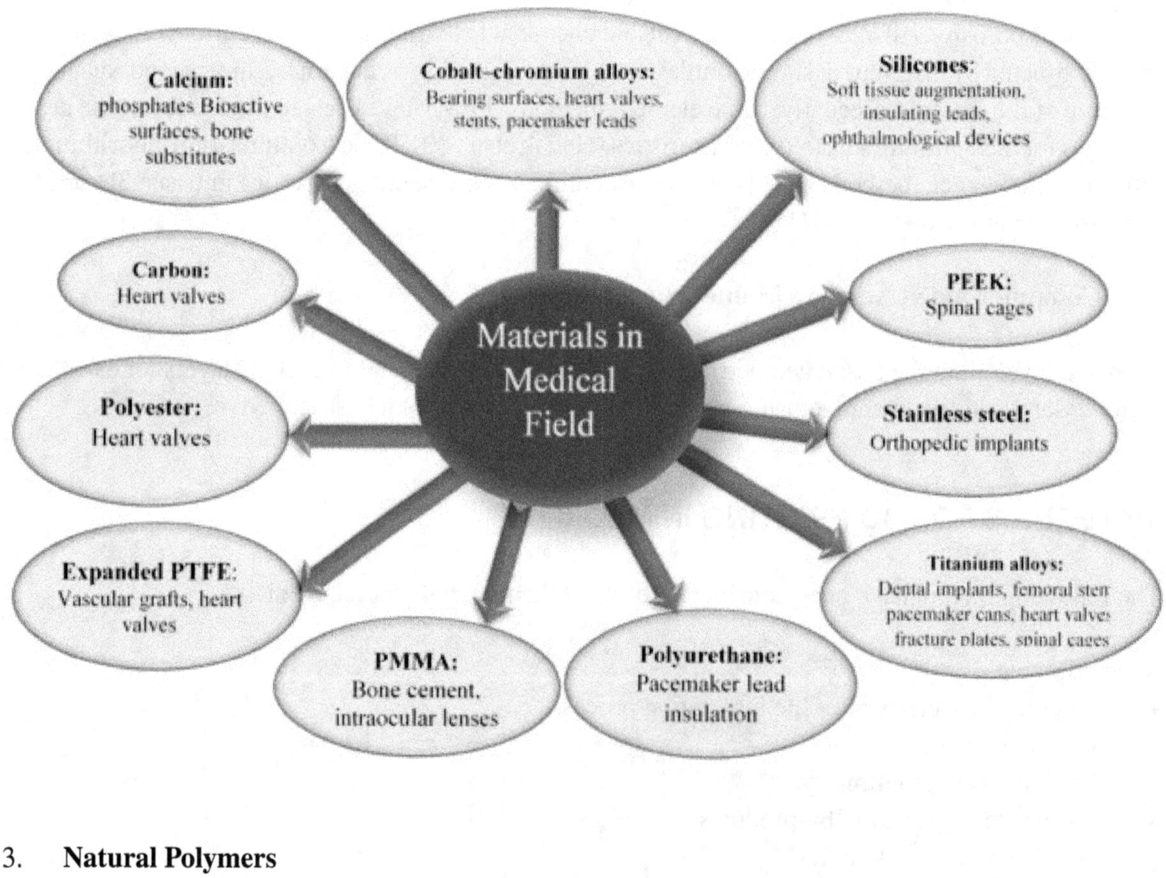

3. **Natural Polymers**

Natural polymers are mostly recommended in the 3D printing of scaffolds due to their natural biodegradability, biocompatibility, and bioactivity. The bacterial cellulose and cellulose from Acetobacter xylinum have been used in the fabrication of cartilage regeneration materials (Hölzl et al., 2016). Collagen and chitosan can also be used as bio-polymers in the tissue engineering field to support cell proliferation and viability. Pure gelatine is a very important material in 3D printing for the fabrication of hepatocyte cells in living environments (Yan et al., 2018).

4. **Synthetic Polymers**

Low-toxicity biodegradable aliphatic polyesters such as PCL or PLGA have been used as source materials in 3D bio-printing processes (Top et al., 2021). The polymer hydrolytic degradation process of aliphatic polyesters has been used in bone regeneration and tissue loss treatments by controlling process time and speed (Oryan et al., 2017). Hydrolysis in enzymatic conditions: principles are applied for the degradation of all types of synthetic polymer scaffolds. These are placed into a living organism, allowing water to penetrate the polymer matrix. Water molecules are used to dilute and break the ester bonds, resulting in polymer chains with cohesive bindings (Heljak et al., 2017, 2018). The printable co-polymers such as hydroxybutyrate, polyglycolic acid, and propylene fumarate are also applied in tissue engineering fields (drug-release applications). Thus, synthetic and copolymers can be used in the TE under in vitro and in vivo conditions for neovascularization and bone ingrowth applications (Daly et al., 2016).

5. **Hydrogels**

Water plays an important role in cross-linking the natural and synthetic polymers called hydrogels. Ethylene glycol and natural gelatin methacrylate (GelMA) are used to prepare the hydrogels. It has low mechanical strength with good biocompatibility, native tissue, and diffusion properties. It is the main material used to fabricate the hybrid tissue scaffolds and soft tissues. The different bio-ink biomaterials, degradation processes, and crosslinking methods are involved in fabricating the various gel states of hydrogel matrixes (Jeznach et al., 2018).

6. **Bio-ceramics**

The 3D printed bio-ceramics materials made from organic salts of phosphate and calcium are good bio- biocompatibility and mechanical strength. It can be used to make various complex structures of bones for specific patients. The ceramic mixed polymer composites are used to create the composite bone structures. Bio-glass, silica, graphene oxide, and zirconium titanate are used in 3D printing in the TE process. The bioactive glass ceramics are applied to fabricate good porosity and high mechanical strength through freeze-drying and sintering processes (Hua et al., 2021; Özcan et al., 2021). Due to their similar chemical structure to human bones, calcium phosphates have been applied in various bone fabrication processes.

Classification In -3D Printing Techniques in Medical FIELDS

The various types of 3D printing techniques applied in the medical field are illustrated in Figure 4. Generally, three types of 3D printing methods have been used in the bioprinting process based on the layer formations: (1) jetting, (2) laser, and (3) laminated types are available. The jet type method is used for liquid based printing materials; laser methods are applied for solid printing materials; the laminated methods have been used for layer formation of gel/ glue state materials. Some 3D printing techniques

Figure 4. Classification of 3D Printing Techniques in Medical Fields

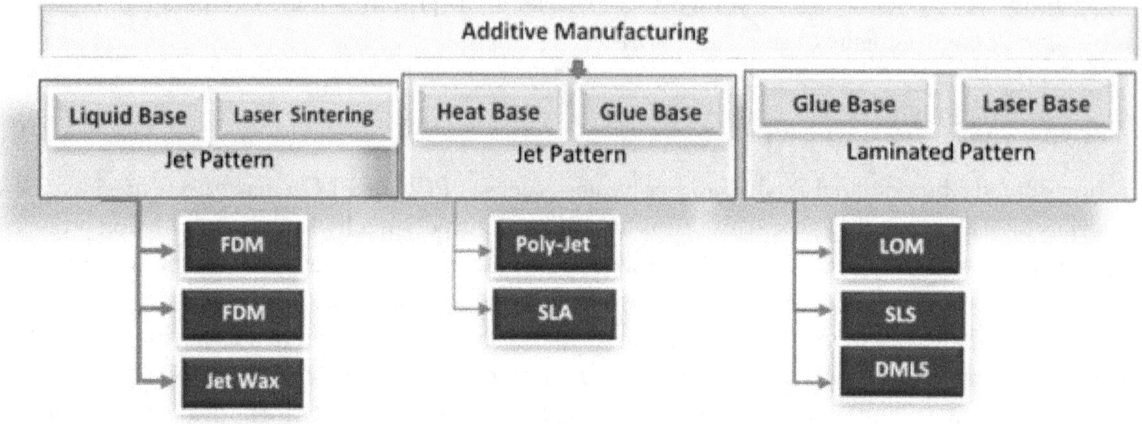

have been used to make the patient-specific parts: computed axial tomography (CT), and cone beam CT (CBCT) are used to make the bones, and magnetic resonance imaging (MRI) is used to make the tissues and organs. In CT methods, the x-ray has been applied to scan the bones, model, and slice them for the 3D printing process. In MRI, magnetic and radio waves are employed to create cross-sectional views of the tissues. The different types of tissues have been detected by the various signals in the MRI. CAD modelling software is used to make precise anatomical models of soft and hard tissues. For the hydrogenous organ manufacturing process, image modalities analysis, mathematical equations and simulations, and computational analysis have been made to provide inclusive 3D models (Klein et al., 2013). The developed models have been sliced into 2D layers and staked into 3D objects to produce the contour or irregular shape of the human parts and various organs (Kang et al., 2016).

3D Printing Approaches in the Medical Field

The important steps should be followed as imaging of organs, designs, and selection of cells and materials. 3D printing approaches are followed in medicine.

(1) **Biomimicry:** It involves selecting suitable materials, functional arrangements of micro-cell-cultural protocols, and integrating biochemicals, medicine, and biophysics using nanotechnology (Ingber et al., 2006).
(2) **Independent self-assembly:** Biological-microstructure and functions based on Extra Cellular Matrix (ECM) components, cell signal and independent cellular organisation are performed (Derby, 2012; Steer & Nigam, 2004).
(3) **Miniature-tissue building blocks:** Tissue and organs are divided into subblocks which comprise the micro-cells and functional elements. It includes the overall organisation of cells, including high resolution and high-resolution, accurate retracements of cells (Kamei et al., 2006).

VARIOUS 3D PRINTING PROCESSES IN TISSUE ENGINEERING

1. Inkjet bioprinting

The schematic representation of the Inkjet Printer in Tissue Engineering is shown in Figure 5. In this process, the mixture of bio-ink (biopolymer gel) and living cells has been stored in the chamber in which the bio-ink mixture is preheated to construct the tissue in the required shape. This method has been rarely used in tissue engineering due to uneven distribution of cells, interpretation by quick solidification, and flow problems. However, it has been applied with low investment and running costs due to a very simple method (Li et al., 2020).

Figure 5. Schematic representation of the Inkjet Printer in Tissue Engineering

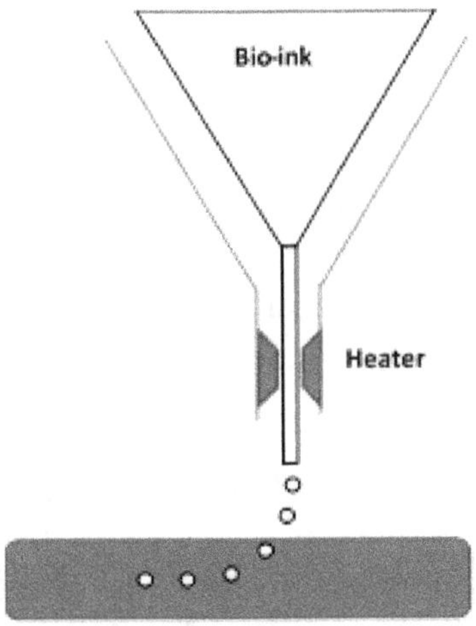

2. Laser assisted bioprinting (LAB)

The schematic representation of Laser Assisted Bioprinting in Tissue Engineering is illustrated in Figure 6. In the LAB method, the bio-ink and donor layer are arranged as shown in the Figure 6. The laser beam is passed to the donor layer. The bio-ink layer is detached from the energy layer by the heat and bubbles are formed between the two layers. The bobble creates a vacuum between the two layers and is ejected. The tissue structure is formed by the droplet-by-droplet stacking principle. This method is applied to form thick/ high-density, and visibility (95%) high-cell tissues with low viscous resistance. However, investment and running costs are higher for the large-scale production process (Hakobyan et al., 2020; Osidak et al., 2020).

Figure 6. Schematic representation of Laser Assisted Bioprinting in Tissue Engineering

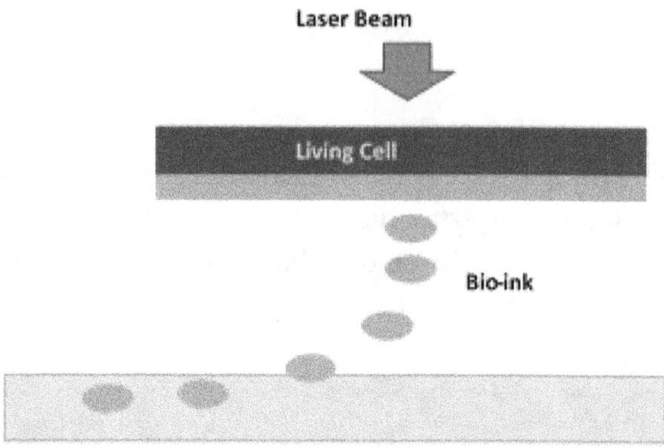

3. Extrusion bioprinting

The schematic arrangement of extrusion bioprinting in tissue engineering is displayed in Figure 7. The gel/solution of bio-ink is ejected from the small needle-shaped nozzle with the help of a high-pressure nozzle. In this process, long strands are used to create the complicated profile of tissues. The uniform density of cells has been produced at the normal atmospheric temperature. Natural biomaterials like hydrogels are printed by this method (Landers & Mülhaupt, 2000).

Figure 7. Schematic representation of Extrusion Bioprinting in Tissue Engineering

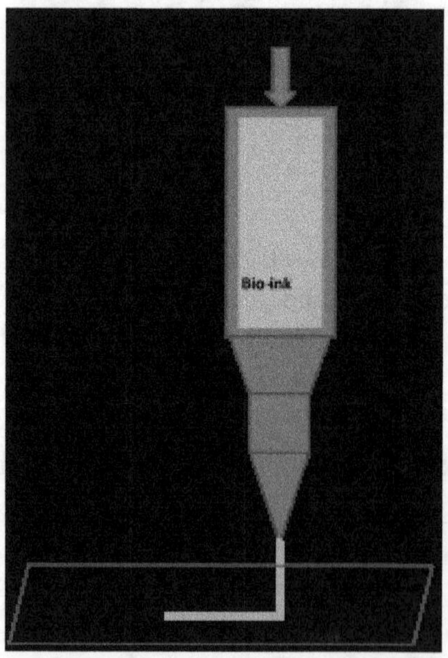

4. Stereolithography

Stereolithography (SLA) is one of the classical techniques using laser beams to form bio-ink polymerization processes. A pictorial view of stereolithography in tissue engineering is represented in Figure 8. After polymerization, the photosensitive hydrogels in the 2D layers are solidified by visible or ultra-violet rays. The above process is repeated to build the tissues to the full required structure by the deposition of layers. The schematic arrangements of the SLA process are illustrated in the Figure 8. The photosensitive hydrogels are namely represented as photo-initiators, Gelatin Methacryloyl (GelMA)(Magalhães et al., 2020), and PEGDA (polyethylene glycol diacrylate).(Jiao et al., 2018). The high-quality tissues have been fabricated by adjusting the SLA process parameters: light intensity, feed rate of printing, 2D layer density, and time of the process(Hong et al., 2005).

Figure 8. Schematic representation of Stereolithography in Tissue Engineering

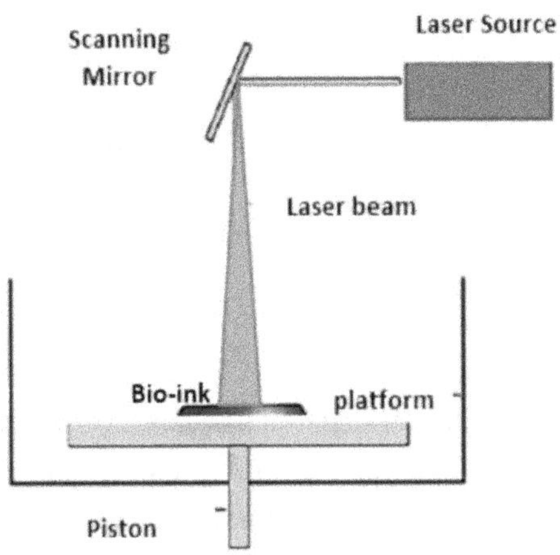

5. Melt-Electro Spinning (MES)

The melt-electro-spinning method is a new technique in the field of 3D tissue engineering and a technological advancement in existing solution electrospinning (SES) without the use of toxic solvents (Lohfeld et al., 2010). The various working element arrangements in Melt-Electro Spinning (MES) processes are shown in Figure 9. The polymer solution plays an important role in the jet form of a solution with a low viscous flow. The layer thickness has been maintained by less than one micrometre of fibres. The schematic arrangements of various MES parts are shown in Figure 9.

Figure 9. Schematic representation of Melt-Electro Spinning (MES) in Tissue Engineering

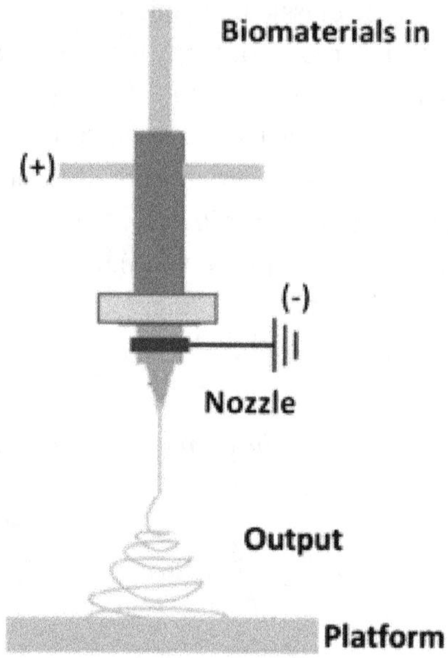

6. Selective Laser Sintering

In this method, the polymeric powders are heated above the polymer glass transition state by the laser beam. The melted powder particles in the gel state are transferred into the fabrication area. During this state, the polymer particles are diffused and placed in the fabricated previous layers. After deposition, the complete structures are moved down to form another layer (Pattanayak et al., 2011). The schematic arrangements of SLS methods for TE are shown in Figure 10.

Figure 10. Schematic representation of Selective Laser Sintering in Tissue Engineering

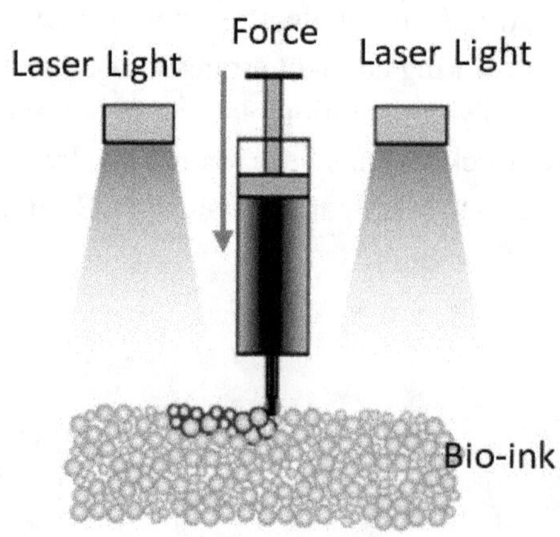

7. Fused Deposition Modelling Method

Figure 11. Schematic representation of Fused Deposition Modelling in Tissue Engineering

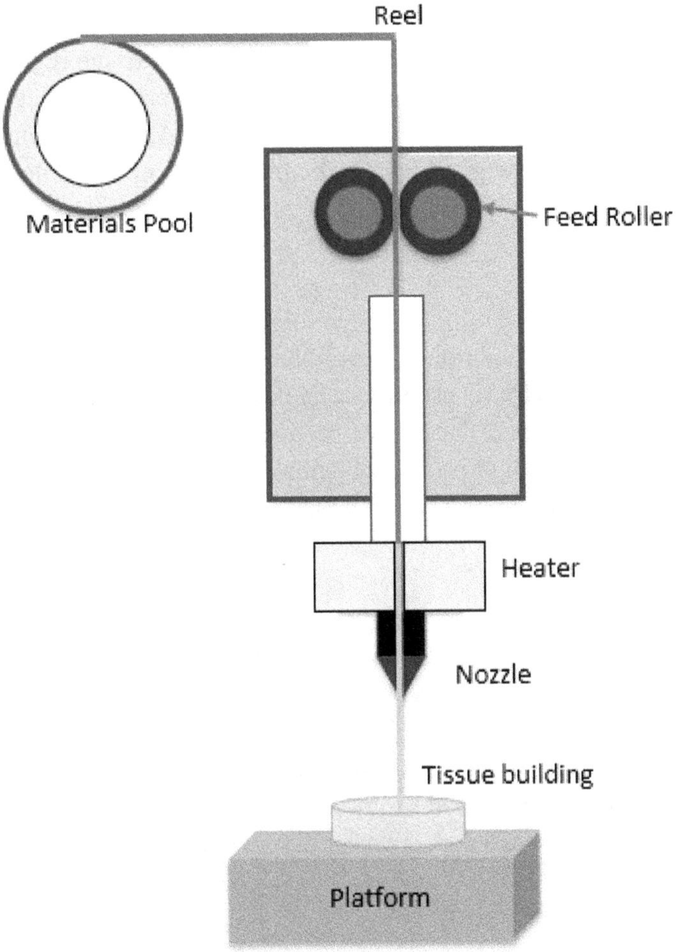

The high-temperature wire shaped polymers are extruded through a convergent nozzle to develop the 3D objects. Important limitations are encountered in the biological applications and fabrication processes using living cells and biomolecules due to high temperature, spatial resolution, and thermal degradation issues. Polycaprolactone (PCL), polylactide (PLA), or polyglycolide (PGA) are used as the input sources of FDM. The polymers are heated for a long time to lose their viscosity to make the parts. In the precision extruding deposition (PED) method, the micro-fibre polymer granules are used to quickly melt and flow through the nozzle (Van Noort, 2012). The FDM process in the tissue fabrication process is shown in Figure 11.

ADVANCED 3D PRINTING TECHNIQUES IN THE TISSUE ENGINEERING FIELD

1. Nervous Tissue

3D printing techniques have been applied to make/ repair the tissue in the central nervous system (CNS), nerve conduits and the peripheral nervous system (PNS). The 3D modelling of brain-blood barrier principles has been applied in pathological and physiological tests and tissue regeneration, drug delivery and tissue engineering. In the nerve regeneration process, the 3D printing of cellularized conduits has been made by cellularized cryo-polymerized gelatine (cryoGelMA) gel with adipose-derived stem cells (ASCs). As per patient-specific tissue generation principles, 3D printing of blood circulation channels has also been fabricated (Kim et al., 2015).

2. Ophthalmology Tissues

In the ophthalmology tissue regeneration process, three-dimensional retina, adult rat retinal ganglion cells, and glia have been fabricated by 3D printing technology. The eye cells have reproduced those parts without affecting phenotypic features. These have been made with collagen-based bio-ink containing encapsulated corneal keratocytes. High-profile modelling software has been employed in this field to make and reproduce soft tissue (Isaacson et al., 2018).

3. Ear

The bionic human ear has been designed with CAD modelling software. The 3D printing of the human ear can be made by a hydrogel matrix containing silver nanoparticles and conductive polymers. Every side of the cochlea-shaped inductive-coiled electrodes is covered by the cartilage tissues using vitro culture. The auditory sensing tissue has been made by a 3D bioprinting technique using lipid tissue and auricular cartilage. Adipose-derived stromal cells in hydrogels are placed in the lipid tissue and auricular cartilage can be used to make the active ear tissues (Lee et al., 2014).

4. Kidney

The extortion 3D printing techniques are applied to make the very fine tissue for the kidney organs. These artificial tissues can be used for drug testing and disease modelling purposes. The tissue scaffolds are made from poly (ethylene glycol) diacrylate, sodium alginate, and calcium sulfate.UVC light and human embryonic kidney cells are used to fabricate the cross-linked scaffolds (Lawlor et al., 2021).

5. Skin

The Laser assisted bioprinting (LOB) method has been applied to develop the 3D-printed skin using cultured collagen type I and Matriderm mixings. The experiments have been performed on murine skin in in-vivo environments. The fibroblasts and keratinocytes suspended in the gelatine-based hydrogel have been used for making multi-layer skin equivalents (SE). These technologies have also been applied to conduct skin disease research activities (Derr et al., 2019).

6. **Cancer Models**

In the current scenario, 3D bioprinting methods were used to develop the cancer disease physical models. Heterogeneity tumours, growth factors, and vasculature and tumour cells are developed based on patient-specific cells. It can be applied to investigate anti-cancer therapy, the structural complexity of tumours and cell-matrix interactions (Datta et al., 2020).

7. **Cartilage and Bone Tissue Regeneration**

Autogenous and allogeneic bone grafting methods have been eliminated by 3D bioprinting of bone tissue regeneration techniques due to some practical issues such as unsuitable donors and cancer problems. During the artificial bone development process, the mechanical strength, size, and shape have been modelled and analysed by CAD software. Ceramics, hydrogels, and polyesters are playing important roles in the cartilage and bone tissue regeneration processes (Oladapo et al., 2019). Composite biodegradable aliphatic polyesters and hydrogels are the best combinations to fabricate the 3D-porous hybrid structure of bone and cartilage tissues. However, the selection of materials and 3D printing techniques is a very important challenge to developing the bi-phase or tri-phase of osteochondral scaffolds (Axpe & Oyen, 2016).

MEDICAL APPLICATIONS

The various medical applications of 3D printing technologies are illustrated in Figure 12. The technology of additive manufacturing has been used in a variety of medical applications, including organ fabrication, disease models, pharmaceutical products, 4D objects, surgical templates, and diagnostic tools.

(1) **Organ constructions:** During the organ bio-fabrication process, the following approaches have been followed (Mironov et al., 2011; Xu et al., 2011):
   a. Selection and utilization of existing 3D printing technology.
   b. Making functional organs in the biological environment.
   c. Implementing automation in the fabrication processes.
   d. Implementing integration process for the simplification.
   e. Biomonitoring system – implementing automatic control system and advanced sensors.
   f. The online monitoring system in the fabrication and testing process.

(2) **Disease modelling**

The development of various disease models is connected to interdisciplinary research activities such as mechanical, medical, electrical, material, tissue, and molecular engineers. Building the new concept and tissue models is important based on the damage to human organs (Moon et al., 2010).

(3) **Pharmaceutical products**

This technology is very important in making personalised medicine with different combinations of doses. The following benefits have been captured while implementing 3D printing technology in pharmaceutical manufacturing systems (Pereira & Bártolo, 2015):

a. Rapid production.
b. Low cost
c. Minimum production time.
d. Mass production of verity products.
e. Inkjet 3DP in commercial applications with low investment.
f. Medical products for the verity of dose concentrations.
g. Developing personalized medicine.
h. Micro-variations of different percentages of chemicals.

Figure 12. Application of 3D printing techniques in various medical fields

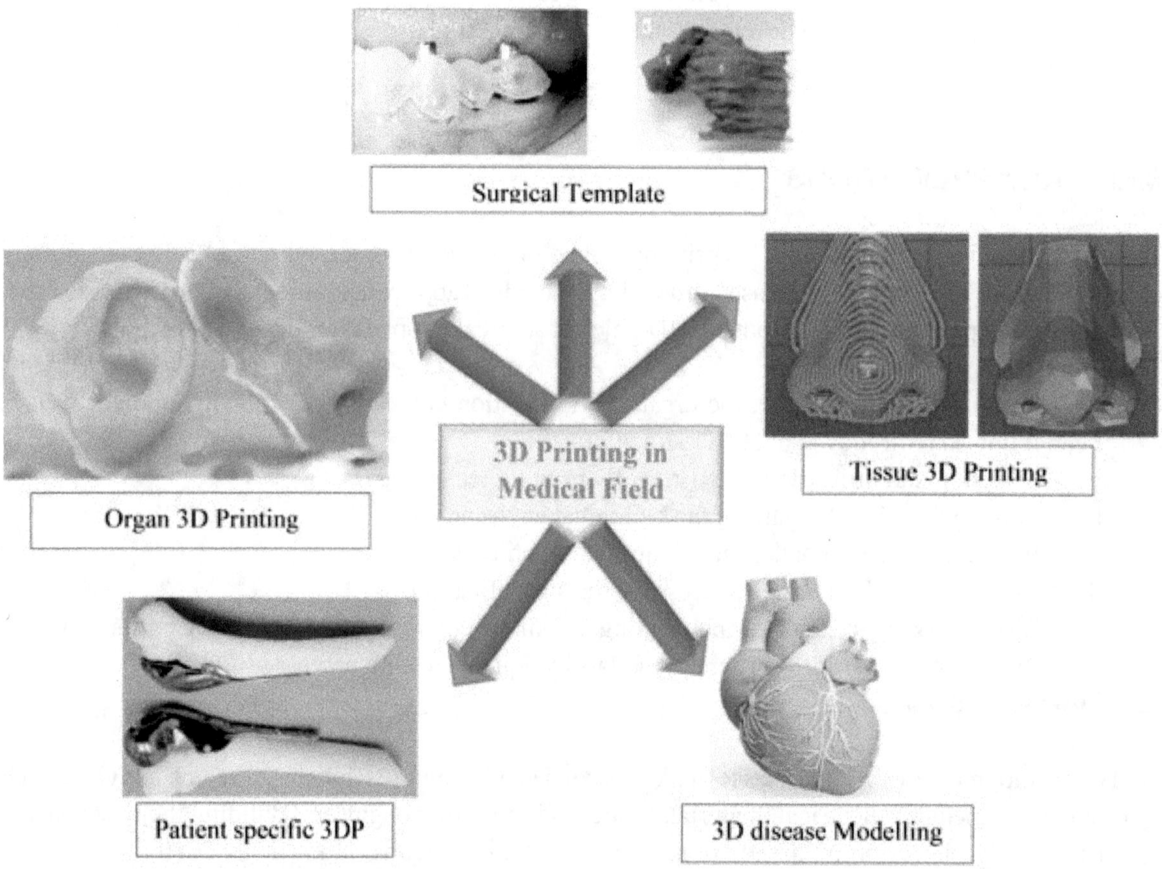

(4) **Surgical templates and diagnostic tools**

The surgical templates have been used as supporting elements to perform the graft implantation surgery, such as drilling support, evaluating surgical locations, and predicting the angle of tool movement, size, and directions. The 3D template guides, supports, and assists the surgical process in the blood supply soft locations to eliminate misalignment and inaccurate placement (Kola et al., 2015). Sometimes, various 3D-printed templates have been used to provide the sequence of surgical operations to doctors. This equipment has been made by the surgical device making industry.

(5) **4D Bioprinting Technology**

In the 4D printing technology, the time of printing is the 4th dimension along with the physical dimensions. In this technology, after blunting and fabrication of organs, how long time is functioning in the biological environment has been considered (Xu et al., 2011).

(6) **Parameter selection for best 3D product**

Parametric optimization of process parameters using design of experiments such as Taguchi (Boopathi & Myilsamy, 2021), mixture (Boopathi, Balasubramani, et al., 2021) and response surface (Kannan et al., 2022; Myilsamy et al., 2021) methodologies. Multi-objective optimization of surface quality, production rate, strength (Boopathi, Thillaivanan, et al., 2021), and mechanical properties (Haribalaji et al., 2021) of additive manufactured products have been improved by TOPSIS (Sampath et al., 2023), MOORA (Boopathi, 2022a), VIKOR (Yupapin et al., 2022) and ABC Techniques. The choose the best additive manufacturing processes for the particular medical applications using conventional and unconventional optimization techniques.

SUMMARY

The additive manufacturing method is an emerging technique in various commercial, industrial, and medical applications. Products are made by adding materials layer by layer based on the model developed by the users. Additive manufacturing (AM) or three-dimensional printing (3DP) techniques are now used in many biomedical applications. In this chapter, the computer-aided tissue modelling process, different fundamental and advanced biomaterials, advanced scaffold manufacturing applications, and additive manufacturing technology has been applied in various sectors: the manufacturing of industrial components, toys, medicine, medical-surgical instruments, and tissue engineering have been illustrated.

REFERENCES

Axpe, E., & Oyen, M. L. (2016). Applications of alginate-based bioinks in 3D bioprinting. *International Journal of Molecular Sciences*, *17*(12), 1976. doi:10.3390/ijms17121976 PMID:27898010

Badekila, A. K., Kini, S., & Jaiswal, A. K. (2021). Fabrication techniques of biomimetic scaffolds in three-dimensional cell culture: A review. *Journal of Cellular Physiology*, *236*(2), 741–762. doi:10.1002/jcp.29935 PMID:32657458

Bongiovanni Abel, S., Montini Ballarin, F., & Abraham, G. A. (2020). Combination of electrospinning with other techniques for the fabrication of 3D polymeric and composite nanofibrous scaffolds with improved cellular interactions. *Nanotechnology, 31*(17), 172002. doi:10.1088/1361-6528/ab6ab4 PMID:31931493

Boopathi, S. (2021). An investigation on gas emission concentration and relative emission rate of the near-dry wire-cut electrical discharge machining process. *Environmental Science and Pollution Research International*, 1–10. doi:10.100711356-021-17658-1 PMID:34837614

Boopathi, S. (2022a). Experimental investigation and multi-objective optimization of cryogenic Friction-stir-welding of AA2014 and AZ31B alloys using MOORA technique. *Materials Today. Communications, 33*, 104937. doi:10.1016/j.mtcomm.2022.104937

Boopathi, S. (2022b). An Extensive Review on Sustainable Developments of Dry and Near-Dry Electrical Discharge Machining Processes. *Journal of Manufacturing Science and Engineering, 144*(5), 1–37. doi:10.1115/1.4052527

Boopathi, S., Balasubramani, V., Kumar, R. S., & Singh, G. R. (2021). The influence of human hair on kenaf and Grewia fiber-based hybrid natural composite material: An experimental study. *Functional Composites and Structures, 3*(4), 045011. doi:10.1088/2631-6331/ac3afc

Boopathi, S., & Myilsamy, S. (2021). Material removal rate and surface roughness study on Near-dry wire electrical discharge Machining process. *Materials Today: Proceedings, 45*, 8149–8156. doi:10.1016/j.matpr.2021.02.267

Boopathi, S., Thillaivanan, A., Pandian, M., Subbiah, R., & Shanmugam, P. (2021). Friction stir processing of boron carbide reinforced aluminium surface (Al-B4C) composite: Mechanical characteristics analysis. *Materials Today: Proceedings, 50*(5), 2430–2435. doi:10.1016/j.matpr.2021.10.261

Daly, A. C., Cunniffe, G. M., Sathy, B. N., Jeon, O., Alsberg, E., & Kelly, D. J. (2016). 3D Bioprinting: 3D Bioprinting of Developmentally Inspired Templates for Whole Bone Organ Engineering. *Advanced Healthcare Materials, 5*(18), 2352. doi:10.1002/adhm.201670100 PMID:27281607

Datta, P., Dey, M., Ataie, Z., Unutmaz, D., & Ozbolat, I. T. (2020). 3D bioprinting for reconstituting the cancer microenvironment. *NPJ Precision Oncology, 4*(1), 1–13. doi:10.103841698-020-0121-2 PMID:32793806

Derby, B. (2012). Printing and prototyping of tissues and scaffolds. *Science, 338*(6109), 921–926. doi:10.1126cience.1226340 PMID:23161993

Derr, K., Zou, J., Luo, K., Song, M. J., Sittampalam, G. S., Zhou, C., Michael, S., Ferrer, M., & Derr, P. (2019). Fully Three-Dimensional Bioprinted Skin Equivalent Constructs with Validated Morphology and Barrier Function. *Tissue Engineering. Part C, Methods, 25*(6), 334–343. doi:10.1089/ten.tec.2018.0318 PMID:31007132

Giannitelli, S. M., Accoto, D., Trombetta, M., & Rainer, A. (2014). Current trends in the design of scaffolds for computer-aided tissue engineering. *Acta Biomaterialia, 10*(2), 580–594. doi:10.1016/j.actbio.2013.10.024 PMID:24184176

Hakobyan, D., Kerouredan, O., Remy, M., Dusserre, N., Medina, C., Devillard, R., Fricain, J. C., & Oliveira, H. (2020). Laser-Assisted Bioprinting for Bone Repair. In *Methods in Molecular Biology* (Vol. 2140, pp. 135–144). Springer., doi:10.1007/978-1-0716-0520-2_8

Haribalaji, V., Boopathi, S., Asif, M. M., Jeyakumar, M., Subbiah, R., & Lewise, K. A. S. (2021). Influences of Friction stir tool parameters for joining two similar AZ61A alloy plates. *Materials Today: Proceedings*, *50*(5), 2547–2553. doi:10.1016/j.matpr.2021.12.074

Heljak, M. K., Kurzydlowski, K. J., & Swieszkowski, W. (2017). Computer aided design of architecture of degradable tissue engineering scaffolds. *Computer Methods in Biomechanics and Biomedical Engineering*, *20*(15), 1623–1632. doi:10.1080/10255842.2017.1399263 PMID:29106807

Heljak, M. K., Moczulska-Heljak, M., Choińska, E., Chlanda, A., Kosik-Kozioł, A., Jaroszewicz, T., Jaroszewicz, J., & Swieszkowski, W. (2018). Micro and nanoscale characterization of poly(DL-lactic-co-glycolic acid) films subjected to the L929 cells and the cyclic mechanical load. *Micron (Oxford, England)*, *115*, 64–72. doi:10.1016/j.micron.2018.09.004 PMID:30253318

Hölzl, K., Lin, S., Tytgat, L., Van Vlierberghe, S., Gu, L., & Ovsianikov, A. (2016). Bioink properties before, during and after 3D bioprinting. *Biofabrication*, *8*(3), 32002. doi:10.1088/1758-5090/8/3/032002 PMID:27658612

Hong, B. T., Shin, K. S., & Kim, D. S. (2005). Ultraviolet-curing behavior of an epoxy acrylate resin system. *Journal of Applied Polymer Science*, *98*(3), 1180–1185. doi:10.1002/app.22130

Hua, L., Lei, T., Qian, H., Zhang, Y., Hu, Y., & Lei, P. (2021). 3D-printed porous tantalum: Recent application in various drug delivery systems to repair hard tissue defects. *Expert Opinion on Drug Delivery*, *18*(5), 625–634. doi:10.1080/17425247.2021.1860015 PMID:33270470

Ingber, D. E., Mow, V. C., Butler, D., Niklason, L., Huard, J., Mao, J., Yannas, I., Kaplan, D., & Vunjak-Novakovic, G. (2006). Tissue engineering and developmental biology: Going biomimetic. *Tissue Engineering*, *12*(12), 3265–3283. doi:10.1089/ten.2006.12.3265 PMID:17518669

Isaacson, A., Swioklo, S., & Connon, C. J. (2018). 3D bioprinting of a corneal stroma equivalent. *Experimental Eye Research*, *173*, 188–193. doi:10.1016/j.exer.2018.05.010 PMID:29772228

Jeznach, O., Kołbuk, D., & Sajkiewicz, P. (2018). Injectable hydrogels and nanocomposite hydrogels for cartilage regeneration. *Journal of Biomedical Materials Research. Part A*, *106*(10), 2762–2776. doi:10.1002/jbm.a.36449 PMID:29726104

Jiao, Z., Wang, C., Yang, Q., & Wang, X. (2018). Preparation and characterization of UV-curable urethane acrylate oligomers modified with cycloaliphatic epoxide resin. *Journal of Coatings Technology and Research*, *15*(2), 251–258. doi:10.100711998-017-9994-6

Kamei, M., Brian Saunders, W., Bayless, K. J., Dye, L., Davis, G. E., & Weinstein, B. M. (2006). Endothelial tubes assemble from intracellular vacuoles in vivo. *Nature*, *442*(7101), 453–456. doi:10.1038/nature04923 PMID:16799567

Kang, H. W., Lee, S. J., Ko, I. K., Kengla, C., Yoo, J. J., & Atala, A. (2016). A 3D bioprinting system to produce human-scale tissue constructs with structural integrity. *Nature Biotechnology*, *34*(3), 312–319. doi:10.1038/nbt.3413 PMID:26878319

Kannan, E., Trabelsi, Y., Boopathi, S., & Alagesan, S. (2022). Influences of cryogenically treated work material on near-dry wire-cut electrical discharge machining process. *Surface Topography: Metrology and Properties*, *10*(1), 015027. Advance online publication. doi:10.1088/2051-672X/ac53e1

Kim, J. A., Kim, H. N., Im, S. K., Chung, S., Kang, J. Y., & Choi, N. (2015). Collagen-based brain microvasculature model in vitro using three-dimensional printed template. *Biomicrofluidics*, *9*(2), 24115. doi:10.1063/1.4917508 PMID:25945141

Klein, G. T., Lu, Y., & Wang, M. Y. (2013). 3D printing and neurosurgery—Ready for prime time? *World Neurosurgery*, *80*(3–4), 233–235. doi:10.1016/j.wneu.2013.07.009 PMID:23871811

Kola, M., Rabah, A., Raghav, D., Sabra, S., Harby, N. M., Khalil, H., & Shah, A. (2015). Surgical templates for dental implant positioning; current knowledge and clinical perspectives. *Nigerian Journal of Surgery: Official Publication of the Nigerian Surgical Research Society*, *21*(1), 1. doi:10.4103/1117-6806.152720 PMID:25838757

Landers, R., & Mülhaupt, R. (2000). Desktop manufacturing of complex objects, prototypes and biomedical scaffolds by means of computer-assisted design combined with computer-guided 3D plotting of polymers and reactive oligomers. *Macromolecular Materials and Engineering*, *282*(1), 17–21. doi:10.1002/1439-2054(20001001)282:1<17::AID-MAME17>3.0.CO;2-8

Lawlor, K. T., Vanslambrouck, J. M., Higgins, J. W., Chambon, A., Bishard, K., Arndt, D., Er, P. X., Wilson, S. B., Howden, S. E., Tan, K. S., Li, F., Hale, L. J., Shepherd, B., Pentoney, S., Presnell, S. C., Chen, A. E., & Little, M. H. (2021). Cellular extrusion bioprinting improves kidney organoid reproducibility and conformation. *Nature Materials*, *20*(2), 260–271. doi:10.103841563-020-00853-9 PMID:33230326

Lee, J. S., Hong, J. M., Jung, J. W., Shim, J. H., Oh, J. H., & Cho, D. W. (2014). 3D printing of composite tissue with complex shape applied to ear regeneration. *Biofabrication*, *6*(2), 24103. doi:10.1088/1758-5082/6/2/024103 PMID:24464765

Li, X., Liu, B., Pei, B., Chen, J., Zhou, D., Peng, J., Zhang, X., Jia, W., & Xu, T. (2020). Inkjet Bioprinting of Biomaterials. *Chemical Reviews*, *120*(19), 10793–10833. doi:10.1021/acs.chemrev.0c00008 PMID:32902959

Lohfeld, S., Tyndyk, M. A., Cahill, S., Flaherty, N., Barron, V., & McHugh, P. E. (2010). A method to fabricate small features on scaffolds for tissue engineering via selective laser sintering. *Journal of Biomedical Science and Engineering*, *03*(02), 138–147. doi:10.4236/jbise.2010.32019

Magalhães, L. S. S. M., Santos, F. E. P., de Maria Vaz Elias, C., Afewerki, S., Sousa, G. F., Furtado, A. S. A., Marciano, F. R., & Lobo, A. O. (2020). Printing 3D hydrogel structures employing low-cost stereolithography technology. *Journal of Functional Biomaterials*, *11*(1), 12. doi:10.3390/jfb11010012 PMID:32098431

Mahmood, M. A., Visan, A. I., Ristoscu, C., & Mihailescu, I. N. (2021). Artificial neural network algorithms for 3D printing. In Materials (Vol. 14, Issue 1, pp. 1–29). doi:10.3390/ma14010163

Mironov, V., Kasyanov, V., & Markwald, R. R. (2011). Organ printing: From bioprinter to organ biofabrication line. *Current Opinion in Biotechnology*, *22*(5), 667–673. doi:10.1016/j.copbio.2011.02.006 PMID:21419621

Moon, S., Hasan, S. K., Song, Y. S., Xu, F., Keles, H. O., Manzur, F., Mikkilineni, S., Hong, J. W., Nagatomi, J., Haeggstrom, E., Khademhosseini, A., & Demirci, U. (2010). Layer by layer three-dimensional tissue epitaxy by cell-laden hydrogel droplets. *Tissue Engineering. Part C, Methods*, *16*(1), 157–166. doi:10.1089/ten.tec.2009.0179 PMID:19586367

Myilsamy, S., Boopathi, S., & Yuvaraj, D. (2021). A study on cryogenically treated molybdenum wire electrode. *Materials Today: Proceedings*, *45*(9), 8130–8135. doi:10.1016/j.matpr.2021.02.049

Oladapo, B. I., Zahedi, S. A., & Adeoye, A. O. M. (2019). 3D printing of bone scaffolds with hybrid biomaterials. *Composites. Part B, Engineering*, *158*, 428–436. doi:10.1016/j.compositesb.2018.09.065

Oryan, A., Kamali, A., Moshirib, A., & Eslaminejad, M. B. (2017). Role of Mesenchymal Stem Cells in Bone Regenerative Medicine: What Is the Evidence? *Cells, Tissues, Organs*, *204*(2), 59–83. doi:10.1159/000469704 PMID:28647733

Osidak, E. O., Kozhukhov, V. I., Osidak, M. S., & Domogatsky, S. P. (2020). Collagen as bioink for bioprinting: A comprehensive review. *International Journal of Bioprinting*, *6*(3), 1–10. doi:10.18063/ijb.v6i3.270 PMID:33088985

Özcan, M., Hotza, D., Fredel, M. C., Cruz, A., & Volpato, C. A. M. (2021). Materials and manufacturing techniques for polymeric and ceramic scaffolds used in implant dentistry. *Journal of Composites Science*, *5*(3), 78. doi:10.3390/jcs5030078

Pattanayak, D. K., Fukuda, A., Matsushita, T., Takemoto, M., Fujibayashi, S., Sasaki, K., Nishida, N., Nakamura, T., & Kokubo, T. (2011). Bioactive Ti metal analogous to human cancellous bone: Fabrication by selective laser melting and chemical treatments. *Acta Biomaterialia*, *7*(3), 1398–1406. doi:10.1016/j.actbio.2010.09.034 PMID:20883832

Pereira, R. F., & Bártolo, P. J. (2015). 3D bioprinting of photocrosslinkable hydrogel constructs. *Journal of Applied Polymer Science*, *132*(48). Advance online publication. doi:10.1002/app.42458

Roopavath, U. K., & Kalaskar, D. M. (2017). Introduction to 3D printing in medicine. In 3D Printing in Medicine (pp. 1–20). Elsevier. doi:10.1016/B978-0-08-100717-4.00001-6

Sampath, B., Sasikumar, C., & Myilsamy, S. (2023). Application of TOPSIS Optimization Technique in the Micro-Machining Process. In Trends, Paradigms, and Advances in Mechatronics Engineering (pp. 162–187). IGI Global.

Steer, D. L., & Nigam, S. K. (2004). Developmental approaches to kidney tissue engineering. *American Journal of Physiology - Renal Physiology, 286*. doi:10.1152/ajprenal.00167.2003

Sun, W., Starly, B., Darling, A., & Gomez, C. (2004). Computer-aided tissue engineering: Application to biomimetic modelling and design of tissue scaffolds. *Biotechnology and Applied Biochemistry*, *39*(1), 49. doi:10.1042/BA20030109 PMID:14556653

Top, N., Şahin, İ., Gökçe, H., & Gökçe, H. (2021). Computer-aided design and additive manufacturing of bone scaffolds for tissue engineering: State of the art. *Journal of Materials Research*, *36*(19), 3725–3745. doi:10.155743578-021-00156-y

Van Noort, R. (2012). The future of dental devices is digital. *Dental Materials*, *28*(1), 3–12. doi:10.1016/j.dental.2011.10.014 PMID:22119539

Williams, D. F. (2008). On the mechanisms of biocompatibility. *Biomaterials*, *29*(20), 2941–2953. doi:10.1016/j.biomaterials.2008.04.023 PMID:18440630

Xu, F., Celli, J., Rizvi, I., Moon, S., Hasan, T., & Demirci, U. (2011). A three-dimensional in vitro ovarian cancer coculture model using a high-throughput cell patterning platform. *Biotechnology Journal*, *6*(2), 204–212. doi:10.1002/biot.201000340 PMID:21298805

Yan, Q., Dong, H., Su, J., Han, J., Song, B., Wei, Q., & Shi, Y. (2018). A Review of 3D Printing Technology for Medical Applications. *Engineering*, *4*(5), 729–742. doi:10.1016/j.eng.2018.07.021

Yupapin, P., Trabelsi, Y., Nattappan, A., & Boopathi, S. (2022). Performance Improvement of Wire-Cut Electrical Discharge Machining Process Using Cryogenically Treated Super-Conductive State of Monel-K500 Alloy. *Iranian Journal of Science and Technology - Transactions of Mechanical Engineering*, 1–17. doi:10.1007/s40997-022-00513-0

Chapter 7
Advance Biodegradable Polymer Composite Materials for Biomedical Additive Manufacturing Applications

Sudheer Kumar
https://orcid.org/0000-0002-8944-0728
JC OrthoHeal Pvt. Ltd., Vadodara, India

Sukhila Krishnan
Sahrdaya College of Engineering and Technology, Thrissur, India

K. Prabakaran
https://orcid.org/0000-0001-5748-0365
KPR Institute of Engineering and Technology, Coimbatore, India

ABSTRACT

In the present period there has been an increasing awareness of the utilization of cost-effective greener resources to design biodegradable polymer composite materials in biomedical additive manufacturing (AM) applications owing to their excellent physical and mechanical properties, lightweight, high strength, stiffness, fatigue, and good corrosion resistance compared to the metal and ceramics materials. On the other hand, the production of bioplastic-based products increased to approximately 30% due to their remarkable role to diminish the greenhouse effect and petroleum-based pollution. Nevertheless, biodegradable polymers are renewable substitutes for petroleum-based AM materials with various advantages such as being biocompatible, biodegradable, bioresorbable, non-toxic, recyclable, and exhibiting better mechanical properties. This chapter discusses the design and preparation of biodegradable polymer composites for AM biomedical applications with different techniques.

DOI: 10.4018/978-1-6684-6009-2.ch007

INTRODUCTION

Additive manufacturing (AM) is a novel method for building 3D components with difficult shape layer-by-layer deposition from liquid and powder with the support of a computer-added model. However, there has been rising awareness and growth of biomaterials areas, owing to the rapid enhancement of medical product requirements reasons by population growth and aging (Madhavadas et al., 2022; Sheoran et al., 2020; Singh et al., 2019). As per the universal industry investigation report, the biomaterials market is projected to touch approximately $ 94 billion in 2018 and is predicted to be over $ 256 billion by 2025, for instance, an enhancement of more than 15% between 2019 and 2025 (Mathew et al., 2022). Figure 1 demonstrated the different materials employed in the AM.

Figure 1. The different materials employed in additive manufacturing

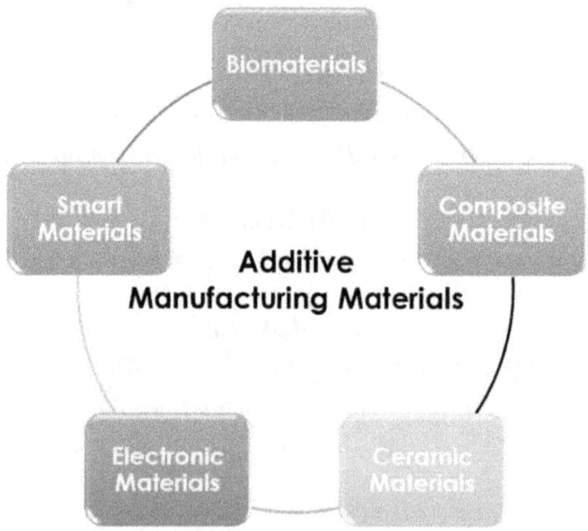

On the other hand, renewable resources were utilized to prepare bio-based 3D printing polymer composite materials in biomedical applications (Durfee & Iaizzo, 2019; Yan et al., 2018). As the annual bioplastic manufacture is increased by 30% estimated (Jamróz et al., 2018), and these bioplastics show a crucial role to diminish petroleum-based plastic pollution and greenhouse emissions effect (Lim et al., 2016). Bio-based polymers are sustainable substitutes for petrochemical-based materials with frequently mentioned benefits instance less dependence on fossil fuel reserves as well as inferior CO_2 footprint, and biodegradability (Gross & Kalra, 2002; Reddy et al., 2012). Consequently, researchers planned an energy-saving system developed for greener 3D printing AM materials, and manufacturing techniques are of a huge reputation around the globe, since of the steady improvement in 3D printing materials utilization.

This chapter aims to endow an updated overview of the development of advanced biodegradable polymer composite materials for AM in biomedical fields. Further, also gives information about the variety of types of biodegradable polymers in assessment for AM initiated by their main points such as physical, chemical, and processing properties in connection to AM processing for medical applications.

ADDITIVE MANUFACTURING IN BIOMEDICAL APPLICATION

The additive manufacturing of polymeric composite materials is currently employed in various biomedical application fields such as surgical training and equipment, 3D anatomical models, prosthetics, implants, and orthopedics. AM techniques provide the freedom to manufacture any design with its cost-effectiveness and reduced material wastages. On the other hand, the product formed by AM exhibited lighter, stronger, and more functional compared to the product made by traditional manufacturing methods. AM techniques bestow more flexibility regarding geometries and processing conditions can be altered after every single product manufacturing. Therefore, AM reveals a very attractive and useful approach for manufacturing sustainable products (Puppi & Chiellini, 2020). Currently, different AM processes are used for product development in biomedical applications as represented in Table 1.

Table 1. The different AM manufacturing process methods and its advantage and limitations (Ambrosi & Pumera, 2016; Wang et al., 2017)

3D printing techniques	Materials form	Process	Advantages	Limitations
Fused diffusion method (FDM) (Resolution range: 50-200 µm)	Filament	Extrusion and deposition	Multi-material capability, low cost, high speed, high tensile strength	Anisotropy nature, nozzle clogging
Selective laser sintering (SLS) (Resolution range: 80-250 µm)	Powder	Laser scanning and heat-induced sintering	Good strength, fine resolution, easy removal of support powder	Very expensive, powdery surface, slow printing
Stereolithography (SLA) (Resolution range: 10 µm)	Liquid photopolymer	Laser scanning and heat-induced curing	The printing resolution is high, biocompatible resins accessible	Materials limitations, high cost, slow printing, cytotoxicity
Binder jetting 3D printing (BJP) (Resolution range: 50-200 µm)	Powder	Binder jetting	Rapid printing, low cost, simple elimination of support powder, flexible to various materials	Coarse-resolution, binder contamination, clogging of binder jet lack of adhesion between layers,
3D plotting (3DP) (Resolution range: 80-250 µm)	Powder	Materials jetting	Low cost, high printing resolution, multiple material capability, compatible with an extensive variety of materials	Low resolution and mechanical properties need post-processing and high porosity
3D (Resolution range: 5-200 µm)	Liquid or plotting paste	Pressurized syringe extrusion, and UV-assisted and heat curing	Materials ability is very soft, high printing resolution	Slow process and low mechanical strength

NATURAL RESOURCE-BASED BIODEGRADABLE POLYMER FOR ADDITIVE MANUFACTURING

The different types of biodegradable polymers employed in the AM such as natural and synthetic polymers and their composite materials. The selection of the polymer is based on the application and material properties. On the other hand, various advantages of polymeric biomaterials such as biocompatible, play a functional role in specific applications because they are degraded in a physiological environment compared to the other petroleum-based AM materials. This chapter endows information about fundamental and scientific descriptions of biodegradable polymers and their composites employed in biomedical AM applications. Figure 2 demonstrated the classification of the biodegradable polymer (Shah & Vasava, 2019) use in biomedical areas with help of various 3D printed techniques such as selective laser sintering (SLS), stereolithography (SLA), binder jetting (BJ), melt and solution extrusion-AM as well as bioprinting, etc. (Puppi & Chiellini, 2020).

Figure 2. Classification of biodegradable polymers used in biomedical AM applications

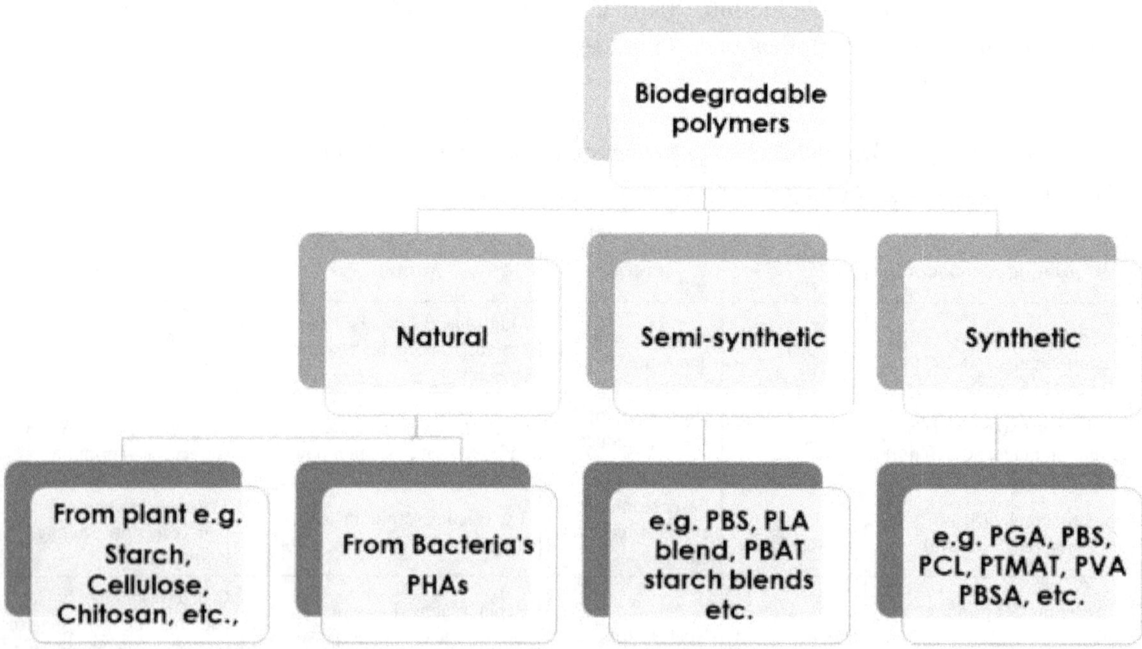

Natural Resource-Based Polymers

Due to the depletion of fossil resources, researchers and academics reduced their dependency on fossil resources and instead rely on natural resources based on polymer because they are sustainable, biodegradable, environmentally friendly, and a better alternative to the present petroleum-based materials as well as reducing the carbon footprint. For that reason, drawing interest in research towards polymers derived from natural resources specially used in biomedical applications because natural polymers reveal various advantages such as inherent biological signaling, encouraging cell adhesion to the surface, remodeling, and cell responsive degradation. Further, the inclusion of the pendant functional group exists in the repeating unit of polysaccharides and proteins endowing several possibilities for chemical derivatization stiff to bioactive functionalization (Morelli et al., 2013). Further, a natural biopolymer and its composites exhibit biodegradability, biocompatibility, and structural similarity and are employed in an extensive variety of applications in biomedical fields.

Naturally-Derived Biomedical Polymers

Polyhydroxyalkanoates (PHAs)

Industrially, PHAs biopolymer is synthesized from various microorganisms by fermentation of sugar and lipids. Owing to their biodegradability, easy processability, and biocompatibility, they are very useful materials for biomedical applications and the degraded product does not provide toxic effects compared to the chemically synthesized aliphatic polyester. From the application point of view, researchers are seeking great attention in PHAs. The degradation of the PHAs is based on two different mechanisms such as enzymatic and nonenzymatic hydrolysis due to the presence of the ester bond. The chemical composition of PHAs is based on the molecular weight, crystallinity, and biodegradation rate depending on the definite physiological atmosphere such as microbial growth conditions suggesting enough opportunity to modify the properties of the obtained material for example mechanical, processing, and biodegradation properties. For that reason, PHAs have been broadly used in the preparation of biodegradable devices e.g., orthopedic screw, skin patches, blood vessels, cartilage, etc. (Raza et al., 2018).

Polyvinyl alcohol (PVA)

PVA is an alternative biodegradable synthetic polymer employed in biomedical areas due to its elasticity, small protein adsorption, as well as higher water solubility. It can be utilized in dissolved supporting materials in 3D printing (extrusion based). Further, after crosslinked PVA looks like tougher hydrogel compared to PEG and it is employed to mimic and provide the mechanical strength of cartilage (Koshut et al., 2020).

Silk fibroin (SF)

SF protein is a natural biopolymer extracted from several insects larvae such as silkworms (bombyx mori) and moth genera respectively and broadly used in tissue engineering applications owing to their specific characteristics such as the low adverse reaction of the immune system, biocompatibility, and proteolytic enzymatic natural degradation in vivo condition. The SF-based scaffold is fabricated by various methods of electrospinning (Liu et al., 2011), freeze-drying (Teimouri et al., 2015), and 3D bioprinting (Wei et al., 2019). Further, it can be used in tissue build-up materials for skin, retina, urethra, blood vessels, bone, cardiac tissue, and cartilage. The SF also provides the adhere and proliferation to human cells such as mesenchymal stem cells, keratinocytes, osteoblast, and fibroblast cells during the scaffold (Chen et al., 2021; Song et al., 2018).

Chitosan

Chitosan is the second most vastly employed linear polysaccharide biopolymer partially derived from the deacetylated chitin, the main constituent of an exoskeleton of crustaceans for example carbs have been used in numerous biomedical purposes for wound dressing, drug delivery, and TE (Ahmed & Ikram, 2016; Croisier & Jérôme, 2013; Hu et al., 2013) due to its huge, biodegradability, biocompatibility, bio renewability, non-toxic, antibacterial nature. Chitosan scaffolds are prepared by the 3D bioprinting method and their hydrophilic character (Ergul et al., 2019) indicates that it is a suitable material for cryogel and hydrogen. As results confirmed that fabricated scaffolds by chitosan displayed positive responses in bone tissue engineering (Saekhor et al., 2019). Cartilage (Kalishwaralal et al., 2018), and cardiovascular tissues (Tian et al., 2012). Figure 3 demonstrated the chemical structure of the naturally derived biodegradable polymers.

Figure 3. Chemical structure of naturally derived biomedical polymers

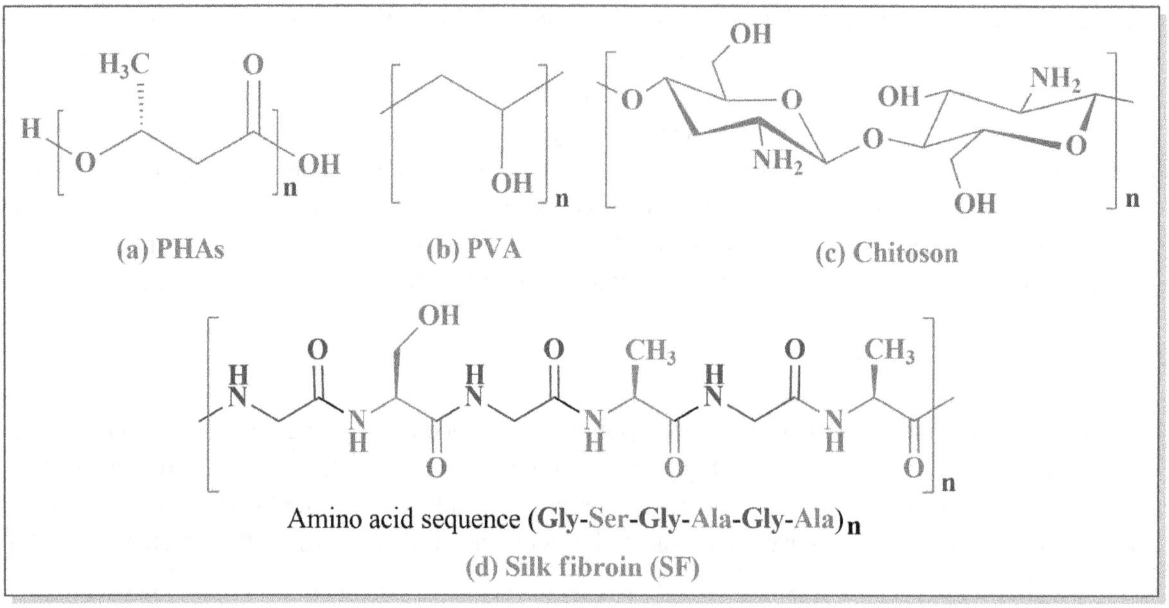

Synthetic Biodegradable Polymers for Biomedical

The synthesized biodegradable polymer demonstrates various advantages due to the presence of natural macromolecules supporting the processing application and industrialization. Another advantage is that we can modify the biodegradation, chemical, mechanical, physical, and functional characteristics of polymers by altering the synthesis or modification procedure condition and finding specific properties based on polymer as per our requirement. An extensive variety of synthetic biodegradable polymers has been synthesized for the pharmaceutical, and medical industries and for instance poly (lactic acid) (PLA), polycaprolactone (PCL), Poly(α-hydroxy acids), poly (glycolic acid) (PGA) and polydioxanone (PDO) (Chen et al., 2021). Conversely, numerous biodegradable polymers are degraded in the human body due to the hydrolysis of ester linkages (-COO-) because of the presence of a lot of biodegradable polyesters group in the backbone. These polymers exhibited two types of degradation hydrolytic and enzymatic in vivo conditions through hydrolysis of the ester. The degradation rate of the polymer is directly associated with the stability of backbone linkage and different other reasons for example degree of crystallinity, pH, material porosity, and water content equilibrium.

Poly (lactic acid) (PLA)

PLA is the most broadly employed synthetic biodegradable aliphatic polyester derived from natural resources such as sugar beet, wheat, sugar, etc. owing to its bioresorbability, biodegradability, easy processability as well as solubility in dissimilar organic solvents (Figure 4a) (Farah et al., 2016). The degradation product of the PLA does not carcinogenic the effect of tissue healing and degradation of PLA is very slow with in vivo conditions in the range of 9 to 15 (Gupta et al., 2007) months up to 3 years (Hutmacher, 2000), depending on the various molecular weight, degree of crystallinity, water diffusion rate and morphology of the polymer. PLA was mostly used in diverse applications for example drug delivery, wound treatment, and scaffold preparation by utilizing different methods such as electrospinning, 3D bioprinting, and fused deposition modeling (FDM). The fabricated scaffold provides suitable proliferation sites for several cells for instant osteoblasts (Howard et al., 2002), adipose stem cells (Narayanan et al., 2016), and mesenchymal stem cells (MSCs) (Gugliandolo et al., 2018). Owing to its biodegradability, as a non-toxic product, and hydrophobic nature, it can be used in the AM for 3D printing into the mesh, pins, and screws for effectiveness as biomedical implants (Kumbar et al., 2014).

Polycaprolactone (PCL)

PCL is the biodegradable polyester polymer developed by the ring-opening polymerization of ϵ-caprolactone (Figure 4b) (Parker, 2001). On the other hand, it is a biocompatible, semicrystalline, hydrophobic, and simple process at low temperatures (Patricio et al., 2013). The degradation of PCL in vivo condition based on the bulk and surface erosion mechanism. The degradation time range of PCL in vivo condition is 9 to 12 months (Hutmacher, 2000) and it's extensively utilized in long-term implant, and drug release applications. The properties of degradation time and other PCL can be modified after blending and copolymerization with other polymers. Further, PCL is used as a very promising biomaterial for implanting and can be degraded with the hydrolysis of the ester linkage. Owing to their crystallinity, hydrophobicity, and very slow degradation rate compared to the PLA, it is suitable for the fabrication of long-term implant materials (Abdo et al., 2017). Thus, PCL fabricated scaffold applies in bone regeneration and tissue repair by electrospinning (Xu et al., 2013) and biocell preparation printing (Domingos et al., 2013). The scaffolds demonstrated better results when employed in vascular (Han et al., 2011), bone (Ekaputra et al., 2009), and nerve tissue (Subramanian et al., 2012) applications. Additionally, PCL is also printed by using FDM and selective laser sintering (SLS), commercial brand monocryl approved by FDA for biomedical use.

Polyethylene glycol (PEG)

PEG is another type of synthetic polymer that is employed in different biomedical applications e.g., TE, drug delivery, and surface modification due to its hydrophilic nature and its linear and branched, neutral polyether structure depicted in Figure 4c (Daly et al., 2020; Liu et al., 2019; Panda et al., 2019). It is soluble in water and organic solvents such as toluene, methanol, and benzene. Further, PEG hydrogels are broadly utilized polymeric materials in the investigation of biomedical applications. Due to their low toxicity, non-immunogenicity, and biocompatibility, PEG hydrogels and their derived polymer are also modified with other biocompatible hydrogel materials and used in scaffold fabrication by 3D bioprinting (Truong et al., 2012). In the manufacture of PEG scaffolds in Tissue engineering (TE), PEG has been cross-linked with several polymers for instant PCL (Remya et al., 2013), PLA (Zhang et al., 2013), PEG acrylate and chitosan (Ma et al., 2010; Roberts & Bryant, 2013), etc. PEG hydrogels showed better results in the preparation of scaffolds for bone, vascular tissue, and cartilage by 3D-printed SLS and digital light projection (DLP) techniques with acrylate resin. The addition of chemical groups like lactate or glycolate can improve, biodegradation and bioabsorption rates (Damodaran et al., 2016).

Polydioxanone (PDO)

The first time Ethicon introduced the biodegradable PDO polymer in the market was in 1981 (Bartholomew, 1981; Goonoo et al., 2015). PDO is synthesized from the ring-opening polymerization of *p*-dioxanone in the presence of an organometallic catalyst depicted in Figure 4d. PDO exhibited various unique characteristics such as shape memory properties and it degraded by hydrolysis, with end product elimination by urine, along with a complete absorption rate of 6 months (Gierek et al., 2018). PDO is mostly used in biomedical such as surgical sutures as well as orthopedics, plastic surgery, cardiovascular, and tissue engineering applications. PDO-manufactured materials were sterilized with ethylene oxide.

Polypropylene fumarate (PPF)

PPF is an unsaturated polymer synthesized for the first time in 1994 by Mikos and Co-workers (Luo et al., 2019) and extensively used in continuous digital light processing (cDLP) 3D printing (Figure 4f) (Childers et al., 2015; Wallace et al., 2014). The degradation of the PPF polymer in the body is based on the Krebs cycle with the main component fumaric acid and food additive (propylene glycol) (Yaszemski et al., 1995). The main application of PPF polymers is in numerous areas such as vascular stents (Fisher et al., 2002), blood vessels (Bracaglia et al., 2014), controlled drug release (Bracaglia et al., 2014), nerve grafts (Choi et al., 2011), cartilage (Wang et al., 2008), and bone tissue engineering (Fisher et al., 2002; Liao et al., 2007). Recently the preparation of the PPF and its copolymers was based on the ring-opening copolymerization technique and control of the molecular mass and distribution (Dean et al., 2005). These properties significantly affect the viscosity of the polymer materials. Instead, the existence of the C-C double bond in the repeating unit provides the crosslinked structure of the covalent polymer system. Thus, PPF is a widely studied, synthetic biodegradable polymer material for photoreactive AM techniques.

Poly(glycolide) (PGA)

PGA is a biodegradable linear polyester synthesized from glycolic acid by polycondensation or ring-opening polymerization the structure is displayed in Figure 4e. The degradation rate of PGA is very high compared to the other synthetic polymers. The solubility parameter of PGA is lower in water and organic solvents, which is ascribed to its high degree of crystallinity. PGA is easily metabolized and removed from the body without harm with the tricarboxylic acid cycle this is approved by US Food and Drug Administration (FDA), and for that reason researchers are interested in using PGA for the development of biomedical devices and products such as different variations of sutures and implants to be used in the fractures and osteotomies by self-reinforcing techniques (Cao & Kuboyama, 2010; Song et al., 2018).

Figure 4. Chemical structure of synthetic biodegradable polymers

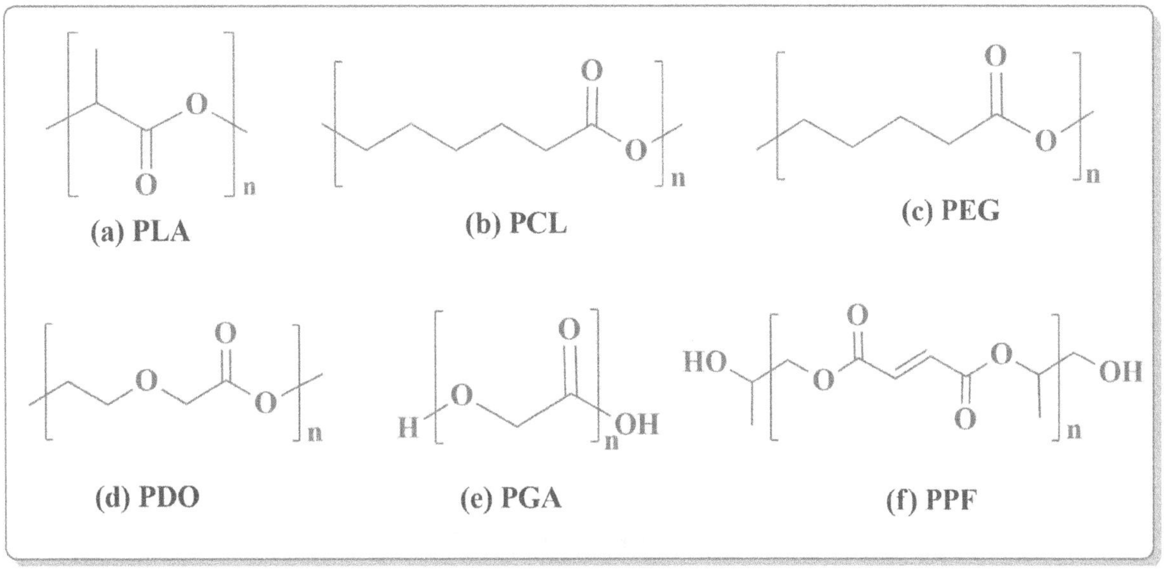

BIODEGRADABLE POLYMER COMPOSITES FOR ADDITIVE MANUFACTURING

The composite materials are a combination of two or more polymeric materials in a macroscopic scale with different compositions and morphology to form new specific materials with dissimilar mechanical and physical properties. The most important benefit of composite materials employed in biomedical applications is that can have a combination of the greatest characteristics of the ingredients. The biodegradable composite materials also demonstrated very useful properties for biomedicals such as biocompatibility, biodegradability, bioactivity, oxygen supply, electrical conductivity, and antimicrobial properties (Guo et al., 2021). However, polymer composites are made from a matrix (polymer) and one or more fillers which bestow chemical, physical, and biological characteristics. The main application of biodegradable composite materials in biomedical applications is depicted in Figure 5.

Backes et al. (Backes et al., 2021) prepared the poly (lactic acid) and PLA/β-tricalcium phosphate (TCP) biocomposites for biomedical AM application by melt blending with the addition of various wt% of TCP (5, 10, and 25) and investigated rheological, mechanical, and thermal properties of prepared composites. After analysis, PLA/5%TCP and PLA/10%TCP revealed similar rheological properties compared to the pure PLA and this is ascribed to the uniform distribution of the TCP into the PLA matrix. While 25%TCP demonstrated lower thermal stability due to the agglomeration. However, the addition of TCP into the PLA e matrix increases biocompatibility and bioactivity characteristic properties. The bioactivity of the developed biocomposite materials was tested by SBF solution for 7 and 21 days after immersing the sample in solution and was confirmed by the surface analysis of PLA/TCP biocomposites by x-ray diffraction (XRD), scanning electron microscopy (SEM) and. The results corroborated the formation of hydroxyapatite. Further, the biocompatibility was also tested by the same procedure for 7 days, and PLA/10%TCP exhibited enhanced relative cell viability compared with the control. Eventually, developed biocomposite materials displayed suitable rheological and thermal properties, filaments for AM, and they were employed to fabricate screw replicas for bone ligament fixation. 3D-printed screws

Figure 5. Application of biodegradable polymer composites applications in biomedical sectors

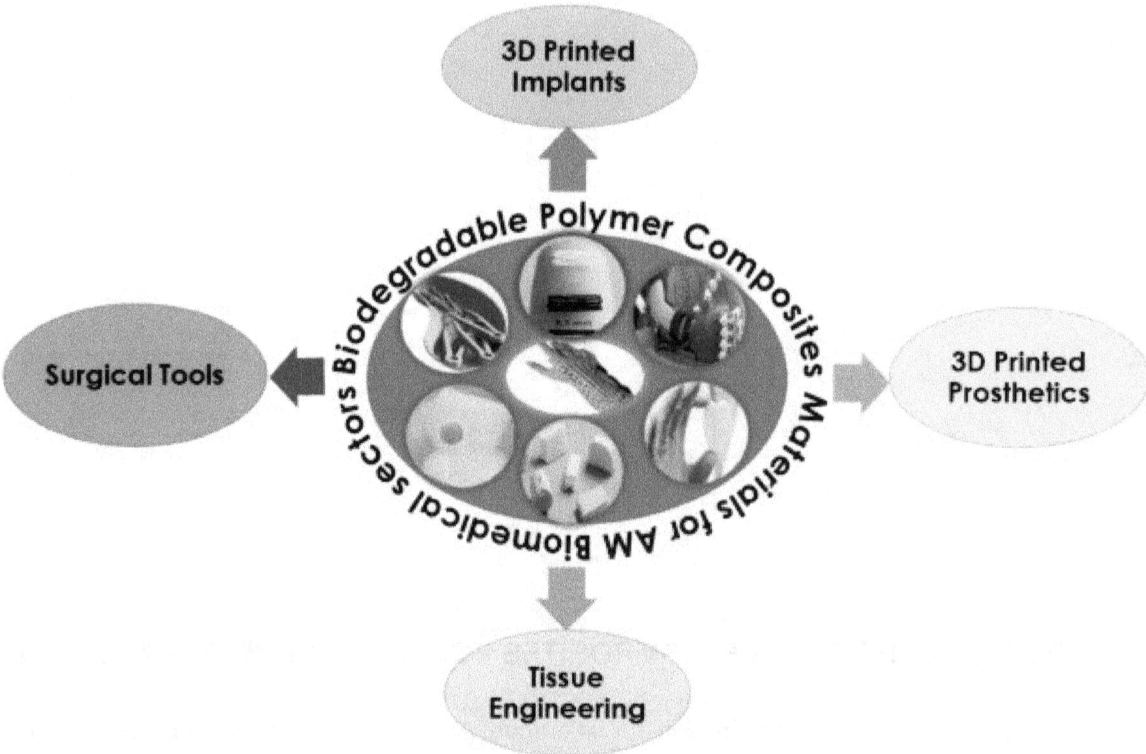

revealed tremendous printability and accuracy. Hence, the prepared PLA/TCP biocomposites can be further utilized in biomedical applications employing AM, specifically guided bone tissue engineering.

Lee et al. (Lee et al., 2019) fabricated PLA/Ti composite scaffold by the addition of various vol% (5 to 20) by fused filament deposition (FDM)-based 3D printing and studied the biological, thermal, and crystallization properties. The thermal stability and degree of crystallization/crystallization temperature of the composite scaffold were reduced by enhancing the titanium content. Alternatively, the glass transition temperature and melting point were parallelly enhanced. Further, the tensile and compressive strength of the developed PLA/Ti scaffold was enhanced with increased inclusion of Ti % up to 10 vol%. These properties of the scaffold are similar to real bone values. The impact strength of the scaffold composites remarkably increased as compared to the virgin PLA. The inclusion of Ti into the pure PLA matrix increases vitro biocompatibility, endorsing the initial attachment, proliferation, and variation of pre-osteoblast cells. Finally, the results corroborated that the addition of Ti into the PLA polymer matrix to prepare PLA/Ti composites scaffold increases mechanical and biological properties and it is a potential applicant for bone TE applications

Polycaprolactone-based bioactive composites scaffold were prepared by the inclusion of bio-derived hydroxyapatite (HA) by 3D printing and studied mechanical properties and vitro bioactivity with pure PCL as reported by Cestari et al. (Cestari et al., 2021). The prepared scaffold demonstrated a compressive elastic modulus of 203 to 316 MPa similar to trabecular bone and higher modulus and lower porosity but pure PCL scaffold exhibited lower modulus (177 MPa), due to the stiffening effect of HA nanoparticles.

As results confirmed, the PLA/bio-HA scaffold composites revealed good bioactivity as compared to the pure PCL, because bio-derived HA promotes improved cell adhesion, proliferation, and activation. The result indicated that bio-derive HA-based materials have better properties to be employed for the preparation of scaffolds with controlled properties such as porosity for suitable 3D printing methods for bone TE.

Neumann et al. (Neumann et al., 2019) also prepared a highly porous biocomposite scaffold with loading dissimilar wt% of inorganic ceramic $CaCO_3$ filler by FDM and mold casting techniques and investigated mechanical, morphological, and degradation properties. The mechanical properties of the scaffold were controlled by the inclusion of $CaCO_3$ into the PCL matrix. $CaCO_3$ is better than the PCL matrix's calcium phosphate (CaP) additive. Therefore, the bioactivity of the employed composite scaffold demonstrated clear mineral formation in vitro conditions and it is a promising material for the area of hard tissue replacement. As a result, 33% of $CaCO_3$ exhibited better processability, buffering effect, and apatite formation in vitro. The $CaCO_3$ demonstrated great potential for the preparation of an open-porous scaffold for hard tissue replacement.

Yeo et al. (Yeo et al., 2021) fabricated the porous PGA/HA composite scaffolds, after the inclusion of various wt% of HA (7.5, 10, 12.5, and 15%) by 3D printing technique. HA is used as a highly bone grafting material for hard tissue regeneration. Subsequently, 12.5 wt% HA loading composites showed maximum compressive modulus, proliferation, mineralization, and biodegradation. The degradation rate of the PGA/HA composite scaffolds was after enhancing the HA. In vivo experiment analysis with rabbits confirmed that PGA/HA composite scaffold revealed higher bone regeneration, greater bone mineral density as well as increased cell growth. Hence, 3D-printed PGA/HA scaffolds bestow a feasible option to endorse bone defects.

Pak et al. (Park et al., 2017) fabricated highly porous PCL/β-TCP composite scaffolds by employing the extrusion 3D printing method after the addition of 0-30 wt% tricalcium phosphate and studied the structure and compressive modulus. In vitro investigation results revealed that the developed scaffolds encourage the cell growth and osteogenic variation of mouse stem cells (mesenchymal).

AM APPLICATIONS FOR THE BIOMEDICAL SECTOR

Additive manufacturing techniques were extensively used in biomedical engineering applications and were developed in 1989 (Chia & Wu, 2015). AM can be employed in medicine to print exact medical products as per patient requirements, enhance the response to the surgery, and enhance surgeons' creativity for required products (Herbert et al., 2005; Mehrali et al., 2018). The 3D printing approach for clinics has grown owing to its remarkable evolution of different kinds of 3D printing methods and it provides the liberty to fabricate any shape and size of customized product (Zhao et al., 2017). In the clinics, they are 4 levels of printing present and cover the printing of surgical implants from 3D printing during the surgery. On the other hand, the fabrication of the product's customized design, such as prosthetics, tissue repair, and scaffold implantable, significantly depends on the quick implementation of the 3D printing technology at clinics. Therefore, 3D printing technology for fabricating any product at the clinic will be much more economical in the future decade.

Orthopaedic Implants

Orthopaedic implants are the technique used for traumatic injuries or joint degeneration treatment. The main work of the implants is to balance or enhance mobility by replacing cartilage, or affixing bone. For that application, the biodegradable polymer is employed in orthopaedic screws, pins, and rods as well as mimic cartilage (Dhandapani et al., 2020). Additionally, biodegradable solves the stress-shielding problems compared to commercial metal implants and decays the mechanical properties eventually (Reis & San Román, 2004). The orthopaedic screws work as a stiff internal fixation device and provide the spiral healing and mobility of damaged tissue during decreasing the fracture gap and bestowing compression diagonally to the fractured bones (Dhandapani et al., 2020). Apart from stress-shielding going on with a metallic screw, the tough incorporation of the screw thread and bone tissue can make difficult the amendment surgery for the duration of removal of the screws.

For this reason, the biodegradable polymer-based screw is a better substitute for metal screws, as their exploit results in deferring required for the second surgery. The fabricated porous orthopaedic screws from biodegradable PLA polymer by using the fused diffusion method (FDM) technique (Figure 6A). The screws with dissimilar sizes were printed to enhance bone ingrowth. According to Dhandapani et al. (Dhandapani et al., 2020) the vitro investigation of the porous screw, the results demonstrated that the improved proliferation, mineralized matrix, and hMSC adhesion were prepared by over 21 days in culture media as contrasted to nonporous screws. Furthermore, the printed screws reveal remarkably enhance vascularization in rat subcutaneous grafting in comparison to control screws.

Figure 6. Schematic representation of biodegradable polymeric orthopedic implants (Dhandapani et al., 2020)

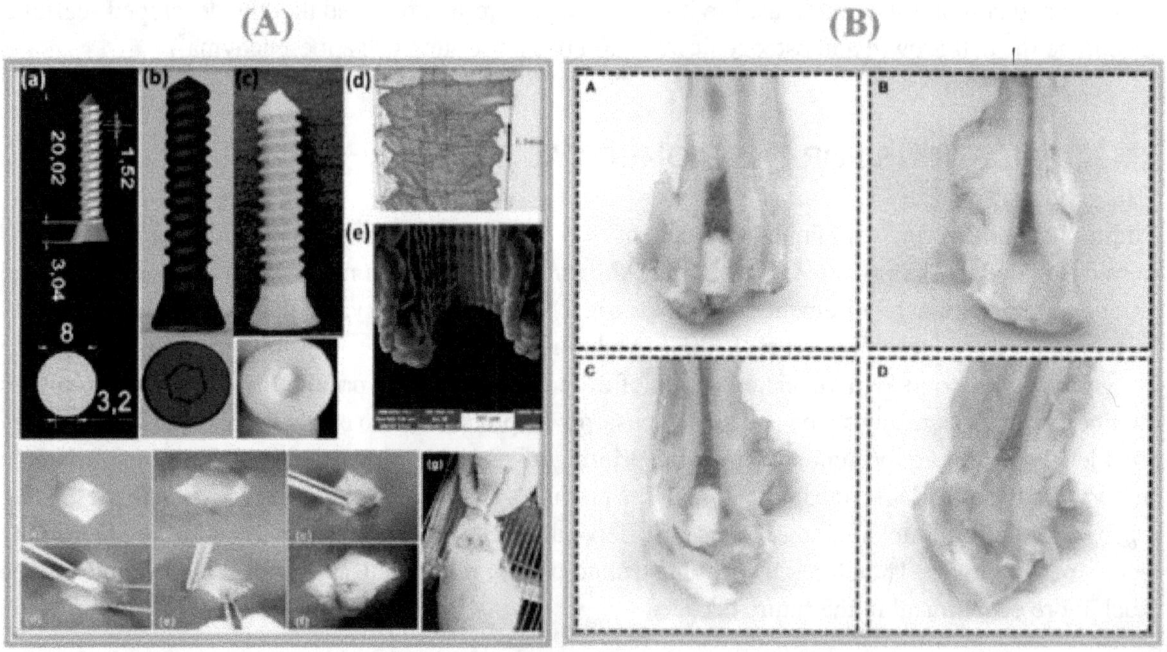

Further, Sanie et al. (Saniei & Mousavi, 2020) also fabricated PLA screws by the FDM technique coated with a PVA hydroxyapatite blend, employing an electrospinning method. On the other hand, PLA screw is frequently utilized in FDM and the medical areas and shows low cell affinity and is helped by PVA-HA coating. The main advantages of these types of screws are nontoxic and more bioactive owing to the electrospun layer. Manavitehrani et al. (Manavitehrani et al., 2017) prepared biodegradable screws from poly (propylene carbonate) and starch blend by melt molding method (Figure 6B), by incorporating glycerol and water as a plasticizer to double the modulus values. After that, the inclusion of bioglass enhanced the bioactivity and modulus. The final results demonstrated that the degradation of PPC–starch–bioglass screws was 45% at six months, much earlier than PLA screws further, tested osseointegration and observed that the bone was entered into the screw at 12 weeks.

3D Printed Prosthetics

In the last decade, various methods included in the preparation of human organ models are very costly and time-consuming (Ten Kate et al., 2017). Conversely, the new 3D printing technology is effectively used to prepare organ models at a very low cost (Mehrali et al., 2018). Currently, doctors also fabricate prosthetics for the patient immediately. The surgeries continuously transfer a risk issue for an unforeseen problem, for that reason doctors required essential whole information about the patient's health situation and surgical implementation. The computer-aided simulation 3D printing techniques resolve this issue

Figure 7. 3D antibacterial filament printed prosthetic finger (Zuniga, 2018)

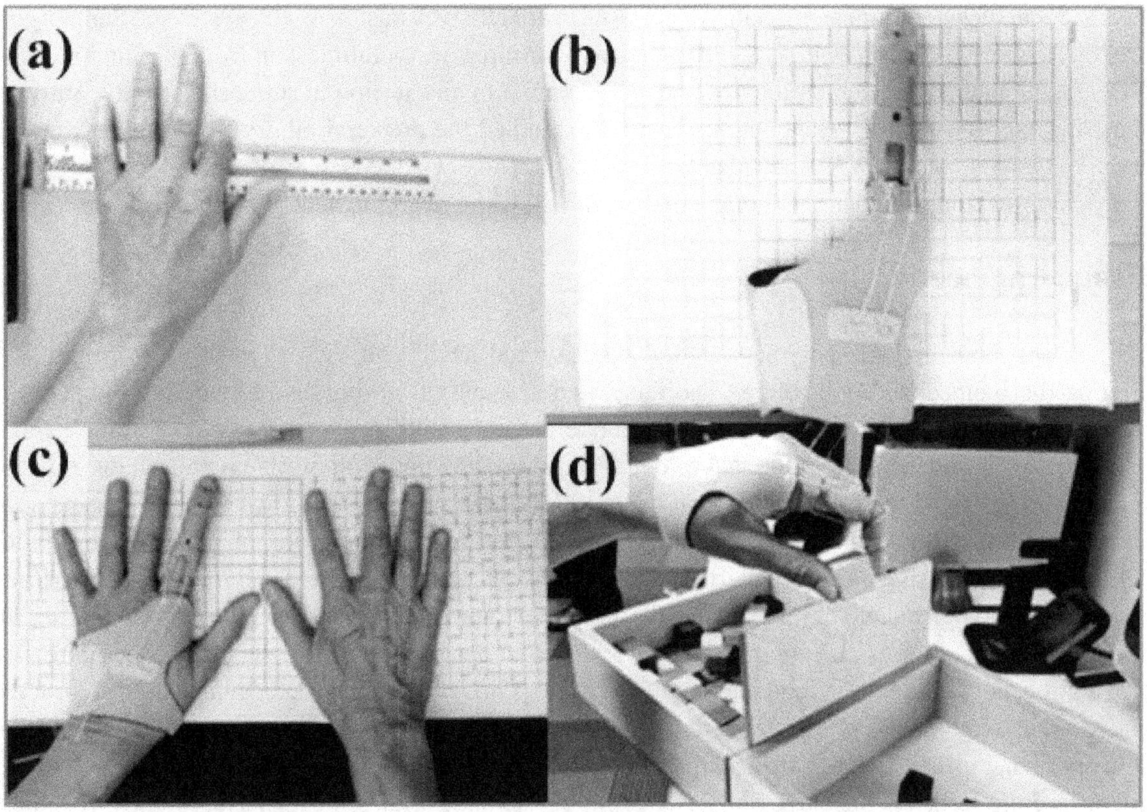

(Rengier et al., 2010). The surgical model was successfully fabricated by 3D printing technology with the required properties low-cost and wide availability. In this way, a simulated model has been prepared for bone reduction clamp, employing online existing software that can be utilized in finger fractures (Fuller et al., 2014).

Correspondingly, Pal et al. (Pal et al., 2021) designed an exoskeletal cortex cast that was employed in support of the trauma zone. The fabricated cortex cast exhibited ultralight, completely ventilated, recyclable, and hygienic conditions. In current years, researchers developed new technologies associated with prosthetic development that have endowed a greener solution to meet the increasing medical needs. AM has been used in various fields for different thermoplastic filament materials, liquid, powder, and hydrogels. In the current scenario, different varieties of biodegradable polymer materials are frequently employed in biomedical to prepare prosthetic and rehabilitation devices. Therefore, the 3D printing method is appropriate for abundant applications in prosthetic manufacturing.

Zuniga et al. (Zuniga, 2018) fabricated a 3D printed prosthetic finger by employing PLA with added copper nanoparticles antibacterial 3D printing filament (PLACTIVE™) depicted in Figure 7. Thus, 3D-printed filament exhibited highly antibacterial properties and was utilized in various medical applications, and had great potential for developing manufacturing devices.

Stents

Currently, stents are broadly utilized in medical surgery for cardiovascular diseases, such as ureteral stents and permanent solutions characteristically designed with bioabsorbable polymer stents on market for the first time in 2015 and approved by FDA and prepared from 3D printing and spray coating, replacing metals stents such as nitinol and stainless steel (Ang et al., 2017; Ormiston & Serruys, 2009; Sangiorgi et al., 2007). Kim et al. (Kim et al., 2019) prepared a 3D-printed drug-eluting stent from polycaprolactone (PCL) its biodegradable and antibiotics and studied them in an exertion to support disruptive salivary gland disease. Boyer et al. (Boyer et al., 2019) also studied the prepared biliary stents by 3D printing with polyvinyl alcohol (PVA) and filled with collagen. The developed stents are effective in revealing hydrophilic swelling character and radiocontrast, which is the main for stent position.

Tissue Engineering

The various biodegradable polymers exhibit porosity and have been used in mostly 3D printing for orthopaedic biomedical applications. The fabricated scaffolds from porous biomaterials support the encouragement of cellular responses and accelerate osteogenesis (Bose et al., 2013). Several cellular responses for instance proliferation, adhesion, penetration, and nutrition diffusion are achieved on the origin of porosity and scaffold composition structure (Hollister, 2005). Recently, very crucial steps were taken by scientists to design many advanced technologies for the fabrication of porous scaffolds for definite structures using various materials such as polymer, glass, ceramics, and metal with increased mechanical, biological, and physicochemical properties (Puppi et al., 2010). The formation of artificial bone in vivo condition is an evolving field of research in various properties such as proper pore size distribution, pore morphology, and interconnectivity respectively which were the main parameters at the macroscopic scale (Perez & Mestres, 2016). Though, the interaction between proteins and absorption delivery of therapeutic agents is the factor at the macroscopic scale that can decide the cellular shape and microstructure surfaces (Babaie & Bhaduri, 2018). A systematic arrangement analysis of the mechanical strength of 3D printed samples from different petroleum-based polymer materials and their composites is compared with the products obtained from additional traditional processing methods.

CONCLUSION

The current chapter presented the details about additive manufacturing biodegradable polymer and its composites materials employed for 3D printing in different biomedical sectors such as implants, surgical tools, prosthetic and orthopedics, tissue engineering, etc. and it is more beneficial compared to the traditional existing materials and manufacturing methods because of their benefits such as less energy consumption and a smaller amount of waste produced, less cost, quick flexibility of design to complex structure. The biodegradable polymer materials do not release any toxic gasses into the environment during the 3D printing process as compared to conventional petroleum-based materials. Therefore, biodegradable polymer materials revealed great potential for use in many biomedical fields and the replacement of traditional polymeric materials.

ACKNOWLEDGMENT

Dr. Sudheer Kumar contributed to this article in his personal capacity. The views expresses are his own and do not necessarily represent the view of the JC Orthoheal Pvt. Ltd. Akota, Vadodara Gujarat.

REFERENCES

Abdo, H. S., Elzatahry, A. A., Alharbi, H. F., & Khalil, K. A. (2017). Electrical conductivity behavior of biopolymer composites. In *Biopolymer Composites in Electronics* (pp. 13–25). Elsevier. doi:10.1016/B978-0-12-809261-3.00002-4

Ahmed, S., & Ikram, S. (2016). Chitosan based scaffolds and their applications in wound healing. *Achievements in the Life Sciences*, *10*(1), 27–37. doi:10.1016/j.als.2016.04.001

Ambrosi, A., & Pumera, M. (2016). 3D-printing technologies for electrochemical applications. *Chemical Society Reviews*, *45*(10), 2740–2755. doi:10.1039/C5CS00714C PMID:27048921

Ang, H. Y., Bulluck, H., Wong, P., Venkatraman, S. S., Huang, Y., & Foin, N. (2017). Bioresorbable stents: Current and upcoming bioresorbable technologies. *International Journal of Cardiology*, *228*, 931–939. doi:10.1016/j.ijcard.2016.11.258 PMID:27912202

Babaie, E., & Bhaduri, S. B. (2018). Fabrication aspects of porous biomaterials in orthopedic applications: A review. *ACS Biomaterials Science & Engineering*, *4*(1), 1–39. doi:10.1021/acsbiomaterials.7b00615 PMID:33418675

Backes, E. H., de Nóbile Pires, L., Selistre-de-Araujo, H. S., Costa, L. C., Passador, F. R., & Pessan, L. A. (2021). Development and characterization of printable PLA/β-TCP bioactive composites for bone tissue applications. *Journal of Applied Polymer Science*, *138*(5), 49759. doi:10.1002/app.49759

Bartholomew, R. S. (1981). PDS (polydioxanone suture): A new synthetic absorbable suture in cataract surgery. *Ophthalmologica*, *183*(2), 81–85. doi:10.1159/000309144 PMID:6795557

Bose, S., Vahabzadeh, S., & Bandyopadhyay, A. (2013). Bone tissue engineering using 3D printing. *Materials Today*, *16*(12), 496–504. doi:10.1016/j.mattod.2013.11.017

Boyer, C. J., Boktor, M., Samant, H., White, L. A., Wang, Y., Ballard, D. H., & Alexander, J. S. (2019). 3D printing for bio-synthetic biliary stents. *Bioengineering (Basel, Switzerland)*, *6*(1), 16. doi:10.3390/bioengineering6010016 PMID:30744131

Bracaglia, L. G., Yu, L., Hibino, N., & Fisher, J. P. (2014). Reinforced pericardium as a hybrid material for cardiovascular applications. *Tissue Engineering. Part A*, *20*(21-22), 2807–2816. doi:10.1089/ten.tea.2014.0516 PMID:25236439

Cao, H., & Kuboyama, N. (2010). A biodegradable porous composite scaffold of PGA/β-TCP for bone tissue engineering. *Bone*, *46*(2), 386–395. doi:10.1016/j.bone.2009.09.031 PMID:19800045

Cestari, F., Petretta, M., Yang, Y., Motta, A., Grigolo, B., & Sglavo, V. M. (2021). 3D printing of PCL/nano-hydroxyapatite scaffolds derived from biogenic sources for bone tissue engineering. *Sustainable Materials and Technologies*, *29*, e00318. doi:10.1016/j.susmat.2021.e00318

Chen, Q., Maalihan, R. D., Ren, J., da Silva, Í. G., Dugos, N. P., Caldona, E. B., & Advincula, R. C. (2021). 3D printing of biomedically relevant polymer materials and biocompatibility. *MRS Communications*, *11*(2), 197–212. doi:10.155743579-021-00038-8 PMID:33936866

Chia, H. N., & Wu, B. M. (2015). Recent advances in 3D printing of biomaterials. *Journal of Biological Engineering*, *9*(1), 1–14. doi:10.118613036-015-0001-4 PMID:25866560

Childers, E. P., Wang, M. O., Becker, M. L., Fisher, J. P., & Dean, D. (2015). 3D printing of resorbable poly (propylene fumarate) tissue engineering scaffolds. *MRS Bulletin*, *40*(2), 119–126. doi:10.1557/mrs.2015.2

Choi, J., Kim, K., Kim, T., Liu, G., Bar-Shir, A., Hyeon, T., & Gilad, A. A. (2011). Multimodal imaging of sustained drug release from 3-D poly (propylene fumarate) (PPF) scaffolds. *Journal of Controlled Release*, *156*(2), 239–245. doi:10.1016/j.jconrel.2011.06.035 PMID:21763735

Croisier, F., & Jérôme, C. (2013). Chitosan-based biomaterials for tissue engineering. *European Polymer Journal*, *49*(4), 780–792. doi:10.1016/j.eurpolymj.2012.12.009

Daly, A. C., Riley, L., Segura, T., & Burdick, J. A. (2020). Hydrogel microparticles for biomedical applications. *Nature Reviews. Materials*, *5*(1), 20–43. doi:10.103841578-019-0148-6 PMID:34123409

Damodaran, V. B., Bhatnagar, D., & Murthy, N. S. (2016). *Biomedical polymers: Synthesis and processing*. Springer.

Dean, D., Wolfe, M. S., Ahmad, Y., Totonchi, A., Chen, J. E. K., Fisher, J. P., Cooke, M. N., Rimnac, C. M., Lennon, D. P., Caplan, A. I., Topham, N. S., & Mikos, A. G. (2005). Effect of transforming growth factor β2 on marrow-infused foam poly (propylene fumarate) tissue-engineered constructs for the repair of critical-size cranial defects in rabbits. *Tissue Engineering*, *11*(5-6), 923–939. doi:10.1089/ten.2005.11.923 PMID:15998232

Dhandapani, R., Krishnan, P. D., Zennifer, A., Kannan, V., Manigandan, A., Arul, M. R., & Sethuraman, S. (2020). Additive manufacturing of biodegradable porous orthopaedic screw. *Bioactive Materials*, *5*(3), 458–467. doi:10.1016/j.bioactmat.2020.03.009 PMID:32280835

Domingos, M., Intranuovo, F., Russo, T., De Santis, R., Gloria, A., Ambrosio, L., & Bartolo, P. (2013). The first systematic analysis of 3D rapid prototyped poly (ε-caprolactone) scaffolds manufactured through BioCell printing: The effect of pore size and geometry on compressive mechanical behaviour and in vitro hMSC viability. *Biofabrication*, *5*(4), 045004. doi:10.1088/1758-5082/5/4/045004 PMID:24192056

Durfee, W. K., & Iaizzo, P. A. (2019). Medical applications of 3D printing. In *Engineering in medicine* (pp. 527–543). Academic Press. doi:10.1016/B978-0-12-813068-1.00021-X

Ekaputra, A. K., Zhou, Y., Cool, S. M., & Hutmacher, D. W. (2009). Composite electrospun scaffolds for engineering tubular bone grafts. *Tissue Engineering. Part A*, *15*(12), 3779–3788. doi:10.1089/ten.tea.2009.0186 PMID:19527183

Ergul, N. M., Unal, S., Kartal, I., Kalkandelen, C., Ekren, N., Kilic, O., & Gunduz, O. (2019). 3D printing of chitosan/poly (vinyl alcohol) hydrogel containing synthesized hydroxyapatite scaffolds for hard-tissue engineering. *Polymer Testing*, *79*, 106006. doi:10.1016/j.polymertesting.2019.106006

Farah, S., Anderson, D. G., & Langer, R. (2016). Physical and mechanical properties of PLA, and their functions in widespread applications—A comprehensive review. *Advanced Drug Delivery Reviews*, *107*, 367–392. doi:10.1016/j.addr.2016.06.012 PMID:27356150

Fisher, J. P., Dean, D., & Mikos, A. G. (2002). Photocrosslinking characteristics and mechanical properties of diethyl fumarate/poly (propylene fumarate) biomaterials. *Biomaterials*, *23*(22), 4333–4343. doi:10.1016/S0142-9612(02)00178-3 PMID:12219823

Fuller, S. M., Butz, D. R., Vevang, C. B., & Makhlouf, M. V. (2014). Application of 3-dimensional printing in hand surgery for production of a novel bone reduction clamp. *The Journal of Hand Surgery*, *39*(9), 1840–1845. doi:10.1016/j.jhsa.2014.06.009 PMID:25042538

Gierek, M., Kuśnierz, K., Lampe, P., Ochała, G., Kurek, J., Hekner, B., & Majewski, J. (2018). Absorbable sutures in general surgery–review, available materials, and optimum choices. *Polish Journal of Surgery*, *90*(2), 34–37. doi:10.5604/01.3001.0010.5632 PMID:29773758

Goonoo, N., Jeetah, R., Bhaw-Luximon, A., & Jhurry, D. (2015). Polydioxanone-based bio-materials for tissue engineering and drug/gene delivery applications. *European Journal of Pharmaceutics and Biopharmaceutics*, *97*, 371–391. doi:10.1016/j.ejpb.2015.05.024 PMID:26614558

Gross, R. A., & Kalra, B. (2002). Biodegradable polymers for the environment. *Science*, *297*(5582), 803–807. doi:10.1126cience.297.5582.803 PMID:12161646

Gugliandolo, A., Diomede, F., Cardelli, P., Bramanti, A., Scionti, D., Bramanti, P., & Mazzon, E. (2018). Transcriptomic analysis of gingival mesenchymal stem cells cultured on 3 d bioprinted scaffold: A promising strategy for neuroregeneration. *Journal of Biomedical Materials Research. Part A*, *106*(1), 126–137. doi:10.1002/jbm.a.36213 PMID:28879677

Guo, Z., Poot, A. A., & Grijpma, D. W. (2021). Advanced polymer-based composites and structures for biomedical applications. *European Polymer Journal, 149*, 110388. doi:10.1016/j.eurpolymj.2021.110388

Gupta, B., Revagade, N., & Hilborn, J. (2007). Poly (lactic acid) fiber: An overview. *Progress in Polymer Science, 32*(4), 455–482. doi:10.1016/j.progpolymsci.2007.01.005

Han, J., Lazarovici, P., Pomerantz, C., Chen, X., Wei, Y., & Lelkes, P. I. (2011). Co-electrospun blends of PLGA, gelatin, and elastin as potential nonthrombogenic scaffolds for vascular tissue engineering. *Biomacromolecules, 12*(2), 399–408. doi:10.1021/bm101149r PMID:21182235

Herbert, N., Simpson, D., Spence, W. D., & Ion, W. (2005). A preliminary investigation into the development of 3-D printing of prosthetic sockets. *Journal of Rehabilitation Research and Development, 42*(2), 141. doi:10.1682/JRRD.2004.08.0134 PMID:15944878

Hollister, S. J. (2005). Porous scaffold design for tissue engineering. *Nature Materials, 4*(7), 518–524. doi:10.1038/nmat1421 PMID:16003400

Howard, D., Partridge, K., Yang, X., Clarke, N. M., Okubo, Y., Bessho, K., & Oreffo, R. O. (2002). Immunoselection and adenoviral genetic modulation of human osteoprogenitors: In vivo bone formation on PLA scaffold. *Biochemical and Biophysical Research Communications, 299*(2), 208–215. doi:10.1016/S0006-291X(02)02561-5 PMID:12437971

Hu, L., Sun, Y., & Wu, Y. (2013). Advances in chitosan-based drug delivery vehicles. *Nanoscale, 5*(8), 3103–3111. doi:10.1039/c3nr00338h PMID:23515527

Hutmacher, D. W. (2000). Scaffolds in tissue engineering bone and cartilage. *Biomaterials, 21*(24), 2529–2543. doi:10.1016/S0142-9612(00)00121-6 PMID:11071603

Jamróz, W., Szafraniec, J., Kurek, M., & Jachowicz, R. (2018). 3D printing in pharmaceutical and medical applications–recent achievements and challenges. *Pharmaceutical Research, 35*(9), 1–22. doi:10.100711095-018-2454-x PMID:29998405

Kalishwaralal, K., Jeyabharathi, S., Sundar, K., Selvamani, S., Prasanna, M., & Muthukumaran, A. (2018). A novel biocompatible chitosan–Selenium nanoparticles (SeNPs) film with electrical conductivity for cardiac tissue engineering application. *Materials Science and Engineering C, 92*, 151–160. doi:10.1016/j.msec.2018.06.036 PMID:30184738

Kim, T. H., Lee, J. H., Ahn, C. B., Hong, J. H., Son, K. H., & Lee, J. W. (2019). Development of a 3D-printed drug-eluting stent for treating obstructive salivary gland disease. *ACS Biomaterials Science & Engineering, 5*(7), 3572–3581. doi:10.1021/acsbiomaterials.9b00636 PMID:33405739

Koshut, W. J., Smoot, D., Rummel, C., Kirillova, A., & Gall, K. (2020). Tensile Fatigue of Poly (Vinyl Alcohol) Hydrogels with Bio-Friendly Toughening Agents. *Macromolecular Materials and Engineering, 305*(3), 1900784. doi:10.1002/mame.201900784

Kumbar, S., Laurencin, C., & Deng, M. (Eds.). (2014). *Natural and synthetic biomedical polymers*. Newnes.

Lee, J., Lee, H., Cheon, K. H., Park, C., Jang, T. S., Kim, H. E., & Jung, H. D. (2019). Fabrication of poly (lactic acid)/Ti composite scaffolds with enhanced mechanical properties and biocompatibility via fused filament fabrication (FFF)–based 3D printing. *Additive Manufacturing*, *30*, 100883. doi:10.1016/j.addma.2019.100883

Liao, E., Yaszemski, M., Krebsbach, P., & Hollister, S. (2007). Tissue-engineered cartilage constructs using composite hyaluronic acid/collagen I hydrogels and designed poly (propylene fumarate) scaffolds. *Tissue Engineering*, *13*(3), 537–550. doi:10.1089/ten.2006.0117 PMID:17319795

Lim, C. W. J., Le, K. Q., Lu, Q., & Wong, C. H. (2016). An overview of 3-D printing in manufacturing, aerospace, and automotive industries. *IEEE Potentials*, *35*(4), 18–22. doi:10.1109/MPOT.2016.2540098

Liu, H., Li, X., Zhou, G., Fan, H., & Fan, Y. (2011). Electrospun sulfated silk fibroin nanofibrous scaffolds for vascular tissue engineering. *Biomaterials*, *32*(15), 3784–3793. doi:10.1016/j.biomaterials.2011.02.002 PMID:21376391

Liu, P., Chen, W., Liu, C., Tian, M., & Liu, P. (2019). A novel poly (vinyl alcohol)/poly (ethylene glycol) scaffold for tissue engineering with a unique bimodal open-celled structure fabricated using supercritical fluid foaming. *Scientific Reports*, *9*(1), 1–12. doi:10.103841598-019-46061-7 PMID:31267014

LogithKumar, R., KeshavNarayan, A., Dhivya, S., Chawla, A., Saravanan, S., & Selvamurugan, N. (2016). A review of chitosan and its derivatives in bone tissue engineering. *Carbohydrate Polymers*, *151*, 172–188. doi:10.1016/j.carbpol.2016.05.049 PMID:27474556

Luo, Y., Le Fer, G., Dean, D., & Becker, M. L. (2019). 3D printing of poly (propylene fumarate) oligomers: Evaluation of resin viscosity, printing characteristics and mechanical properties. *Biomacromolecules*, *20*(4), 1699–1708. doi:10.1021/acs.biomac.9b00076 PMID:30807696

Ma, G., Yang, D., Li, Q., Wang, K., Chen, B., Kennedy, J. F., & Nie, J. (2010). Injectable hydrogels based on chitosan derivative/polyethylene glycol dimethacrylate/N, N-dimethylacrylamide as bone tissue engineering matrix. *Carbohydrate Polymers*, *79*(3), 620–627. doi:10.1016/j.carbpol.2009.09.015

Madhavadas, V., Srivastava, D., Chadha, U., Raj, S. A., Sultan, M. T. H., Shahar, F. S., & Shah, A. U. M. (2022). A review on metal additive manufacturing for intricately shaped aerospace components. *CIRP Journal of Manufacturing Science and Technology*, *39*, 18–36. doi:10.1016/j.cirpj.2022.07.005

Manavitehrani, I., Fathi, A., Wang, Y., Maitz, P. K., Mirmohseni, F., Cheng, T. L., & Dehghani, F. (2017). Fabrication of a biodegradable implant with tunable characteristics for bone implant applications. *Biomacromolecules*, *18*(6), 1736–1746. doi:10.1021/acs.biomac.7b00078 PMID:28535038

Mathew, J., Das, J. P., Tp, M., & Kumar, S. (2022). Development of poly (butylene adipate-co-terephthalate) PBAT toughened poly (lactic acid) blends 3D printing filament. *Journal of Polymer Research*, *29*(11), 1–15. doi:10.100710965-022-03320-y

Mehrali, M., Bagherifard, S., Akbari, M., Thakur, A., Mirani, B., Mehrali, M., & Dolatshahi-Pirouz, A. (2018). Blending electronics with the human body: A pathway toward a cybernetic future. *Advancement of Science*, *5*(10), 1700931. doi:10.1002/advs.201700931 PMID:30356969

Morelli, A., Puppi, D., & Chiellini, F. (2013). Polymers from renewable resources. *Journal of Renewable Materials*, *1*(2), 83–112. doi:10.7569/JRM.2012.634106

Narayanan, L. K., Huebner, P., Fisher, M. B., Spang, J. T., Starly, B., & Shirwaiker, R. A. (2016). 3D-bioprinting of polylactic acid (PLA) nanofiber–alginate hydrogel bioink containing human adipose-derived stem cells. *ACS Biomaterials Science & Engineering*, *2*(10), 1732–1742. doi:10.1021/acsbiomaterials.6b00196 PMID:33440471

Neumann, R., Neunzehn, J., Hinueber, C., Flath, T., Schulze, F. P., & Wiesmann, H. P. (2019). 3D-printed poly-ε-caprolactone-CaCO 3-biocompositescaffolds for hard tissue regeneration. *Express Polymer Letters*, *13*(1), 2–17. doi:10.3144/expresspolymlett.2019.2

Ormiston, J. A., & Serruys, P. W. (2009). Bioabsorbable coronary stents. *Circulation: Cardiovascular Interventions*, *2*(3), 255–260. doi:10.1161/CIRCINTERVENTIONS.109.859173 PMID:20031723

Pal, A. K., Mohanty, A. K., & Misra, M. (2021). Additive manufacturing technology of polymeric materials for customized products: Recent developments and future prospective. *RSC Advances*, *11*(58), 36398–36438. doi:10.1039/D1RA04060J PMID:35494368

Panda, B. P., Krishnamoorthy, R., Bhattamisra, S. K., Shivashekaregowda, N. K. H., Seng, L. B., & Patnaik, S. (2019). Fabrication of second generation smarter PLGA based nanocrystal carriers for improvement of drug delivery and therapeutic efficacy of gliclazide in type-2 diabetes rat model. *Scientific Reports*, *9*(1), 1–15. doi:10.103841598-019-53996-4 PMID:31758056

Park, S. H., Park, S. A., Kang, Y. G., Shin, J. W., Park, Y. S., Gu, S. R., & Shin, J. W. (2017). PCL/β-TCP composite scaffolds exhibit positive osteogenic differentiation with mechanical stimulation. *Tissue Engineering and Regenerative Medicine*, *14*(4), 349–358. doi:10.100713770-017-0022-9 PMID:30603491

Parker, G. (2001). Encyclopedia of materials: Science and technology. Academic Press.

Patricio, T., Domingos, M., Gloria, A., & Bártolo, P. (2013). Characterisation of PCL and PCL/PLA scaffolds for tissue engineering. *Procedia CIRP*, *5*, 110–114. doi:10.1016/j.procir.2013.01.022

Perez, R. A., & Mestres, G. (2016). Role of pore size and morphology in musculo-skeletal tissue regeneration. *Materials Science and Engineering C*, *61*, 922–939. doi:10.1016/j.msec.2015.12.087 PMID:26838923

Puppi, D., & Chiellini, F. (2020). Biodegradable polymers for biomedical additive manufacturing. *Applied Materials Today*, *20*, 100700. doi:10.1016/j.apmt.2020.100700

Puppi, D., Chiellini, F., Piras, A. M., & Chiellini, E. (2010). Polymeric materials for bone and cartilage repair. *Progress in Polymer Science*, *35*(4), 403–440. doi:10.1016/j.progpolymsci.2010.01.006

Raza, Z. A., Abid, S., & Banat, I. M. (2018). Polyhydroxyalkanoates: Characteristics, production, recent developments and applications. *International Biodeterioration & Biodegradation*, *126*, 45–56. doi:10.1016/j.ibiod.2017.10.001

Reddy, M. M., Misra, M., & Mohanty, A. K. (2012). Bio-based materials in the new bio-economy. *Chemical Engineering Progress*, *108*(5), 37–42.

Reis, R. L., & San Román, J. (2004). *Biodegradable systems in tissue engineering and regenerative medicine*. CRC Press. doi:10.1201/9780203491232

Remya, K. R., Joseph, J., Mani, S., John, A., Varma, H. K., & Ramesh, P. (2013). Nanohydroxyapatite incorporated electrospun polycaprolactone/polycaprolactone–polyethyleneglycol–polycaprolactone blend scaffold for bone tissue engineering applications. *Journal of Biomedical Nanotechnology, 9*(9), 1483–1494. doi:10.1166/jbn.2013.1640 PMID:23980497

Rengier, F., Mehndiratta, A., Von Tengg-Kobligk, H., Zechmann, C. M., Unterhinninghofen, R., Kauczor, H. U., & Giesel, F. L. (2010). 3D printing based on imaging data: Review of medical applications. *International Journal of Computer Assisted Radiology and Surgery, 5*(4), 335–341. doi:10.100711548-010-0476-x PMID:20467825

Roberts, J. J., & Bryant, S. J. (2013). Comparison of photopolymerizable thiol-ene PEG and acrylate-based PEG hydrogels for cartilage development. *Biomaterials, 34*(38), 9969–9979. doi:10.1016/j.biomaterials.2013.09.020 PMID:24060418

Saekhor, K., Udomsinprasert, W., Honsawek, S., & Tachaboonyakiat, W. (2019). Preparation of an injectable modified chitosan-based hydrogel approaching for bone tissue engineering. *International Journal of Biological Macromolecules, 123*, 167–173. doi:10.1016/j.ijbiomac.2018.11.041 PMID:30423397

Sangiorgi, G., Melzi, G., Agostoni, P., Cola, C., Clementi, F., Romitelli, P., & Colombo, A. (2007). Engineering aspects of stents design and their translation into clinical practice. *Annali dell'Istituto Superiore di Sanita, 43*(1), 89–100. PMID:17536159

Saniei, H., & Mousavi, S. (2020). Surface modification of PLA 3D-printed implants by electrospinning with enhanced bioactivity and cell affinity. *Polymer, 196*, 122467. doi:10.1016/j.polymer.2020.122467

Shah, T. V., & Vasava, D. V. (2019). A glimpse of biodegradable polymers and their biomedical applications. *e-Polymers, 19*(1), 385-410.

Sheoran, A. J., Kumar, H., Arora, P. K., & Moona, G. (2020). Bio-medical applications of additive manufacturing: A review. *Procedia Manufacturing, 51*, 663–670. doi:10.1016/j.promfg.2020.10.093

Singh, A. V., Ansari, M. H. D., Wang, S., Laux, P., Luch, A., Kumar, A., & Nussberger, S. (2019). The adoption of three-dimensional additive manufacturing from biomedical material design to 3D organ printing. *Applied Sciences-Basel, 9*(4).

Song, R., Murphy, M., Li, C., Ting, K., Soo, C., & Zheng, Z. (2018). Current development of biodegradable polymeric materials for biomedical applications. *Drug Design, Development and Therapy, 12*, 3117–3145. doi:10.2147/DDDT.S165440 PMID:30288019

Subramanian, A., Krishnan, U. M., & Sethuraman, S. (2012). Fabrication, characterization and in vitro evaluation of aligned PLGA–PCL nanofibers for neural regeneration. *Annals of Biomedical Engineering, 40*(10), 2098–2110. doi:10.100710439-012-0592-6 PMID:22618802

Teimouri, A., Azadi, M., Emadi, R., Lari, J., & Chermahini, A. N. (2015). Preparation, characterization, degradation, and biocompatibility of different silk fibroin based composite scaffolds prepared by freeze-drying method for tissue engineering application. *Polymer Degradation & Stability*, *121*, 18–29. doi:10.1016/j.polymdegradstab.2015.08.004

Ten Kate, J., Smit, G., & Breedveld, P. (2017). 3D-printed upper limb prostheses: A review. *Disability and Rehabilitation. Assistive Technology*, *12*(3), 300–314. doi:10.1080/17483107.2016.1253117 PMID:28152642

Tian, H., Tang, Z., Zhuang, X., Chen, X., & Jing, X. (2012). Biodegradable synthetic polymers: Preparation, functionalization and biomedical application. *Progress in Polymer Science*, *37*(2), 237–280. doi:10.1016/j.progpolymsci.2011.06.004

Truong, V., Blakey, I., & Whittaker, A. K. (2012). Hydrophilic and amphiphilic polyethylene glycol-based hydrogels with tunable degradability prepared by "click" chemistry. *Biomacromolecules*, *13*(12), 4012–4021. doi:10.1021/bm3012924 PMID:23134321

Wallace, J., Wang, M. O., Thompson, P., Busso, M., Belle, V., Mammoser, N., & Dean, D. (2014). Validating continuous digital light processing (cDLP) additive manufacturing accuracy and tissue engineering utility of a dye-initiator package. *Biofabrication*, *6*(1), 015003. doi:10.1088/1758-5082/6/1/015003 PMID:24429508

Wang, S., Kempen, D. H., Simha, N. K., Lewis, J. L., Windebank, A. J., Yaszemski, M. J., & Lu, L. (2008). Photo-cross-linked hybrid polymer networks consisting of poly (propylene fumarate) and poly (caprolactone fumarate): Controlled physical properties and regulated bone and nerve cell responses. *Biomacromolecules*, *9*(4), 1229–1241. doi:10.1021/bm7012313 PMID:18307311

Wang, X., Jiang, M., Zhou, Z., Gou, J., & Hui, D. (2017). 3D printing of polymer matrix composites: A review and prospective. *Composites. Part B, Engineering*, *110*, 442–458. doi:10.1016/j.compositesb.2016.11.034

Wei, L., Wu, S., Kuss, M., Jiang, X., Sun, R., Reid, P., & Duan, B. (2019). 3D printing of silk fibroin-based hybrid scaffold treated with platelet rich plasma for bone tissue engineering. *Bioactive Materials*, *4*, 256–260. doi:10.1016/j.bioactmat.2019.09.001 PMID:31667442

Xu, H., Cui, W., & Chang, J. (2013). Fabrication of patterned PDLLA/PCL composite scaffold by electrospinning. *Journal of Applied Polymer Science*, *127*(3), 1550–1554. doi:10.1002/app.37505

Yan, Q., Dong, H., Su, J., Han, J., Song, B., Wei, Q., & Shi, Y. (2018). A review of 3D printing technology for medical applications. *Engineering*, *4*(5), 729–742. doi:10.1016/j.eng.2018.07.021

Yaszemski, M. J., Payne, R. G., Hayes, W. C., Langer, R. S., Aufdemorte, T. B., & Mikos, A. G. (1995). The ingrowth of new bone tissue and initial mechanical properties of a degrading polymeric composite scaffold. *Tissue Engineering*, *1*(1), 41–52. doi:10.1089/ten.1995.1.41 PMID:19877914

Yeo, T., Ko, Y. G., Kim, E. J., Kwon, O. K., Chung, H. Y., & Kwon, O. H. (2021). Promoting bone regeneration by 3D-printed poly (glycolic acid)/hydroxyapatite composite scaffolds. *Journal of Industrial and Engineering Chemistry*, *94*, 343–351. doi:10.1016/j.jiec.2020.11.004

Zhang, W., Chen, B., Zhao, H., Yu, P., Fu, D., Wen, J., & Peng, X. (2013). Processing and characterization of supercritical CO2 batch foamed poly (lactic acid)/poly (ethylene glycol) scaffold for tissue engineering application. *Journal of Applied Polymer Science*, *130*(5), 3066–3073. doi:10.1002/app.39523

Zhao, H., Yang, F., Fu, J., Gao, Q., Liu, A., Sun, M., & He, Y. (2017). email protected]: From Medical Models to Organ Implants. *ACS Biomaterials Science & Engineering*, *3*(12), 3083–3097. doi:10.1021/acsbiomaterials.7b00542 PMID:33445353

Zuniga, J. M. (2018). 3D printed antibacterial prostheses. *Applied Sciences (Basel, Switzerland)*, *8*(9), 1651. doi:10.3390/app8091651

Chapter 8
An Overview of 3D-Printed Smart Polymers and Composites

Amithkumar Gajakosh
Dayananda Sagara College of Engineering, India

Suresh Erannagari
BMS College of Engineering, India

R. Suresh Kumar
BMS College of Engineering, India

Thyagaraj N. R.
SJC Institute of Technology, India

Mallaradhya H. M.
SJC Institute of Technology, India

Rudresha S.
JSS Academy of Technical Education, India

ABSTRACT

This chapter highlights recent developments and provides an overview of the rapid application of fused deposition modelling (FDM) for polymeric smart and composites. The review is divided into sections that describe the processing conditions and characteristics of FDM components made of polymer and its composites as well as shape memory polymers/composites. The chapter covers a wide range of applications, including PVDF structures and components. While FDM adoption has been rapid in this field, more coordinated efforts in the areas of smart polymer feedstock synthesis, process tuning, and testing are required. This chapter provides an overview of 3D printed smart polymer materials and composites as well as their properties, performance, and potential applications. Additionally, this chapter discusses the motivation for future 3D printing research.

DOI: 10.4018/978-1-6684-6009-2.ch008

INTRODUCTION

The emergence of 3D printing has simultaneously given rise to a new generation of enthusiasts and transformed it into a pervasive technique with several applications in a variety of fields. This encompasses the aircraft, automobile, bio-medical, consumer goods, agriculture, clothing, architecture, modeling, metalworking, defense, and pharmaceutical sectors among others (Gibson et al., 2014, F C Godoi et al., 2018, A. Paolini et al., 2019, L.E. Murr et al., 2016). The innovative emergence and permeation of 3D-printing into residences, low to higher-level educational institutes, and the private organizations in the past ten years has not been achieved overnight for an advanced technology that once demands extensive machinery which was for a longer period considered an instrument for replicating by the design team, performers, and engineers. A series of circumstances came together to make this ascent feasible in part.

Since 2010, numerous innovative makers (including Ultimakers, Markforgeds, XYZprintingss, Zortax, FlashForges, RepRaps, etc.) have had the flexibility to create reasonably priced 3D printers thanks to the exhalation of a ground-breaking invention regarding the technical expertise of fused deposition modeling (a subset of 3D-printing). This is one of the obvious forces behind the 3d-printing transformation into a common technology (T.D. Ngo et al., 2018). The innovation is, however, determined to continue gathering traction as we enter the new decade, according to recent findings. This is because there are numerous additional enabling variables. is a non-exhaustive summary of these characteristics contains (U.M. Dilberoglu et al., 2017, N.A. Gershenfeld, 2005, B. Berman, 2012, J.-Y. Lee et al., 2017, K.B. Mustapha et al., 2019)

- Industry 4.0 and digitalisation are both boosting the demand for embedded smart materials, opening the door for the quick creation of agile production hubs, and making it easier to integrate smart manufacturing systems.
- A wider variety of materials (polymer, metallic, ceramic, wood, composite materials, etc.) are now available for 3D printing techniques, making it more affordable than conventional production methods when working with difficult geometries.
- Enablement of component customisation for low to medium scale manufacture of printed electronic components multi-material parts than others.
- Rapid iterations of product files and administration enabled by seamless interface with design systems, which speeds up innovations.
- Reductions in tool expenditures.
- Manufacturing of compact and integrated systems into a single step.
- Lowering the risk associated with inventories and eliminating waste.

What 3D printing represents has been described by numerous researchers. 3D printing M was formally defined as a "solid freeform production technology for the straightforward automated manufacturing of customised goods and components in very small to moderate-sized batches without relying on particular molds and tooling" by (M. Monzon et al., 2015) in their definition. Although there are many additional definitions of the technologies, it generally entails the slicing-by-slicing manufacture of actual parts from computer-aided design virtual objects that are transformed by a particular 3D printing machine. This has become now well accepted that there are multiple approaches to achieving this continuous layer-by-layer part production. So, 3D printing has been defined as a catch-all phrase that, in accordance with the ISO standards, groups together procedures like vat polymerisation, materials extrusion, powdered bed fusion, materials and binders jetting, sheets lamination, etc.

The underlying technique used by each of the 3D printing techniques varies and each one focuses on a specific category of materials. For example, thermosetting resins are the focus of the vat polymerisation technique, which uses SLA as its most notable case. The core of the material extrusion-based technology is the employment of the heat extruded technique to melt a filament, which is then permitted to solidify. This group includes fused deposition modeling (FDM), which makes use of melting solid thermo-plastic filament to create molten forms that are successively deposited layer by layer to create the required form during crystallization. On the other side, techniques like powdered bed fusion use the use of elevated-energy resources to solidify powder materials (selective laser sintering (SLS) and selective laser melting (SLM) are some examples). In contrast, materials jetting relies on the polymerisation of jets of feedstock components to function similarly to inkjet printing techniques. Binder jetting, on the contrary, uses targeted spraying of liquid binders to speed up the conversion of particles into solid. Sheet lamination is the second classification of 3D printing. Although sheet lamination has a much lower acceptance rate than other 3D printing techniques, it primarily entails the swift and methodical application of binders to a variety of sheet-based components for the creation of parts (Z.X. Khoo et al., 2015, Z. Liu et al., 2019, B. Redwood et al., 2017).

Fascinatingly, amongst all of these methods, fusion deposition modeling has created a distinct niche for itself in the field of 3D printing research. Scott created the first fusion deposition modeling in the year 1988. The easiness of the printing with the fusion deposition modeling route, the accessibility of inexpensive and less-maintenance fused deposition modeling machinery, some of which can be purchased for much less than 400 dollars, and the propagation of a broad range of printable components are all reasons why the method has recently attracted significant attention. For example, among the commonly existing materials for fusion deposition modeling include ABS, polylactic acid (PLA), polyamide (PA), and polycarbonate (PC) composites. However, the performance of the fused deposition model parts that are commercially accessible is typically constrained. There's been an increase in the utilization of material structures created in different labs as a consequence, which you'll note is extremely typical for the creation of expensive smart products. As previously mentioned, in the fused deposition modeling process, the actual item is printed utilizing the geometrical data from the STL data of the computed-aided drafting file to direct the depositing of a quasi-liquid thermo-plastic polymer (hot, extrusion, the dispensation of them in a pasty state, and then chilled to create the required shape). However, polymers must have a relatively low melting point and the ability to undergo phase change through rapid cooling in order to be used with fusion deposition modeling (X. Wang et al., 2017, S. Kumar et al., G. Spinelli et al., 2018).

3D PRINTING OF SMART MATERIALS

Increasing the utility of smart materials depends on their capacity to be manufactured. Both smart materials and the processes used to create them come in a diversity. However, the majority of the conventional approaches come within the pricy, drawn-out, and time-taking semiconductor-based solutions foundry manufacturing methods. There's been an intensified quest for alternate affordable ways of making smart materials and systems out of smart components, which underlies the growing needs for less price sensing and intelligent systems for the upcoming generations' energy creation and rapid industrialization. Particularly, the growing complexity of 3D printing machines and the accessibility of various material feedstocks over recent times have increased interest in 3d printing as a substitute for the conventional techniques of producing smart components. These material structures, which are considered to be stimuli-responsive, respond to outside forces in a predicted and repetitive way. As a result, they can be used as actuators, energy collectors, shape-shifting devices, and sensors. Smart material structures have a variety of coupled characteristics that enable them to display characteristics that span several distinct physical regimes, such as shapes changing, self-actuating, healing, diagnosing, and sensing (commonly 2 for a particular group). Mechanical, electric, biochemical, heat transfer, and electromagnetic are a few types of diverse physical regimes used in the research of smart components (K. Schroeter, 2008, M. E. Morales Rodriguez, 2017, R. Bogue, 2014).

For the creation of intelligent polymeric structures, 3D printing has various benefits. The capacity to modify the molecular chain of smart polymer components such that they have adjustable mechanical, electromagnetic, or chemical characteristics is one such strategic benefit (J. Oliveira et al., 2018, D.I. Woodward et al., 2015, S.J. Leigh et al., 2012). Following the establishment of 3D printing, there have been numerous articles that cover a wide range of 3D printing-related topics. Some of it is more general in their overview, while others are focused on particular features of every classification of 3D printing. However, the majority of these studies concentrated on the application of 3D printing to passive parts. One of the more insightful summaries of the topic was provided by (S.C. Ligon et al., 2017). The use of various 3D printing techniques for the creation of various material categories is taken into account by the researchers, amongst other aspects. The research, nevertheless, had much less investigation of the interface between fused deposition modeling and smart components. A few researchers have also examined how additive manufacturing or 3D printing is used to create polymeric reinforcement composite structures (P. Parandoush and D. Lin, 2017, G.D. Goh et al., 2019, V.K. Balla et al., 2019, I. Fidan et al., 2019, S.F. Kabir et al., 2020). (C. Weller et al., 2015) studied the business structure and financial effects of additive manufacturing. The impact of process parameters on the characteristics of fused deposition modeling-built components was summarised by (S. Kumar et al.). In their examination, (E.A. Papon et al. 2020) looked at the physics, mathematical analysis, molten flow, layer-by-layer depositing, crystallization, binding, and their effects on the microstructure of parts manufactured by fused deposition modeling. The analysis of (Liu et al., 2019) concentrated on the application of fused deposition modeling to the production of polylactic acid parts, but it paid no attention to the impacts of shape memory. (P.K. Penumakala et al., 2020) reviewed his earlier research on thermoplastic composite structures and fused deposition modeling. The impact of methodology-induced flaws resulting from the fused deposition modeling on the mechanical characteristics of fiber reinforcement composite structures was examined in a study by (S. Wickramasinghe et al., 2020). They also described different ways to remediate these flaws in order to improve the functional qualities of fused deposition modeling-made components.

The increasing importance of additive manufacturing and fused deposition modeling has been noted in every one of the previous studies. The themes in almost all of them, though, are limited to traditional passive components. Smart materials are just briefly described in all those where they are referenced. For example, (A. Subash and B. Kandasubramanian, 2020) only included a few of the studies at the intersection of fused deposition modeling and smart polymer components when they reviewed the 3D printing of shape memory polymer components. The research of (Z.X Khoo et al., 2015), which examined the usage and implications of AM for the creation of smart material components several years back, must also be mentioned in the context of this article. Nevertheless, a lot of the discussions in (J. Gardan's, 2019) debate focused on a few of the previous research on the use of additive manufacturing in smart material structures. However, considering that such an evaluation is just several years old, it is understandable that there aren't many specifics concerning new developments in fused deposition modeling for smart material structures. 4D-printing utilizing fused deposition modeling was recently reviewed by (J. Carrell et al., 2020). Although polymer-based piezo-electric, as well as other forms of shape memory polymeric materials, are partly obscured, the limitation of the coverage is restricted to specific shape memory polymeric materials. Nevertheless, there is a substantial body of scientific research on the application of fused deposition modeling to the production of intelligent and useful components. In the universe of materials for functionally graded engineering fields, smart materials occupy a distinct place. Numerous research and industrial studies have started to be interested in them (M.A.C. Stuart et al., 2010).

The aforementioned concept serves as the foundation for this research, which aims to evaluate the outcomes of several recent investigations at the interface of fused deposition modeling and a number of smart polymers that have recently experienced growing utilization. The manufacturing of these smart polymer components and the fused deposition modeling-based manufacturing of operating structures from them are covered in the study. The performance information of the fused deposition modeling-printed smart materials is presented, together with a comprehensive study of the fused deposition modeling processing parameters provided for these smart polymers. It should be mentioned that 4D printing has recently become another name for 3D printing of smart material components in the existing studies (J.-Y. Lee et al., 2017). This happens frequently in 3D printing production, especially when a substance with a shape memory influence is used to create a product that can change shape over time. However, we continue to employ additive manufacturing and only occasionally 4D printing in the upcoming studies. The reality that a scarcity of reasonably priced feedstock materials continues to have been one of the obstacles to the widespread adoption of fused deposition modeling for smart materials and functionally graded materials is another significant theme that has evolved over the past several years. Nevertheless, as this will be demonstrated in the succeeding chapters, a number of authors have focused on this task, and the outcomes were reported and dispersed among publications in many domains. This article moderately aims to highlight the advancement of how fused deposition modeling has flourished in the manufacturing of polymer-based smart materials and their composites by distilling and outlining some of these seemingly unrelated developments. The phrase "smart polymers and their composites" is admittedly rather wide. The subchapter is therefore focused deeply on a variety of investigations on the fused deposition modeling manufacturing of polymer and its composites, as well as a selection of shape memory polymers and their composite materials.

FDM FOR MANUFACTURING DIFFERENT SMART MATERIALS AND COMPOSITES

Piezoelectric Poly Polymer Materials

A discussion of investigations on the 3D printing of piezoelectric polymer materials opens this section's detailed review. For a very long time, polymeric materials were only used for construction and insulation purposes in the realm of electronics since they were thought to be electrically passive. With the development of multi-functional polymer materials, nevertheless, a new utilization dimension emerged (A. Holmes-Siedle et al., 1983). Due to their strength, higher energy conversion rates, less density, mobility, bio-compatibility, biochemical, and thermal thermodynamic stability, piezoelectric polymer materials which are mostly controlled by fluoropolymers, have attracted a lot of study attention since this development (C. Chang et al., 2010, H. Kawai 1969, N.A. Shepelin et al., 2020). These polymeric materials have a wide variety of uses, which have been well defined through a number of studies. These uses include greater frequency transmitters, tissue regeneration, energy harvesting, wearable electronic components, robots, vibrations measurement, pressure transducers, storage systems, and less-density devices (N.A. Shepelin et al., 2020, Y.-S. Lee et al., 2011, R.M. Habibur et al., 2018, Y. Jiang et al., 2010, Y. Xin et al., 2016., T.S. Roopa et al., 2018, T. Sharma et al., 2012, T. Wang et al., 2016). Poly (vinylidene fluoride) (PVDF), nylon 9, poly (b-hydroxy-butyrate), PLLA, Poly (vinylidene fluoride)-TrFE, sulfonated tetra-fluoro-ethylene, polyimides, and polyureas are a few examples of piezoelectric polymer materials (J. Harrison and Z. Ounaies, 2002). Poly (vinylidene fluoride) stands out as the most well-known piezoelectric polymeric material among the examples given above. It is the fluoropolymer with the highest structural shear modulus (D.W Grainger, 2020).

PVDF is offered extensively in the following forms: filaments, granule, coatings (or film), solutions, and powders. The limitations imposed by the present manufacturing and processing techniques are one of the obstacles to the wide-scale application of PVDF. Cryogenic ball milling, hot compression, L-B deposition, solvent castings, and electro-spraying are a few of the traditional methods of manufacture. In general, these techniques are the best for creating thin film coatings and plane geometry. They cannot, nevertheless, create actual piezoelectric systems with a large enough working surface, homogeneity, complicated forms, and sensitivity. Since additive manufacturing has become more popular, numerous researchers tried to use it to fabricate piezoelectric PVDF and its composite structures. Some of this research is covered in the following subsections.

FDM for manufacturing of uniform PVDF structures

Various researchers have shown how to fabricate PVDF polymeric components using fused deposition modeling in both uniform and composite forms. (C. Lee and J.A Tarbutton, 2014) examined the manufacturing of piezoelectric materials from PVDF rods in a method that involves fused deposition modeling and is known as EPAM. In this procedure, a strong electrical field is created between the nozzle's tip of fused deposition modeling equipment, which serves as the anodic one, and the printing bed, which serves as the cathodic one. The electrical field's existence aids in a chain realignment that changes the PVDF filament's state from the alpha-dominating phases to the beta-dominating crystal structures. The researchers studied the impacts of 0 MV/m, 1MV/m, 2 MV/m, and 3 MV/m and used FTIR to analyze the crystalline nature of the built components at an extruding and a bed temperatures of 503 K and 373 K. Electric breakage was observed to occur at 3 MV/m electrical field strength, suggesting that this was a number that was too strong for poling the building polymeric components. Nevertheless, under mechanical stress, the 2 MV/m showed maximum charge and current discharges and caused the greatest stimulation of the beta-phase. The impact of variations in the printing process conditions on the electro-mechanical reaction of the manufactured specimens, nevertheless, was not taken into account in this investigation. The construction of a shock sensor is mentioned as yet another development of the team's work (J. Tarbuttona et al., 2017, E.D. Burnham-Fay et al., 2017, M. Marandi and J. Tarbutton, 2019).

(D.A. Porter et al., 2017) created thicker PVDF films utilizing fused deposits at various printing rates and an in-situ poling of 1000V supplied at the high-temperature side of the extruding equipment. The experiment looked at how the layer's thickness and building orientation affected how the produced specimens responded. Tension and piezoelectric reaction experiments, as well as FTIR analysis, were used to evaluate the specimens. The highest Elastic constant (0.4GPa to 0.5GPa) was achieved for the

PVDF-built components utilizing the zero-degree building angle during the tension testing, which was carried out at a strain rate of 0.5mm/min. The quantity of the beta-phase in the manufactured specimens is discovered to be influenced by printing rate, high-temperature side, and bed temperature. (N. Momenzadeh et al., 2020) examined a wider range of deposition parameters and more rigorously examined the impact of fused deposition modeling process conditions on the manufacturing of PVDF with the Reprap Prusa i3 fused deposition modeling equipment. It is shown that a number of the chosen printing process conditions result in subpar bonding and printing quality while a constrained set of process conditions yields portions of adequate standard. This limited range was used to develop a sequence of ten-layer samples with a thickness of 1.5mm and a 70 percent infill, which were then examined for microstructure characteristics using a scanning electron microscope and FTIR. Additionally, using a universal tension testing setup and a fixed strain rate of 1.25mm/min, they performed tension studies in accordance with ASTM standards. According to reports, keeping heating bed temperatures between 363 K and 383 K and a printing rate between 10mm/s and 25mm/s will improve dimensional accuracy. The print specimens have the highest recorded elastic constant, which is 772MPa. The estimated elastic constant for the bulk PVDF, which is 1.7GPa, is significantly lower than this number. However, the author's assumption of a 70 percent infill quantity makes the lesser elastic constant plausible (S. Mahmood et al., 2017, J. Fan et al., 2020). It should be emphasized that this work offers no description of in situ poling throughout the fused deposition modeling route. In PVDF sheets with a thickness of 250um to 300um were created using fused deposition modeling for pressure sensor applications. In contrast to earlier research in which the filament was made in-house, the researchers depended on the usage of a commercialized PVDF filament named (FluorX, 2021). With an extruded temperature of 523 K and a printing rate of 10mm/s, printable PVDF films were created. Although the impact of post-manufacturing poling process conditions (such as voltage, temperatures, and time) was examined, the physico - mechanical characteristics are not discussed. The study's key finding is that, at fixed temperatures, the poling voltage and piezoelectric constant correlated positively. At 750V and 393 K, a maximal piezoelectric constant of 32pC/N was attained.

FDM for manufacturing of composite PVDF structures

Studies devoted to improving their electrical, dielectric, and mechanical characteristics by the addition of filler materials are now increasing, in addition to using PVDF as the only feedstock. Scientists are working on this approach to integrate the special qualities of PVDF, such as its higher dielectric constant, outstanding chemical resilience, non-toxicity, and manufacturing ease, with the beneficial characteristics of other polymers. Reduced graphene oxide, carbon nanotubes, and ferroelectric ceramics like barium titanate are typical filler materials that were explored and published (I. Fidan et al., 2019, H. Kim et al., 2017). Nevertheless, while the dielectric constant of the resulting composite structure is improved by the reinforcing of PVDF with barium titanate, the ceramic particles agglomerate in a larger proportion within the PVDF's matrix. It may later result in the formation of cavities and micro-cracks (Y. Jin et al., 2016). Despite this negative aspect, barium titanate (BTO) has been widely used to enhance the PVDF's characteristics. For example, (H. Kim et al., 2017) investigated the manufacturing of smart materials utilizing PVDF reinforcement with barium titanate filler material using a combined fused deposition modeling and poling technique. According to reports, adding barium titanate filler materials makes the PVDF's -phase content higher, which improves the piezoelectric sensitivity. The existence of the barium titanate filler material is said to produce an increase in tensile strength of roughly 45 percent according to mechanical tests of the resulting polymer composite structures under tensile and fatigue applied loads. The report indicates that the solvent casting of PVDF/barium titanate specimens had a higher beta-phase concentration than the fused deposition modeling-made samples in a subsequent investigation (H. Kim et al., 2018). Contrary to their additive manufactured components, the PVDF/barium titanate composite samples that were solvent cast suffer from increased aggregation, micro-pores, and cracks. However, compared to the film made by the solvent-casting approach, the piezoelectric reactions in the fused deposition modeling manufactured sample display greater output current. The issue, however, is that it was shown to cause fractures when a larger weight proportion of barium titanate (12 percent to 15 percent) was added.

Three distinct formulation techniques such as sonication, ball, and extrusion were used by (S. Bodkhe et al., 2017) to investigate the printing ability of PVDF when barium titanate nanoparticles were added. Once more, the barium titanate nanoparticles were used in the investigation to faster the nucleation of the piezoelectric beta-polymorph in the PVDF. The researchers look at how the PVDF composite's piezoelectric characteristics were affected by process variables and filler material content. The findings imply that PVDF components with 10 percent barium titanate nanoparticles got high dielectric and piezoelectric characteristics through the fused deposition modeling method. When the built sample was characterized, it had a piezoelectric coefficient of 18 pC/N, which is practically in line with that of conventional PVDF sensors. Despite the use of additional characterization techniques like X-ray diffraction, scanning electron microscope, and FTIR, little is known about the mechanical characteristics of the built components that have been produced. The manufacture of PVDF-based sensor sheets for the identification of volatile organic compounds was examined by the researchers in (Z. Kennedy et al., 2017). A PVDF nanocomposite structure with greater conductance and strong chemical stability was created by reinforcing the PVDF with multi-wall carbon nanotubes. The formulations of the PVDF and multi-wall carbon nanotubes mix in various proportions, solution casting aided by DMF solution, and formulating the PVDF multi-wall carbon nanotube filaments feedstock using a plunge extrusion machine at 573 K are all procedures that went into creating the nanocomposite filaments. Samples were fused deposition modeling-printed utilizing the composite filaments at a rate of 25mm/s for characterization studies based on X-ray diffraction, thermographic analysis, and resistance testing. Temperatures for the extrusion and bed were kept at 593 K and 333 K, respectively. No mechanical evaluation was mentioned in this work because the chemical resistance-related sensors are the main emphasis. Nevertheless, the specimens' chemical-resistive ability was shown and found to be acceptable. Barium titanate nanofibers were combined with PVDF by (N. Phatharapeetranun et al., 2017). Utilizing electro-spinning, calcination, and solgel synthesis are all necessary for the production of barium titanate/PVDF. In addition to using X-ray diffraction to check the crystalline nature of the additive manufactured parts, the morphological and rheological characteristics of the produced feedstock were examined. The nozzle's dimension (2.5mm and 1.2mm) is the primary fused deposition modeling processing conditions taken into account in this work. It is used to investigate its impact on the dispersion of the nanofibers at various concentrations. According to reports, it is possible to significantly improve the dielectric characteristics compared to pure PVDF by more than one order of magnitude.

Shape Memory Polymers (SMPs)

Presenting a different family of smart materials, shape memory polymers, a subcategory of sensory-responsive polymers with a sensory-sensitives switch that encourage a programmed shape-changing propensity (Q. Zhao et al., 2015). Elastic polymers network that exhibits extremely higher elastic deformations serves as examples of shape memory polymer behavior. They vary considerably from shape memory alloys (A. Lendlein and S. Kelch, 2002). For starters, unlike shape memory alloys, a large proportion of shape memory alloys have less density, are biodegradable, recover slowly, and typically have variable stiffness. Additionally, coding a transient shape in shape memory alloys is possible at a specific temperature, whereas recovery requires high thermal loads above the martensite transition temperatures (C, Liu et al., 2007). The shape memory impact, however, can be activated by a wide variety of stimuli thanks to the support provided by many types of shape memory polymers. Temperature, current, magnetic field, and lighting are only a few examples. However, for the sake of concision, the explanation of shape memory polymers in this introduction will focus on those in which the shape memory effect is activated thermally. For example, the studies by (C. Liu et al., 2007, J. Leng et al., 2011, Q. Zhao et al., 2015), among others, provide a thorough detailed investigation into the many mechanisms behind the behavior of shape memory polymers.

According to a general definition, a shape memory polymer exhibits a shape memory impact when it can acquire and maintain a temporary shape after coding (which includes physical influences in addition to external influences) and can return to its normal shape after being exposed to an external stimulation (H. Meng and G. Li, 2013). The polymers may exhibit this shape-changing ability in the manner of a reversible change in concentration, stretching, compression, bending, and twisting. Regardless of the type, this behavior is said to be closely related to the relationship between the molecular structures, methodology, methodology, and coding parameters of the shape memory polymers (J. Carrell et al., 2020, A. Lendlein and S. Kelch, 2002). The shape memory polymer may demonstrate a 1-way (dual-shape or multiple-shapes memory effects) or 2-way shape memory effects, according to a number of findings from earlier studies (Y. Osada and A. Matsuda, 1995, P. Rastogi et al., 2019). The one-way shape memory effect involves a shape-changing behavior in only a single direction (and is typically irreversible), as excellently discussed and emphasized by Zhao et al. (Q. Zhao et al., 2015). They showed programmed and reversible switching between two different shapes in the case of the shape memory polymers with 2-way shape memory effects. The most straightforward and typical behavior of shape memory polymers is the dual-shape shape memory effect. For example, a dual-form shape memory effect in a thermoresponsive shape memory polymer initially undergoes a specialized production procedure to give the polymer a fixed shape (Fig.1). It is then designed to take on a transient shape (with changes in temperature and external stresses). The polymer returns to its actual shape after stimulation (such as heat for the thermally-responsive shape memory polymer) (Q. Meng and J. Hu, 2009). The shape memory polymers with a multi-shape shape memory effect, on the other hand, exhibit a sophisticated transition behavior. Importantly, they possess the capacity to recall many transient shapes and switch smoothly from one temporary shape to a regular shape (Q. Zhao et al., 2015).

Regarding the theoretical justification for shape memory polymer behavior, a deeper examination of the fundamental molecular mechanism in the thermo-responsive dual-way shape memory effect is instructive (again, a detailed account can be obtained in the studies of shape memory polymers (Q. Zhao et al., 2015, C. Liu et al., 2007, J. Leng et al., 2011). According to reports, the shape memory polymers have a structure that is enhanced by the interaction of two/more phase-segregated subdomains (Fig.1) (68, 71).

As a result, the strong segment (or net points) that has the first highest thermal transformation temperature governs the fixed shape. Physically/chemically cross-linked microscopic phases create the net points. Similar to this, the molecular switch is made up primarily of molecular chains for whom the mobility can be activated or hindered, and is located in the domain with the second-highest transforma-

Figure 1. Programming and new shape establishment for a thermal responsive shape memory polymer

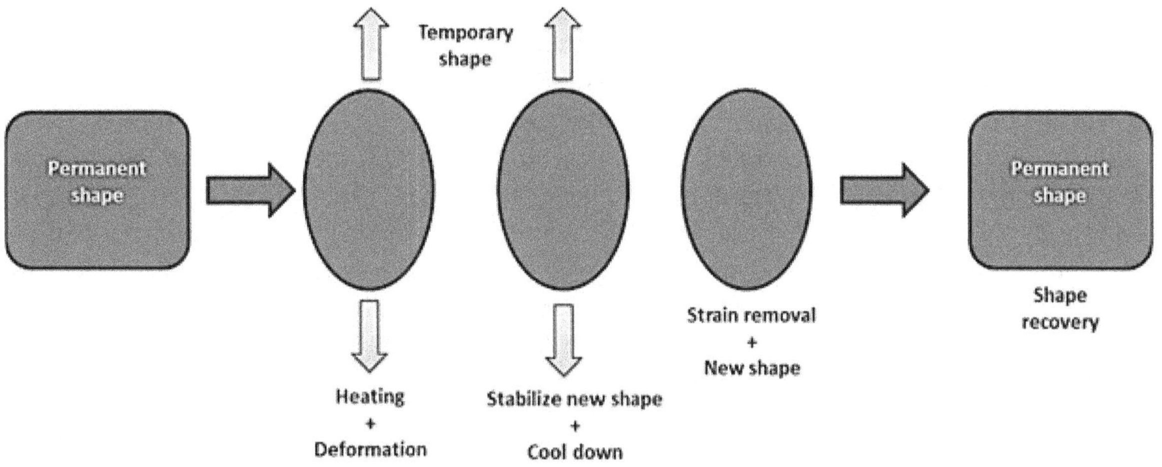

Figure 2. Effect of interaction between subdomains

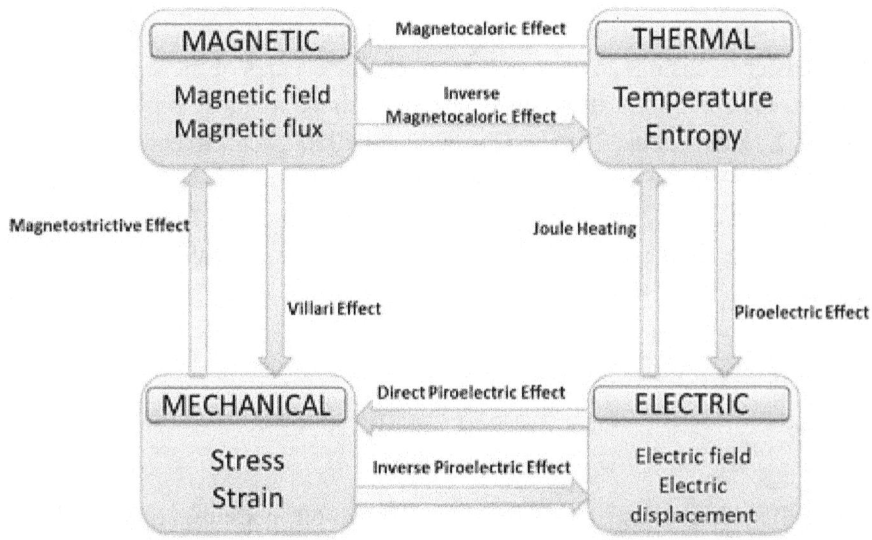

tion temperature. (A. Lendlein and S. kelch 2002) assert that the fixing of a transient shape in a polymers network system with a shape memory effect is facilitated by the second domain acting as the molecular switch. As a result, in real life, whenever a shape memory polymer is exposed to a temperature, the part exhibits a high elastic constant if the temperature is less second-highest transition temperature. It is important to note that the glass transformation temperature or the melting point temperature equals the second-highest transition temperature for some shape memory polymers (Q. Zhao et al., 2015). On the other side, the softer segment lacks stiffness when the temperature is greater than the second-highest transformation temperature, resulting in a portion with a lower elastic constant. In the latter case, when deformation is applied and the component is cooled to a temperature less than the second highest transition temperature, the component can be designed to take on a temporary shape. Importantly, the transient shape is fixed when the temperature is less than the second highest transformation temperature. Nevertheless, the portion regains its fixed shape after the temperature will become greater than the second-highest transformation temperature. The discharge of the tension energy held in the soft section is what theoretically causes the shape memory effect (Q. Meng and J. Hu, 2009, Q. Zhao et al., 2015, M Behlet al., 2010). For multi-way shape memory polymers, a similar principle holds true, with the exception that more than one thermal transformation temperature is required so that more than one shape can be momentarily fixed (Y. Osada and A. Matsuda, 1995). However, it's this capability that makes it possible to program components made from shape memory polymers to take on various transient shapes for various uses. Transportable hinges (C. Liu et al., 2007), micro-electromechanical equipment (D Santiago et al., 2014, C. Schmidt et al., 2011), self-assembly mechanisms, vasculature stents (Y.C. Jung and J.W. Cho, 2010, C.M. Yakacki et al., 2009), orthodontics (J. Cui et al., 2010, X. Lan et al., 2009), bone flaw filler materials (C.-C. Fu et al., 2009, C.-C. Fu, et al., 2010), are just a few examples of the many fields in which shape memory polymers have been applied.

As per earlier investigations (S.M. Felton, et al., 2013, G. M. Baer, et al., 2009), polyurethane, polylactic acid, poly (ethylene vinyl acetate), and polycaprolactone are some of the polymers that have been reported to show shape memory effects. Such shape memory polymers have historically been produced using techniques including solvent casting and injection molding. It is also acknowledged that additive manufacturing processes like fused deposition modeling, which enable the creation of complicated structures and make it easier to customize the mechanical characteristics of these materials, provide greater control over the process conditions. The future growth of shape memory polymer will be shaped by all of these benefits (Z.X. Khoo, et al., 2015, J. Carrell et al., 2020, S.-Y. Gu et al., 2016, I. Navarro-Baena et al., 2016). The basic available processing is additive manufacturing, with the difference that additive manufacturing-made structures have the potential to exhibit relieving stiffness with time (S. Tibbits, 2014, Yaxin Zhang Fang Huang Ercong Zhang Liang Zhang, 2023, Keshav Lalit Ameta et al., 2022, Soyeon Park et al., 2022, S.Santhosh et al., 2022). Nevertheless, certain recent research that deals with additive manufacturing and shape memory polymers frequently refers to 4D printing.

SUMMARY

The advantages of 3D printing are offered by fused deposition modeling, a particular technology that falls under the category of material extruding technique. It stimulates flexibility in design, makes it easier to produce complicated components without using pricey molds, cuts down the lead-time for product creation, and enables for product customization for individual ownership. Amongst different 3D printing techniques, FDM offers additional significant benefits in the form of affordable, less-maintenance 3D printers and affordable feedstock material. Additionally, it enables the fine-tuning of component qualities to produce products with compositions inspired by nature.

This assessment acknowledged the widespread use of fused deposition modeling and described how it was developing new uses and prospects for producing useful and intelligent parts. Various parts of the discussion have illustrated and emphasized the idea that fused deposition modeling has evolved into a multipurpose instrument in the advancement of smart materials, which is based on numerous novel research published in recent times. The overview has particularly emphasized the broad range of fused deposition modeling applications, including its use to create: (a) simple beams; (b) energized smart composites; (c) responsive meta-materials; (d) pressure sensor systems; (e) light-sensitive smart parts; (f) thermo-sensitive smart elements; (g) PH-sensitive smart parts; (h) grips and soft robots; and (i) responsive solar thermal panels. It is noteworthy that different degrees of achievement were identified with fused deposition modeling in addressing the necessity of operating with polymer piezoelectric materials (specifically PVDF and its composite parts) and many kinds of shape memory polymers and of their composites. This is based on the core publications that were extracted. As fused deposition modeling gains traction, it is anticipated that its use will expand even more and that the operational capabilities of the produced specimens will improve. However, there are still issues that must be resolved in order to optimize the effect of methodology, despite the position of fused deposition modeling as an easily accessible technology for the manufacturing of smart products. Constructing on the concepts presented in the surveyed publications, a few of the key issues are determined to circle around: (a) simulation of the fused deposition modeling ; (b) extending the range of materials used as feedstocks and filaments; (c) improving the fused deposition modeling surface and microstructure; and (d) intensifying the rigorous testing of fused deposition modeling-printed specimens to satisfy the demands of the extreme conditions where usable devices are probable to be used in the future. Below, each of these points is further developed.

Simulation of FDM Process

There isn't an overall, commonly used computer-assisted tool for simulating the fused deposition modeling method as of the time of this writing. Because few smart polymer material components are costly to get or make, it is important to minimize wastage while designing and producing complicated components with them. The reducing waste process will be significantly aided by the use of sophisticated simulation software to identify any faults and defects in such materials before the fused deposition modeling workflow is started.

Extending the Range of Materials Used as Feedstocks and Filaments

The variety of materials that are readily available is another potential obstacle to the growth of fused deposition modeling that has to be addressed. PLA and ABS have historically dominated the list of materials that may be utilized with fused deposition modeling, which is proven to work primarily with thermo-plastic materials. Approximately 70 percent of the market for commercial filaments is still dominated by these materials. Such materials are without a doubt effective at manufacturing parts that are appropriate for systems where structural support, strength, and stability are essential criteria.

Improving the Fused Deposition Modeling Surface and Microstructure

Many process variables that must be carefully chosen to guarantee appropriate and adequate microstructural features, surfaces, and elastic and dimensional accuracy attributes are believed to have a significant impact on the quality of the fused deposition modeling method. Extruding temperature, feed rate, printing rate, layer thickness, bed temperature, raster angle, infill patterns, and infill concentration are just a few of the numerous variables that have been found to have an impact on the quality of the finished fused deposition modeling-made parts.

Characterization

The standardization of techniques for gathering essential engineering data from examination of fuses deposition modeling-fabricated functioning specimens is another crucial barrier that would have a considerable influence on the direction of developing smart materials parts with fused deposition modeling. Such information is required, amongst other things, for strength forecast, mathematical methods, constitutive modeling, and materials informatics.

MOTIVATION

In summary, overcoming a few of the challenges raised above would go a long way toward bridging the gap between the advent of smart material parts in practical workplaces and research lab demonstration as researchers attempt to push the limits of what is achievable using fused deposition modeling. This study is intended to help readers gain a better grasp of the fast-changing research environment at the intersection of 3D printing in general, smart polymeric materials, shape memory alloys, and fused deposition modeling in particular.

REFERENCES

Gibson, I., Rosen, D. W., & Stucker, B. (2014). *Additive manufacturing technologies*. Springer.

Godoi, B. Prakash, & Zhang. (2018). Fundamentals of 3D food printing and applications. Academic Press.

Paolini, A., Kollmannsberger, S., & Rank, E. (2019). Additive manufacturing in construction: A review on processes, applications, and digital planning methods. *Additive Manufacturing*, *30*, 100894. doi:10.1016/j.addma.2019.100894

Murr, L. E. (2016). Frontiers of 3D printing/additive manufacturing: From human organs to aircraft fabrication. *Journal of Materials Science and Technology*, *32*(10), 987–995. doi:10.1016/j.jmst.2016.08.011

Ngo, T. D., Kashani, A., Imbalzano, G., Nguyen, K. T., & Hui, D. (2018). Additive manufacturing (3D printing): A review of materials, methods, applications and challenges, Compos. *B Eng.*, *143*, 172–196. doi:10.1016/j.compositesb.2018.02.012

Dilberoglu, U. M., Gharehpapagh, B., Yaman, U., & Dolen, M. (2017). The Role of Additive Manufacturing in the Era of Industry 4.0. *Procedia Manufacturing*, *11*, 545–554. doi:10.1016/j.promfg.2017.07.148

Gershenfeld, N. A. (2005). *Fab: The coming revolution on your desktop–from personal computers to personal fabrication*. Basic Books.

Berman, B. (2012). 3-D printing: The new industrial revolution. *Business Horizons*, *55*(2), 155–162.

Lee, J.-Y., An, J., & Chua, C. K. (2017). Fundamentals and applications of 3D printing for novel materials. *Appl. Mater. Today*, *7*, 120–133.

Mustapha, K. B., Saad, M., & Abakr, Y. A. (2020). Flatwise Compression and Buckling Characterizations of Adhesive-Free Additively Manufactured Defected Architected Structures. In *Proceedings of International Conference of Aerospace and Mechanical Engineering 2019*. Springer.

Monz'on, M., Ortega, Z., Martínez, A., & Ortega, F. (2015). Standardization in additive manufacturing: Activities carried out by international organizations and projects. *International Journal of Advanced Manufacturing Technology*, *76*(5–8), 1111–1121.

Khoo, Z. X. (2015). 3D printing of smart materials: A review on recent progresses in 4D printing. *Virtual and Physical Prototyping*, *10*(3), 103–122.

Liu, Z., Wang, Y., Wu, B., Cui, C., Guo, Y., & Yan, C. (2019). A critical review of fused deposition modeling 3D printing technology in manufacturing polylactic acid parts. *International Journal of Advanced Manufacturing Technology*, *102*(9), 2877–2889.

Redwood, Sch¨offer, & Garret. (2017). *The 3D Printing Handbook: Technologies, Design and Applications*. 3D Hubs B.V.

Wang, X., Jiang, M., Zhou, Z., & Gou, J. D. (2017). 3D printing of polymer matrix composites: A review and prospective. *Compos. B Eng.*, *110*, 442–458. doi:10.1016/j.compositesb.2016.11.034

Stratasys. (n.d.). https://www.stratasys.com/

Kumar, Singh, Singh, & Batish. (n.d.). Fused filament fabrication: A comprehensive review. *J. Thermoplast. Compos. Mater.* doi:10.1177/0892705720970629

Spinelli, G. (2018). Morphological, rheological and electromagnetic properties of nanocarbon/poly (lactic) acid for 3d printing: Solution blending vs Melt mixing. *Materials (Basel)*, *11*(11), 2256.

Schroeter, K. (2008). Printed sensors: Enabling new applications. *Sensor Review*.

Morales Rodríguez. (2017). *Printed Sensors for Energy Applications: Challenges and Implications*. Academic Press.

Bogue, R. (2014). Smart materials: A review of capabilities and applications. *Assembly Automation*.

Oliveira, J., Correia, V., Castro, H., Martins, P., & Lanceros-Mendez, S. (2018). Polymer-based smart materials by printing technologies: Improving application and integration. *Addit. Manuf.*, *21*, 269–283. .addma.2018.03.012 doi:10.1016/j

Woodward, D. I., Purssell, C. P., Billson, D. R., Hutchins, D. A., & Leigh, S. J. (2015). Additively-manufactured piezoelectric devices. *Physica Status Solidi. A, Applications and Materials Science*, *212*(10), 2107–2113. https://doi.org/10.1002/pssa.201532272

Leigh, S. J., Bradley, R. J., Purssell, C. P., Billson, D. R., & Hutchins, D. A. (2012). A simple, low-cost conductive composite material for 3D printing of electronic sensors. *PLoS One*, *7*(11), e49365.

Ligon, S. C., Liska, R., Stampfl, J. R., Gurr, M., & Mülhaupt, R. (2017). Polymers for 3D printing and customized additive manufacturing. *Chemical Reviews*, *117*(15), 10212–10290.

Parandoush, P., & Lin, D. (2017). A review on additive manufacturing of polymer-fiber composites. *Composite Structures*, *182*, 36–53.

Goh, G. D., Yap, Y. L., Agarwala, S., & Yeong, W. Y. (2019). Recent progress in additive manufacturing of fiber reinforced polymer composite. *Adv. Mater. Technol.*, *4*(1), 1800271.

Balla, V. K., Kate, K. H., Satyavolu, J., Singh, P., & Tadimeti, J. G. D. (2019). Additive manufacturing of natural fiber reinforced polymer composites: Processing and prospects, Compos. *B Eng.*, *174*, 106956.

Fidan, I., & .... (2019). The trends and challenges of fiber reinforced additive manufacturing. *International Journal of Advanced Manufacturing Technology*, *102*(5–8), 1801–1818.

Kabir, S. F., Mathur, K., & Seyam, A.-F. M. (2020). A critical review on 3D printed continuous fiber-reinforced composites: History, mechanism, materials and properties. *Composite Structures*, *232*, 111476.

Weller, C., Kleer, R., & Piller, F. T. (2015). Economic implications of 3D printing: Market structure models in light of additive manufacturing revisited. *International Journal of Production Economics*, *164*, 43–56. https://doi.org/10.1016/j.ijpe.2015.02.020

Papon, E. A., & Haque, A. (2020). Review on process model, structure-property relationship of composites and future needs in fused filament fabrication. *J. Reinf. Plast. Compos.*, *39*(19–20), 758–789. 0731684420929757 doi:10.1177/

Penumakala, P. K., Santo, J., & Thomas, A. (2020). A critical review on the fused deposition modeling of thermoplastic polymer composites. *Compos. B Eng.*, *201*. doi:10.1016/j.compositesb.2020.108336

Wickramasinghe, S., Do, T., & Tran, P. (2020). FDM-based 3D printing of polymer and associated composite: A review on mechanical properties, defects and treatments. *Polymers*, *12*(7), 1529.

Subash, A., & Kandasubramanian, B. (2020). 4D printing of shape memory polymers. *Eur. Polym. J.*, *134*. .eurpolymj.2020.109771 doi:10.1016/j

Gardan, J. (2019). Smart materials in additive manufacturing: State of the art and trends. *Virtual and Physical Prototyping*, *14*(1), 1–18.

Carrell, J., Gruss, G., & Gomez, E. (2020). Four-dimensional printing using fused-deposition modeling: A review. *Rapid Prototyping Journal*.

Stuart, M. A. C. (2010). Emerging applications of stimuli-responsive polymer materials. *Nature Materials*, *9*(2), 101–113.

Holmes-Siedle, A., Wilson, P., & Verrall, A. (1983). PVdF: An electronically-active polymer for industry. *Materials & Design*, *4*(6), 910–918.

Chang, C., Tran, V. H., Wang, J., Fuh, Y.-K., & Lin, L. (2010). Direct-write piezoelectric polymeric nanogenerator with high energy conversion efficiency. *Nano Letters*, *10*(2), 726–731.

Kawai, H. (1969). The piezoelectricity of poly (vinylidene fluoride). *Japanese Journal of Applied Physics*, *8*(7), 975.

Shepelin, N. A. (2020). Printed recyclable and self-poled polymer piezoelectric generators through single-walled carbon nanotube templating. *Energy & Environmental Science*, *13*(3), 868–883. https://doi.org/10.1039/C9EE03059J

Lee, Y.-S., Collins, G., & Arinzeh, T. L. (2011). Neurite extension of primary neurons on electrospun piezoelectric scaffolds. *Acta Biomaterialia*, *7*(11), 3877–3886.

Habibur, R. M., Yaqoob, U., Muhammad, S., Uddin, A. S. M. I., & Kim, H. C. (2018). The effect of RGO on dielectric and energy harvesting properties of P(VDF-TrFE) matrix by optimizing electroactive β phase without traditional polling process. *Mater. Chem. Phys., 215*, 46–55. .matchemphys.2018.05.010 doi:10.1016/j

Jiang, Y., Shiono, S., Hamada, H., Fujita, T., Higuchi, K., & Maenaka, K. (2010). *Low-frequency energy harvesting using a laminated PVDF cantilever with a magnetic mass* (Vol. 375378). Power MEMS.

Xin, Y. (2016). The use of polyvinylidene fluoride (PVDF) films as sensors for vibration measurement: A brief review. *Ferroelectrics*, *502*(1), 28–42.

Roopaa, T. S., Narasimha Murthy, H. N., Praveen Kumar, V. V., & Krishna, M. (2018). Development and Characterization of PVDF Thin Films for pressure sensors. *Materials Today: Proceedings*, *5*(10, Part 1), 21082–21090. https://doi.org/ 10.1016/j.matpr.2018.06.503

Sharma, T., Je, S.-S., Gill, B., & Zhang, J. X. (2012). Patterning piezoelectric thin film PVDF–TrFE based pressure sensor for catheter application. *Sensors and Actuators. A, Physical*, *177*, 87–92.

Wang, T. (2016). Electroactive polymers for sensing. *Interface Focus*, *6*(4), 20160026.

Harrison & Ounaies. (2002). Piezoelectric polymers. *Encyclopedia Polym. Sci. Technol., 3.*

Grainger, D. W. (2020). 1.3.2C - Fluorinated Biomaterials. In W. R. Wagner, S. E. Sakiyama-Elbert, G. Zhang, & M. J. Yaszemski (Eds.), *Biomaterials Science* (4th ed., pp. 125–138). Academic Press.

Lee, C., & Tarbutton, J. A. (2014). Electric poling-assisted additive manufacturing process for PVDF polymer-based piezoelectric device applications. *Smart Materials and Structures*, *23*(9), 095044.

Tarbuttona, J., Leb, T., Helfrichb, G., & Kirkpatrickb, M. (2017). Phase Transformation and Shock Sensor Response of Additively Manufactured Piezoelectric PVDF. *Procedia Manufacturing*, *10*, 982–989. https://doi.org/10.1016/j.promfg.2017.07.089

Burnham-Fay, E. D., Le, T., Tarbutton, J. A., & Ellis, J. D. (2017). Strain characteristics of additive manufactured polyvinylidene fluoride (PVDF) actuators. *Sensors and Actuators. A, Physical*, *266*, 85–92.

Marandi, M., & Tarbutton, J. (2019). Additive manufacturing of single-and double-layer piezoelectric PVDF-TrFE copolymer sensors. *Procedia Manufacturing*, *34*, 666–671.

Porter, D. A., Hoang, T. V., & Berfield, T. A. (2017). Effects of in-situ poling and process parameters on fused filament fabrication printed PVDF sheet mechanical and electrical properties. *Addit. Manuf.*, *13*, 81–92.

Momenzadeh, N., Miyanaji, H., Porter, D. A., & Berfield, T. A. (2020). Polyvinylidene fluoride (PVDF) as a feedstock for material extrusion additive manufacturing. *Rapid Prototyping Journal.*

Mahmood, S., Qureshi, A., Goh, K. L., & Talamona, D. (2017). Tensile strength of partially filled FFF printed parts: Experimental results. *Rapid Prototyping Journal.*

Fan, J., Gonzalez, D., Garcia, J., Newell, B., & Nawrocki, R. A. (2020). The Effects of Additive Manufacturing and Electric Poling Techniques on PVdF Thin Films: Towards 3D Printed Functional Materials, in: Smart Materials. Adaptive Structures and Intelligent Systems, 84027.

FluorX. (n.d.). https://go.arkema.com/fluorx-filament

Kim, H. (2017). Increased piezoelectric response in functional nanocomposites through multiwall carbon nanotube interface and fused-deposition modeling three-dimensional printing. *MRS Communications, 7*(4), 960–966.

Jin, Y., Xia, N., & Gerhardt, R. A. (2016). Enhanced dielectric properties of polymer matrix composites with BaTiO3 and MWCNT hybrid fillers using simple phase separation. *Nano Energy, 30*, 407–416.

Kim, H., Torres, F., Villagran, D., Stewart, C., & Lin, Y. (2017). 3D printing of BaTiO3/PVDF composites with electric in situ poling for pressure sensor applications. *Macromolecular Materials and Engineering, 302*(11), 1700229.

Kim, H., Fernando, T., Li, M., Lin, Y., & Tseng, T.-L. B. (2018). Fabrication and characterization of 3D printed BaTiO3/PVDF nanocomposites. *Journal of Composite Materials, 52*(2), 197–206.

Bodkhe, S., Turcot, G., Gosselin, F. P., & Therriault, D. (2017). One-step solvent evaporation-assisted 3D printing of piezoelectric PVDF nanocomposite structures. *ACS Applied Materials & Interfaces, 9*(24), 20833–20842.

Kennedy, Z. (2017). 3D-printed poly (vinylidene fluoride)/carbon nanotube composites as a tunable, low-cost chemical vapour sensing platform. *Nanoscale, 9*(17), 5458–5466.

Phatharapeetranun, N., Ksapabutr, B., Marani, D., & Bowen, J. R. (2017). 3D-printed barium titanate/poly-(vinylidene fluoride) nano-hybrids with anisotropic dielectric properties. *Journal of Materials Chemistry. C, Materials for Optical and Electronic Devices, 5*(47), 12430–12440.

Zhao, Q., Qi, H. J., & Xie, T. (2015). Recent progress in shape memory polymer: New behavior, enabling materials, and mechanistic understanding. *Progress in Polymer Science, 49*, 79–120.

Lendlein, A., & Kelch, S. (2002). Shape-memory polymers. *Angewandte Chemie International Edition, 41*(12), 2034–2057.

Liu, C., Qin, H., & Mather, P. (2007). Review of progress in shape-memory polymers. *Journal of Materials Chemistry, 17*(16), 1543–1558.

Leng, J., Lan, X., Liu, Y., & Du, S. (2011). Shape-memory polymers and their composites: Stimulus methods and applications. *Progress in Materials Science, 56*(7), 1077–1135.

Meng, H., & Li, G. (2013). A review of stimuli-responsive shape memory polymer composites. *Polymer, 54*(9), 2199–2221.

Osada, Y., & Matsuda, A. (1995). Shape memory in hydrogels. *Nature, 376*(6537), 219–219.

Rastogi, P., Njuguna, J., & Kandasubramanian, B. (2019). Exploration of elastomeric and polymeric liquid crystals with photothermal actuation: A review. *European Polymer Journal*, *121*, 109287.

Meng, Q., & Hu, J. (2019). A review of shape memory polymer composites and blends. *Compos. A Appl. Sci. Manuf.*, *40*(11), 1661–1672. doi:10.1016/j.compositesa.2009.08.011

Zhao, Q., Qi, H. J., & Xie, T. (2015). Recent progress in shape memory polymer: New behavior, enabling materials, and mechanistic understanding. *Progress in Polymer Science*, *49–50*, 79–120. https://doi.org/10.1016/j.progpolymsci.2015.04.001

Behl, M., Razzaq, M. Y., & Lendlein, A. (2010). Multifunctional shape-memory polymers. *Advanced Materials*, *22*(31), 3388–3410.

Liu, C., Qin, H., & Mather, P. T. (2007). Review of progress in shape-memory polymers. *Journal of Materials Chemistry*, *17*(16). https://doi.org/10.1039/b615954k

Santiago, D., Ferrando, F., & De la Flor, S. (2014). Influence of Holding Time on Shape Recovery in a Polyurethane Shape-Memory Polymer. *Journal of Materials Engineering and Performance*, *23*(7), 2567–2573. https://doi.org/10.1007/s11665-014-0997-5

Schmidt, C., Chowdhury, A. M. S., Neuking, K., & Eggeler, G. (2011). Studies on the cycling, processing and programming of an industrially applicable shape memory polymer Tecoflex® (or TFX EG 72D). *High Performance Polymers*, *23*(4), 300–307. https://doi.org/10.1177/0954008311405245

Jung, Y. C., & Cho, J. W. (2010). Application of shape memory polyurethane in orthodontic. *J. Mater. Sci. - Mater. Med.*, *21*(10), 2881–2886. s10856-008-3538-7 doi:10.1007/

Yakacki, C. M., Griffis, J., Poukalova, M., & Gall, K. (2009). Bearing area: A new indication for suture anchor pullout strength? *Journal of Orthopaedic Research*, *27*(8), 1048–1054. https:// doi.org/10.1002/jor.20856

Cui, J., Kratz, K., & Lendlein, A. (2010). Adjusting shape-memory properties of amorphous polyether urethanes and radio-opaque composites thereof by variation of physical parameters during programming. *Smart Materials and Structures*, *19*(6), 065019. https://doi.org/10.1088/0964-1726/19/6/065019

Lan, X., Liu, Y., Lv, H., Wang, X., Leng, J., & Du, S. (2009). Fiber reinforced shape-memory polymer composite and its application in a deployable hinge. *Smart Materials and Structures*, *18*(2), 024002.

Fu, C.-C. (2009). Tunable Nanowrinkles on Shape Memory Polymer Sheets. *Advanced Materials*, *21*(44), 4472–4476. https://doi.org/10.1002/adma.200902294

Fu, C.-C. (2010). Bimetallic nanopetals for thousand-fold fluorescence enhancements. *Applied Physics Letters*, *97*(20), 203101. https://doi.org/ 10.1063/1.3495773

Felton, S. M. (2013). Self-folding with shape memory composites. *Soft Matter*, *9*(32), 7688–7694.

Baer, G. M. (2007). Fabrication and in vitro deployment of a laser-activated shape memory polymer vascular stent. *Biomedical Engineering Online*, *6*(1), 43. https:// doi.org/10.1186/1475-925X-6-43

Baer, G. M. (2009). Thermomechanical properties, collapse pressure, and expansion of shape memory polymer neurovascular stent prototypes. *Journal of Biomedical Materials Research. Part B, Applied Biomaterials*, *90*(1), 421–429.

Gu, S.-Y., Jin, S.-P., Gao, X.-F., & Mu, J. (2016). Polylactide-based polyurethane shape memory nanocomposites (Fe3O4/PLAUs) with fast magnetic responsiveness. *Smart Materials and Structures*, *25*(5), 055036.

Zhang, Z.-X., He, Z.-Z., Yang, J.-H., Huang, T., Zhang, N., & Wang, Y. (2016). Crystallization controlled shape memory behaviors of dynamically vulcanized poly (l-lactide)/ poly (ethylene vinyl acetate) blends. *Polymer Testing*, *51*, 82–92.

Navarro-Baena, I., Sessini, V., Dominici, F., Torre, L., Kenny, J. M., & Peponi, L. (2016). Design of biodegradable blends based on PLA and PCL: From morphological, thermal and mechanical studies to shape memory behavior. *Polymer Degradation & Stability*, *132*, 97–108.

Tibbits, S. (2014). 4D printing: Multi-material shape change. *Architec. Des.*, *84*(1), 116–121.

Yaxin, Z. F. H. E. Z. L. Z. (2023, January). Effect of the support bath on embedded 3D printing of soft elastomeric composites. *Materials (Basel)*, *331*(15), 133475.

Ameta, Solanki, Haque, Singh, Devi, & Chundawat. (2022). Critical appraisal and systematic review of 3D & 4D printing in sustainable and environment-friendly smart manufacturing technologies. *Sustainable Materials and Technologies*.

Santosh, Nithyanandh, Ashwath, & Lalith Kishore. (2022, November). Comparison of internal friction measurements on Ni-Ti reinforced smart composites prepared by additive manufacturing. *Journal of Alloys and Compounds*, *924*(30), 166027.

Chapter 9
Applications of Image Processing Techniques in Fused Deposition Modeling

Manasa R.
Dayananda Sagar College of Engineering, India

Karibasappa K.
Graphic Era University (Deemed), India

Abhishek M. B.
Dayananda Sagar College of Engineering, India

Vibha T. G.
Dayananda Sagar College of Engineering, India

Bindu H. M.
Dayananda Sagar College of Engineering, India

ABSTRACT

3D printing uses fused plastics and composite filaments to layer components. This method produces parts for household tasks, industrial applications, and rapid prototyping. Developing a printing model requires determining extruder speed, extrusion height, and bed temperature (slicing). Poorly defined parameters can result in poor dimension precision, surface integrity, mechanical properties, and finally printing flaws in plastic and composite sections, limiting or prohibiting their use. During the first stage of 3D printing, a component's first layer is created. This chapter primarily reviews research on the use of image processing methods at different phases of the fused modeling process. Additionally, a review of the broad literature on the junction of FDM and image processing has been presented. The advantages and disadvantages of image processing for the FDM process, as well as existing barriers, have been discussed.

DOI: 10.4018/978-1-6684-6009-2.ch009

INTRODUCTION

AM, also referred to as 3D, has increased immensely in popularity over the years as suggested by L. Lu, J. Zheng et al 2015 and J. Jiang, X. Xu 2019. The authors I. Campbell et al 2012 and J. Jiang et al 2019 discuss that a range of industries, which include automobiles, aerospace, and medicine, have adopted it because of its unrivaled ability to lower various design metrics, such as time-to-market, the ability to produce high-complexity as well as performance products rapidly, and the ability to mass customize. C. Kousiatza et al 2016 have told that Fused Deposition Modelling is one of the most frequently used techniques in classifying additive manufacturing processes due to the high speed, low machinery price, and durable part materials. A typical FDM process entails melting thermoplastic material, extruding plastic via a nozzle, and layering it on a worktable is given by D. Espalin et al 2014. Its observed by Q. Y. Lu et al 2018 and W. Devesse et al 2017 that the uncertainty associated with FDM printing technology continues to be a significant problem and results in geometric errors or surface defects. Provided the long and complex production cycle connected with FDM having to print, it is essential to identify flaws and failures slightly earlier in the print process to avoid wasting time and money as seen in Y. Wang et al 2019 and Y. Wang et al 2020. Numerous researchers like Y. Tlegenov 2018 have been done to forecast printer effectiveness using specific printing parameters such as part temperature, nozzle condition etc. These techniques, however, necessitate the development of precise mathematical modeling techniques to link abnormal printer parameters to specific defects. Due to the complexity of 3d - printed mechanisms, developing a comprehensive framework to identify most of the problems in practice is extremely difficult. Unlike the indirect monitoring mentioned previously, operators inspect printing components directly for defects. This way, any flaws in the appearance can be found right away. Since the parts' appearance now reflects their deviations first from the design model, their quality can be determined by their appearance. On the other hand, constantly monitoring innumerable printing processes is extremely difficult.

There are now many automated systems for monitoring the printing process parts that don't need to be monitored by humans. Its because new computer vision techniques have made it easier for these tasks to be done. Gonzalez and Woods, 2002 showed that Image processing is capturing data using imaging devices, including cameras and scanners, and converting it to a computer or even device-readable format by Russ, 2016. This field emerging image sensor manages the object's dimensions and detects its color using image processing techniques. Ultrasound, electroscope, and computer-based images were also constructed from various components retrieved from the image source to accomplish a specific task Samtaş and Gülesin, 2011.

Image processing includes three stages: transferring images to electronic media, analyzing them, and printing the results Neumann et al., 2011. The initial step, analysis, is designed to minimize noise from the image (image blur, lack of sharpness, and poor image quality). For example, low-level operations like image filtering and high-level operations like image analysis are used to achieve this Bellaire et al., 1998. Along with the image processing steps, two types of image processing are used: analog and digital image processing. Digital and analog image processing consist of the following stages: pre-processing, during which data must pass; development and visualization; but rather information and extraction Lin et al., 2018. Yuanbin Wang et al 2020, Alberto Boschetto et al 2015, Ketai He1 2018, Koray Ozsoy 2021, Yi WU 2018, Hongyao Shen et al 2020, Matheus Godoy Fonseca Carmo 2020 have discussed that Images can be processed to improve their color and light quality, enhance their clarity, store and transfer them more efficiently, and even think of all the people in them. Established research mainly focuses on part geometry and infill pattern defects. Generally, this method needs to take a moment and recreate the

Figure 1. Typical workflow

printing part's 3D model and its affiliated point cloud. This is a lengthy and computationally intensive operation. It's because FDM processes build up layers that give off a lot of information about the part's quality through the way it spends its time by Paschalis 2021. While detecting defects using the visual image would be efficient, this will never be thoroughly investigated.

FEATURES OF CAMERA

Camera specifications are also critical to capture errors during printing and enhance the quality of printing. It should be **Compatible.** There are only a few camera specifications listed. The camera sensor type most frequently used in FDM is CMOS as Complex metal oxide semiconductor. Resolutions of 3000 and 2000 pixels are possible. A minor resolution of 5 Megapixels, a frame rate of 14-15 frames per second, color, and a USB interface are required for FDM cameras. Cameras must be plug-and-play compatible with any software chosen. Even if it is a high-quality camera, you'll need a way to view its video feed to monitor your print. OctoPrint, a system that connects a single-board computer like a Raspberry Pi to your Printer and can control the parts of your machine from afar, is probably the most popular choice. You can connect your camera to your host computer's USB port or camera interface and view it through your online OctoPrint interface. If your 3D Printer is close to a desktop or another stationary computer, you can connect it directly to a computer e.g., a nearby laptop, PC, or a spare Raspberry Pi board. Using a webcam monitoring program such as at Home Video Streamer, TeamViewer, or Splash top, you could remotely watch your 3D Printer using your camera stream. While the popular microcontroller is compatible with most USB webcams, the Raspberry Pi Camera Module V2 is the only camera made exclusively for a Raspberry Pi. Its size is unrivaled, with dimensions of 25 x 24 x 9 mm and a weight of about three grams. Supporting 1080p full HD video at 30 frames per second, the resolution is also quite good fps. The Raspberry Pi's popularity makes it an excellent choice for a 3D printer camera. Many in the community use this camera, with many 3D printed stands and mounts available. Some were designed to directly mount on a specific printer's bed, which can give you a front-row seat to your print's progress. Some of the other examples are Creality CRCC-S7, PlayStation Eye https://www.hubs.com

APPLICATIONS OF IMAGE PROCESSING FOR FDM

3D printer applications in image processing have recently garnered popularity and attention in the medical field Rengier, 2010. Von, 2008 have suggested many ways that rapid prototyping can be used. It can be used to look at and show images in three dimensions and make three-dimensional structures of different tissue types by pressing live cells & biomaterials together simultaneously. Build live vessels with a 3D printer and quickly detect deformation Armillotta, 2006. Like existing 3D printing machines, this study used a CMOS fields scanning camera to acquire photographs following sample fabrication. Using the Computer language and also the Opencv but also Diffimg libraries, the various sample images produced from the camera were compared to the corresponding image, and the surface quality error rate was identified. Thus, the functionality and capacity of 3d printing techniques will be enhanced by enabling continuous printing without user interference. Firstly, the picture of each layer during manufacturing is caught with a camera. Secondly, the contours of these pictures are extracted with image processing methods. Finally, each layer's profile errors are calculated, the data is analyzed with the methods of profile control, and the control status is acquired. It can be concluded that all defects in a counter profile can be successfully detected through this method. It needs to consider the structure of the product and features, which significantly simplifies the complexity of the process and has a good application prospect in the product development stage Yi WU,2018.

SURFACE QUALITY TESTING BASED ON IMAGE PROCESSING

This investigation used an autonomous 3D printing apparatus competent in continuous writing to conduct printing tests at various operating speeds. Opencv, Qt, Diffimg, and Numpy libraries were used to scan images of manufactured samples. The database was set up with input images of samples that needed to be created before the packaged software could be prepared. The RGB as Red Green Blue input images of the image have been converted to sequence variables using the Image Diffimg Library taken from each section after the sample manufacturing was completed. The error % rate was calculated by comparing each pixel's similarity ratio. The index of an RGB colour image pixels is used to refer to them. The pixel similarity ratio is calculated beginning with the first-pixel value and finishing with the last pixel value. The percentage error rate was obtained by dividing the determined pixel similarity ratios by the total number of pixels. Following the creation of the reference image, a camera was used to photograph the part's surface. The 3D automatic equipment runs at 75, 100, 125, 150, and 175 mm/s. Three times at each speed, the technique yielded 15 20x20x20 mm sample prints. For each speed, the manufacturing times appear to be 46, 38, 30, 23, and 18 minutes. For each sample, the time necessary to transfer the sample out from the Printer was 27 seconds—the photos of the samples that were created. Here, the component surface deteriorates as the axes' speed increases to motor acceleration.

The reference images and sample images' pixel values are extracted separately. The mathematical values obtained are in the RGB color space and consist of three channels for each pixel. This distinction was between source images and the printable sample image regarding surface quality. Ideas are transformed into matrix arrays according to their size in the image processing technique. The required computations were performed to use the Numpy library, which may be used in conjunction with the OpenCV library to allow quicker array processing. Comparing the digitized images was accomplished through the use of the diffimg library and numerical calculation methods. The error images are produced when the reference image is compared to the sample images. Each press was used to inspect the surface quality of the three samples. The Phyton programming language was used to compare the reference image and the images of a manufactured sample. By comparison, the mean percentage error rate for every group was calculated.

The researchers discovered that a prototype of an autonomous 3D printing machine capable of printing indefinitely, functional testing, improved surface tests, and dimension precision tests had been built. A camera for assessing surface conditions is also included in the conveyor-based device. To use an image analysis tool, the superior surface errors of the extracts were found to be 3.3–16.72 percent. The samples' dimensional accuracy erroneous values were determined to be between 0.1 and 2%.

The conclusions of this project, which included the design and manufacture of reliable 3D printing equipment based on an image processing technology, are summarised as follows:

- The use of 3D printing technology necessitates the presence of humans. Removing components off the platform is essential before starting a new manufacturing process. • Shifting concerns on printed parts via Printer are overcome through printed testing. This study would have aimed to design and manufacture a device with such a constant conveyer belt system that could be mass produced. In reality, the system's usefulness has already been shown.
- The model was made to boost the usage of 3D printing machines in the industry and to act as a model for future research.
- A practical application that could be used in this same industry and serve as a foundation for future research has been completed.
- By enabling continuous printing without user intervention, 3D printing technologies will increase manufacturing volume, functionality, and capacity. Future research shows that the 3D printer system can be monitored throughout sample production. The error rate is computed using photos taken after sample production and images captured in real-time while producing using image processing algorithms.

GEOMETRY DEVIATION ANALYSIS BASED ON BORDER SIGNATURE METHOD

Each layer of a manufactured object may be mapped out by providing a two-dimensional profile of pixels. To determine whether the part is genuinely out of tolerance in terms of dimensions, a comparison between the as-built profiles and their corresponding designer (CAD) concept should be performed, as previously stated. In the past, the likely-to-fit parameterized function method was frequently used to characterize and calculate profile error. On the other hand, developing a proper function for a single complex profile is hard, if not impossible. To address this problem, this chapter proposes to convert the profiles into a part of distance and angle using the border signature technique. This technique applies to both simple and complex shapes. It analyses the variation first from a nominal value more efficiently and accurately. It was demonstrated how to use the border signature method to portray geometry deviation analysis. This article reveals a polar coordinate concerning the first layer due to fixed relation between printed layers. O (X0, Y0) is chosen as the origin, where signifies the angle, d denotes the radius, and z denotes the layer height within the polar coordinate system.

The remaining layers then are predicted into this polar coordinate system's polar coordinate system. The projection correlation between layer 1 but also layer I I = 2, 3, was presumed to be, whereas the dashed line corresponds to the ideal design profile. Thus, at an angle of 180, the precise profile point on layer I like I = 1, 2 is Pi,zi,di, whereas the ideal profile point is P). At this point, the geometric difference between every layer of a part can be defined as the radius difference between the corresponding points on both the desired and real profiles at same angle. Calculate the angles and radius of the boundary pixels of each layer and plot them in a Cartesian coordinate system with the angle as that of the horizontal axis and the radius as the vertical axis. Thus, the profile error of layer I at an angle can be calculated as 180 180. The radius of the ideal profile at this angle, denotes the radius of a real profile layer I at this angle. Generate a profile defect point cloud for every layer of the printed object using the analysis described above.

It's worth noting that the machine vision system's image data is represented in pixel units. To evaluate the physical size of a measured part, calibration of a CCD dimension sensor has been required. To simplify the calculations, it is essential to avoid the effects of translation, torsion, and lens distortion as much as possible. The image equivalent then becomes a constant: d = L N mm pixel, where N is the total number of pixels equal to the physical length L in mm.

Figure 2. Deviation calculation with image processing details

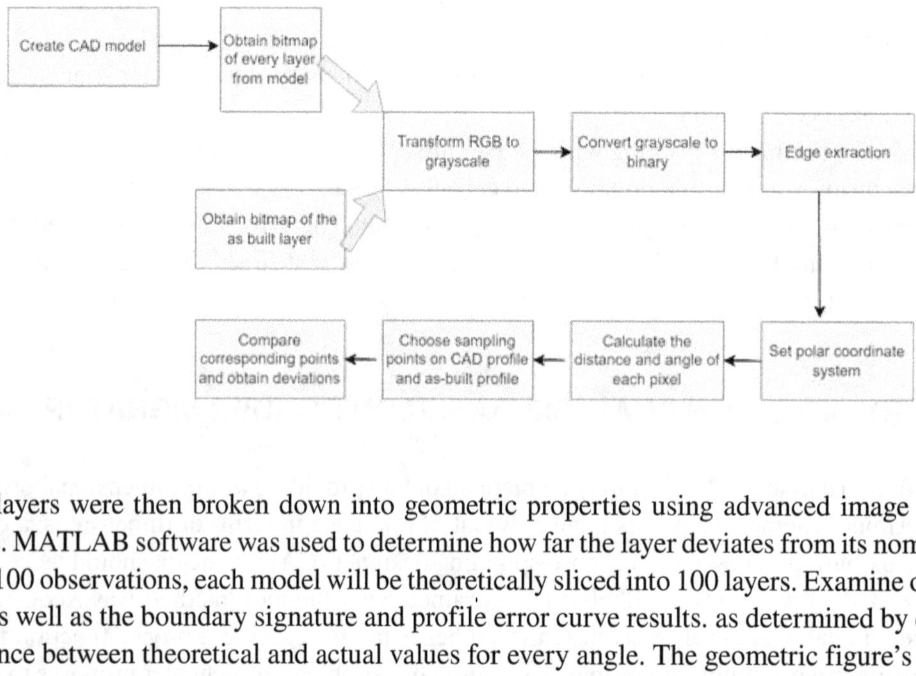

These layers were then broken down into geometric properties using advanced image processing techniques. MATLAB software was used to determine how far the layer deviates from its nominal value. To obtain 100 observations, each model will be theoretically sliced into 100 layers. Examine one layer of model 1, as well as the boundary signature and profile error curve results. as determined by calculating the difference between theoretical and actual values for every angle. The geometric figure's center will be marked by a red dot for differentiation and was selected as the origin. Angle values are displayed on the horizontal axis, while distance values are displayed on the vertical axis. There are approximately 13 pixels per inch in this experiment. 0.5 to 0.5 mm is a typical range for random errors in the contour when printing typically. Material phase change and machine accuracy lead to the majority of printing process errors... The sampling angle interval is important because the smaller the gap, the more exact the interpretation of profile information is. In this study, the sampling angle period is set to one. Depending on the intended minimum standard of the detected fault, the data samples may be positioned differently in other applications.

We used a Flash Cast Designer to make these models, which you can read about here. It was used during the fused deposition process model to get image data from some parts. It was connected to the computer by an image sensor called a CCD. This article created these models to use a Flash Cast Creator 3D printer. Images were captured using an image sensor connected to the computer during the fused deposition modeling process. The geometric properties of a part's layers were then extracted using image processing techniques. This same layer's deviations first from nominal value have been calculated and used the border signature technique in MATLAB software.

The procedure for analyzing geometric manufacturing errors and performing quality control on the FDM process. Use non-contact computer vision-based measurement instead of conventional measurement methods like CNN. You can get more than just a few critical excellent features and can only use something after the manufacturing procedure is finished. This enables real-time retrieval of the manufacturing process's complete geometry. The profile-based quality control monitors Every layer, maintaining control over the entire FDM process. Three distinct designs within each flaw, including dislocation, staircase, and gradual change. The remaining two are standard components that come in a variety of shapes. The experiments demonstrate the method's efficacy in detecting anomalies. Whenever a defect occurs, it's indeed possible to find it inside the part body but instead obtain real-time information about the associated error.

IMAGE PROCESSING TECHNIQUES FOR DIMENSIONAL ACCURACY

People used this study's image processing to ensure that an image's dimensions were correct. The Mean Filter, K-means algorithm, the Gaussian Filter, cluster algorithm, and other methods were used to ensure that an image was accurate in terms of its dimensions. Excellent filtering, also known as mean filtering, is a linear filter that removes noise. Mathematical, harmonic, or geometric averages are taken from each pixel and used to make a new value for the next one. This method is called "mean filtering." A median is a non-filtering technique for reducing image noise. The median filtering technique makes use of a kernel matrix of user-defined size. Gaussian filtering uses two-dimensional blurring for noise removal and details from images.

K-means algorithm: K-means is a clustering algorithm based on unsupervised learning. Indeed, K-means is a technique for segmented clustering that assigns each piece of data to a unique cluster. There are I data objects in the data set, and the K-means algorithm split it up amongst the m data sets that were given as input parameters. The K-means algorithm initiates with a random point called m. (clusters in number). Each data item is clustered according to its proximity to the nearest center point. A cluster center's value is calculated by averaging the pts assigned to it, a process that is reproduced until the centers' values stabilize.

Morphological operations. Morphological operations are set-theoretic methodologies that generate geometric patterns from the geometric patterns of the objects in an image. Morphology operations can be carried out on binary means white and black and greyscale images. Morphological processes can be classified into two broad categories: image dilation and erosion.

The Chan Vese is a segmentation algorithm. It is possible to separate objects in an image using segmentation, a standard image processing tool. A region-based segmentation technique proposed by the Chinese is based on active credit and level sets. With an open set of variables and two different constants, c1 and c2, we can address the problem of operationally minimal separation and the given curve. The first term compresses this same contour, whereas the derivative expands the contour in this method. These two forces are in equilibrium when the study area reaches its limit.

Detection of image edge points: Edge points can be present in an image using image processing techniques. This allows us to figure out the borders of the objects in the picture. Methods for detecting edge points include the Robert and Prewitt edge models, the Sobel and Gauss edge models, and the Laplace and Canny edge models. In aspects of edge detection and noise reduction, the Canny edge significantly achieves the others. Most of the time, Robert edge detection is often seen in hardware applications because of its speed and can be used easily which are very important. By making use of integer filter, sobel edge detection technique can be seen that image can be bent vertically and horizontally.

Three different image processing methods are utilized to figure out how much the sample part made with the additive manufacturing method was off in terms of its dimensions and area. The Mean, Median, and Gaussian filters are used to calculate the differences as well as errors between real part and images. The Gaussian filter, on the other hand, does have a greater error rate. To smooth an image and make it easier to see the lines and edges, the Mean and Median filtration methodologies are used.

The geometrics part of the bottom layer image is compared to 3 different processing methods of images with average width, length, as well as area. Median filtering was used to reduce measurement error in length and width for tool geometry dimensional control. The Mean filtering process was used to obtain the lowest possible error in the area measurement. With almost the exact bottom layer actual image of such a type of part geometry and one calibrated with an average of the three images taken, several different ideas could be produced and compared. These two things work better together just to keep the edges safe. They look at three different ways to process images and see how their average width, length, and area compare to the part geometries in the middle layer of images. Has been used for dimension control of geometrical to use median and Gaussian filter methods. These methodologies have been used to get the minimum error when measuring width and length.

Additionally, the median filtering technique was used to gather results with the least amount of area measurement error. Gaussian filtering was also used to ensure the lowest possible measurement error. Again, the median filter method produces excellent results for dimension control of length, width, and area in lower layers of part geometry. After the part geometry's middle layer, the dimensional control margins of error significantly increased. This is because the part's geometry is discrete and small, reducing its sensitivity. Analyses of relevant academic literature for the study were conducted, and data for comparison between the studies were provided. The powder bed's dimensional accuracy was achieved by Aminzadeh using image processing techniques.

In the building chamber, an 8 Megapixel camera monitored the manufacturing process. There are 2160 4096 pixels in the camera's images. All images are filtered before analysis with a median filter of 7 7. The point-to-point error was obtained for each image's boundary. This error was found to be 69 pixels, or 476 microns, or 0.48 mm across all images. On a part with complex geometry, the investigation used a 3x3 mean filter and three image processing algorithms. Following that, a three-by-three median filter based on Kmeans was used. Methods such as Kmeans and morphology were also employed. When all layer images are examined in a study, there must be a minimum of 0.2mm (length) between points. This is 0.06mm for the Mean filter. This is 0.06mm for the Median and Gaussian. As per Xiong and Zhang41, the maximum difference in width between corrected and uncorrected values is approximately 2%.

The Gaussian filter, the Sobel operator, and the Hough transformation have been used to extract the characteristic points of the bead geometry. The results of the tests illustrate that detection precision can meet the dimension control requirements for additive manufacturing. The digital image task has become more difficult in this study due to the image resolution, illumination, and low contrast. Despite these issues, the average width and length errors again for dimensional control are kept to a minimum across all layers. Python is a programming language that is free and open-source, and it was used to create the interface program. Interface programs allow the user to take a screenshot of the layer as it currently appears. Then, during manufacturing, the layer image can be filtered to achieve dimension accuracy. The obtained data can be used to calculate the dimensional accuracy error rate.

DEFECT ANALYSIS OF FDM PRODUCTS

Defect detection is a rapidly growing field of study. The two types of defect detection methods that are currently available are based on machine vision for example, machine vision-based monitoring systems and laser scanning for example, laser scanning-based monitoring systems. Statistical process control and machine learning can detect defects in the fused deposition modeling process. To begin, images of every layer are captured during the manufacturing process using a camera. Second, contours are extracted from these images using image processing techniques. In the end, statistical process control methodologies are used to look at these data and figure out how well the process is being controlled. FDM is a straightforward and efficient method of additive manufacturing. As a result, the FDM is used to verify the method's feasibility.

The former uses cameras to take images, while those can determine an object's height, which is impossible to monocular vision. To detect overfill and underfill defects on the upper surface of deposited parts during the additive manufacturing process, compare the existing point cloud to pre-sliced stereolithography STL model. Surface defect detection can be improved by using a stereo vision measurement system that can simultaneously acquire surface grayscale and depth images without additional information registration or calibration. With the help of measured reflectance datasets, the researchers developed a Gaussian process GP model that accurately represents the nonlinear reflectance behavior of various materials. There are two types of defect detection algorithms: traditional and deep learning-based. The significant difference between these two techniques is that deep learning employs convolutional networks to instantaneously extract features, although conventional machine learning methods necessitate manually designing correlating features engineering.

Many researchers are working on deep learning techniques for specific applications. Using bright field illumination, a surface image acquisition module was created. Tiny Defect Net is a model that uses deep learning to figure out where and what kind of defects are. A deep learning architectural style for surface crack detection and segmentation. Industrial 4.0 application of deep neural networks to classify optical defects. Typically, deep learning methods necessitate a vast amount of data. The UCR convolutional deep neural network detect common and uncommon surface defects on aluminum profiles. The Multilayer Perceptron Network (FPN) is used to improve model performance by augmenting data and developing algorithms for sensing X-ray image deficiencies in automobile die-cast aluminum parts. Although small datasets can be more efficient for some machine learning methods, for deep learning models, many labels must be provided to get the best results. Automated inspection system with five-plane array CCD cameras and four LED lights. It was decided to use the SVM algorithm in classify defects based on features extracted from candidate defect regions.

A computer program helps you figure out the location, shape, number, and size of surface defects on complex parts—visual and numerical instruments for analyzing a rough surface by imaging its micro-roughness and utilizing the parameter value. For image impulse noise, two new adaptive filtering media techniques were shown. Most current research focuses on finding defects just on the upper side of FDM. Outer surface detection benefits from FDM's even layers. Whenever the layer gaps between the layers are asymmetrical, defects are much more likely to happen. Our prior work showed how the cracks, layer upon layer on the outer surface, are always the same. This allowed us to offer a multi-view and the all vision detection method for the outer surface. This system can detect defects on the object's external surface online, which also keeps tabs on the object's 3-D printing progress. But it only works on parts with such a slight surface change. When looking at someone's face, it's hard to tell if their eyes are normal or if they have a defect because of how quickly the outer surface of their face changes from one angle to another. This section discusses a technique for detecting shape defects in robot-based three-dimensional printing.

After projecting and capturing images, both contours are detected using image processing technology predicated on the laminate structure characteristics of FDM. By comparing the contour variables of the image taken from the platform to those projected from a theoretical model, defects and their associated parts can be found, and they can be found by comparing them. To determine whether a feature or a fault is present, one must compare the parameters of two different contours and decide whether they are similar enough—this system's processing of components.

If you want to know how an industrial CCD camera works, you need to see how the original point and camera work together to determine the projection point. In the image, the coordinates of a' are determined by the real-world coordinates of a and the camera parameters, such as the camera's center of the field of view. As a result, a monocular camera can be used to determine the mapping relationship. A theoretical projection picture is created by projecting earlier extracted feature points onto the image. Finally, image processing will define the projected image's contours.

In the FDM manufacturing process, each step impacts the part's precision. For example, incorrect parameter settings, changes in the environment, and problems with the machines can affect the quality of the features that are made.

There are five types of defects in FDM manufacturing: filament stacking as well as pulling, warpage deformation, abnormal mutation, shape anomaly, and material interference, and start/stop error. Each surface has a beginning/end point due to the slicing. If the speed of the nozzle and the speed of the silk at the start and stop point are not compatible, the starting and halting material would be both too much and not enough. Even though these flaws are distinct, they all share one trait. The defect layer's shape differs from the standard layer's, affecting the manufacturing section's outer contour. A direct visual inspection can evaluate if the manufacturing is abnormal by inspecting each layer's contour shape for abnormalities.

Machine vision and a contour control map are used to detect defects during FDM manufacturing to avoid interfering with the FDM molding process. As a result of image processing, the manufacturing process's cross-sectional region and contour are determined. At last, the line profile error of each layer section is analyzed and determined using a contour control chart to evaluate the manufacturing process's irregularity.

Image-Based Parameter Extraction

FDM manufacturing is done with the help of a camera. It takes pictures of each layer as it is being made. The images have a lot of information, but they also have a lot of things that can mess things up, like noise and other items in the way. Image processing can extract contour parameters from each layer section, allowing for statistical analysis. The following describes the image processing procedure:

First, take a view of the current surface section. Then transform the color image into a grayscale image & extract each of the color components from it, including a frequency distribution graph and more analysis of the color image with the most contrast. This experiment uses red material, with the most transparent part being the red part of the material). To achieve a binary value, the red component map has been binarized utilizing threshold segmentation, and then the resulting grayscale image is selected. Closed and open operations process the binary image, which retains a few slight noise points and breaks. Recognize the current layer's contours to determine its cross-sectional profile.

Calibration of Size

CAD and 3D software use vector graphics, but a camera captures a bitmap picture. To understand the difference between vector and bitmap images, it is necessary to know how they differ. The data in the two images above are too diverse to be compared directly. It is essential to convert vector images into bitmaps before performing a comparison. Calibration is used to decide the camera's internal and external parameters. It is possible to determine these parameters by comparing the object's reference point x,y,z and its image point XY,y,z coordinates u, v. An X-Y ruler and an N-pixel pixel ruler are used to specifying the number of pixels needed to represent a given length L in pixels. The ratio L to N denotes each pixel's actual size S. This allows for capturing at least each valid image for each layer cross-section while the experiment is running. The light levels significantly affect the image's quality throughout the investigation.

When processing images and getting information, a fixed light source should be used to keep natural light from interfering with the experiment. People who read this article will learn how omnidirectional lighting is used to ensure the object being measured is illuminated from many angles, preventing unevenness in the image because of shadows. The LED light strip is often used as a light source in the experiment to avoid uneven illumination during photography, which reduces the precision of information extraction.

FDM PROCESS DEFECT RECOGNITION BASED ON CONTOUR CONTROL CHART

Real-time image processing is used using a camera to extract forming section profile parameters from each layer's forming section even during the FDM molding process. The profile of an ideal model is compared to a profile of the actual model. To determine if the model isn't right when it's being moulded, figure out how much each layer's profile error is and create the joint control diagram. Because of the height difference between the lower and upper layers, it can be seen that the data for pts 27 to 30 exceed the limit, indicating an abnormal shape inside the cross section's cylinder and also that the 26th layer is just too thin 0.18 mm due to height difference between the lower and upper layers.

Although the outcome of the detecting defects is deferred by one layer, the delay could be ignored since a single-layered defect has so little influence in 3D printing—the joint control diagram for the variable section model. The information is categorized in 39-45 layers, as can be seen. If it exceeds the decision boundary and an abnormal section can be identified during the FDM forming. The transition layer between layers 36 and 38 is standard.

Figure 3. Image processing Process

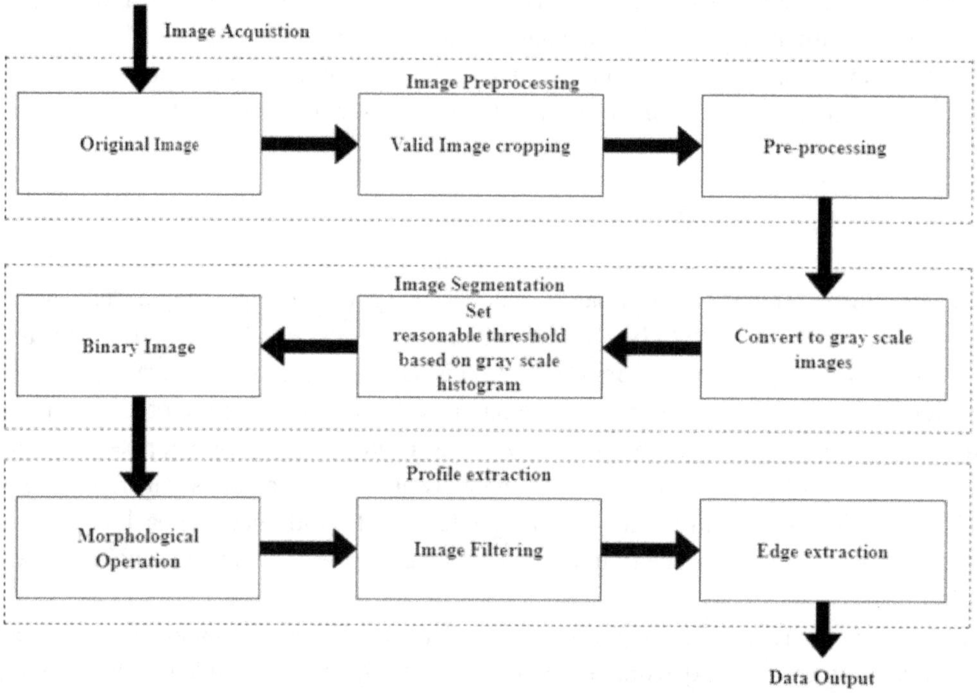

IMAGE PROCESSING FOR MANUFACTURING

Each layer of the manufacturing process is photographed using a digital camera for quality control purposes. Images should then be processed to extract contour and area information. This data is used as the starting point for more statistical analysis. Counter and area information should be obtained for each layer using appropriate image processing methods. Due to the interference from the outside world, the image will have a noisy signal. Three steps are required to obtain the layer counter. During the pre-processing stage, the image's noise signal must be eliminated. Second, using global threshold segmentation, binary images can be obtained. Finally, using feature extraction, extract this same section and layer counters. The image processing procedure is depicted in Figure 3.

The image was taken at the beginning of the production process. Operation speed is boosted by selecting the most effective area of the background, as there are many ineffective areas. It is first necessary to extract the original image's grayscale and RGB components. Figure 3 shows the image's grayscale and RGB components. As can be seen in this experiment, the R component exhibits the most significant contrast. When getting information about each layer, the R component is chosen.

Image processing techniques can be used to calculate the section's area and the number of layers in the section. When you compare a counter and area values between every layer, you can look at models with the same section. The variable cross-section model, on the other hand, relies on the previous design as a reference. Defects can be identified by comparing the differences between the real and design sections. So, when making a model with a variable cross-section, you should figure out the region of the reference image in advance. Size Calibration is required once Machine Vision is used to detect defects. Orthogonally rulers were taped to a machine's platform to serve as a reference for determining the contour size in the images. During manufacturing, rulers could be captured with each layer section. Count the pixels on the rulers to find the contour's actual size. Then, using the following formula, one can determine the exact length in one pixel Calibration valuation, S. L denotes the measured distance. N is the length in pixels that is being measured. S=L/N

This process, known as 3D printing or Fused Deposition Modeling (FDM), is used to create parts by layering plastic fusion filament. This process generates components suitable for domestic use, rapid prototyping, and final applications. Preparedness the having to print model slicing includes defining extruder speed, height from the bed, and bed temperature. Incorrectly defined parameters can lead to poor dimensional accuracy, surface quality, mechanical resistance, and eventually multiple printing faults in the parts, limiting or preventing their use. During the first layer of a piece's construction, a critical point in the process of 3D printing occurs. When printing parameters are incorrectly defined, strange geometrical patterns emerge. That's why this project is being undertaken: to find out why. Monolayer parts were printed using polylactic acid PLA filament. We used a Graber i3 3D printer for home use. Using three different variations of the extruder height relative to the bed, three other pieces were printed. A DCP-L2540DW model scanning was used to scan the printed components at a resolution of display with a resolution. Matlab® software was used to analyse the captured images and ascertain the pieces' geometrical properties. To better understand how incorrect parameter definitions affect the geometric defects in three-dimensional model processing, this study is the first step.

Matlab® version R2017b's image processing as well as computational view toolbox included ImageViewer®, which was used for the measurements. Each image was independently accessed, and the Means of measuring Distance feature was utilised. The above tool determines the amount of pixels in two dimensions between two identified dots X and Y. In addition to dimensions, the data included layer thickness and diagonal length. The diagonal measurements revealed extruder alignment issues, while the thickness of the internally and externally layers revealed table height variations. Following the information extraction process, images of the pieces with the specified dimensions were saved inside the Tagged Image File Format TIFF to create high-resolution graphic representations. All the data extracted was put into cells in the Matlab® technology, which made it easy to compare the data from every repetition by condition. As a final step, we compared the results of the printed conditions to the representative measures we defined earlier in the printing process.

VISUAL QUALITY CONTROL IN 3D PRINTING

This article describes our in-situ monitoring method for assessing the surface quality of FDM 3D prints. Images were taken and analyzed of the layers that had been deposited throughout the process. Grayscale value changes on deposited surfaces enabled overfill and underfill detection. An infrared camera with a long-range is used to keep an eye on the temperature of parts made with an FDM 3D printer and link it to their mechanical properties and quality. Scanners were used to identify and correct printing errors by analyzing feedback signals produced during the printing process—a three-dimensional digital image similarity system 3D-DIC for monitoring and quantifying printed part surface quality online. Specifically, 3D Digital Image Association was used to capture stereo images and spatially correlate them to retrieve surface properties and generate 3D geometry. A method for detecting manufacturing errors in 3D printing. According to the findings, a multi-camera structure with 5 monitor features, each equipped with such a Raspberry Pi camera, will be used to inspect the product quality in this study. The Dot Net framework and C# were used to process the images. The imagery was captured from eight predetermined locations, as each camera was placed at five different angles, as part of a process that allowed a more accurate comparison of what was captured and the last 3D printed piece. A correlation was performed pixel by pixel.

For instance, while the experimental procedure can detect premature process terminations that result in inadequate product or filament injection failures through into Printer's nozzle, it cannot evaluate vital geometrical characteristics and structural defects, which are the most common FDM errors. It has been proved that a novel method for 3D restructuring doesn't require the camera to be moved entirely around the printed object or use multiple static cameras. Additionally, by cropping the images once they are acquired, the printable part of the image can be kept separate from the surroundings and other undesired information. Layer by layer, the 3D model is constructed, necessitating online monitoring and error detection. Finally, we evaluated the procedure using an Immersive Virtual reality AR prototype of the printed object. As per a recent study, the geometric accuracy of printed parts can be calculated using image-sensing processes and artificial intelligence algorithms.

This article outlines a method of achieving in-process quality control on an FDM 3D printer using the standard supervised learning algorithm SVM classifier and a single camera. This work categorized the condition of the printing process as 'defective' or 'good' using a machine learning technique and an image caught during the construction procedure.

The experimental setup for this work consists of a commercial 3D printer known as the 'Ultimaker 3 Extended,' an ADXL345 accelerometer, and an RGB-Depth camera known as the 'Asus Xtion Pro Live.' The Ultimaker 3 Extended 3D printer features the highest construct buildings of 215x215x300 mm, a layer resolution of 20 to 200 microns, and an XYZ print accuracy of 12.5, 12.5, 2.5 micrometers. In the experimental procedures, polylactic acid PLA was used as an eco-friendly feedstock material. The process parameters described in Table 1 were selected and maintained throughout the experiments. Because of this, when the print head attained a checkpoint during the FDM process, the accelerometer sent a signal back to the optical monitoring program, which allowed the optical tracking system to take images of that point.

The Asus Xtion Pro Live sensor captures color images as well as data about the depth of each pixel. The integrated infrared depth sensor of the camera provides depth information for the scene, while the RGB sensor renders the scene in color. After registering these data sources, an RGB image of the location and information about the scene's pixels-per-pixel depth are obtained. Only the depth sensing capabilities of the camera will be used in this article. The laboratory equipment mentioned previously is used to monitor the 3D printed objects throughout the process. It is worth noting that the experimental setup proposed here is easily adaptable to any type of 3D Printer and a variety of additive manufacturing technologies. The developed software was written in C++, and the Point Cloud Library PCL was used to manage the 3D points; as a result, a pipeline was developed that automates the in-situ monitoring of the FDM 3D printing procedure. Finally, numerous experiments were conducted in various lighting conditions to capture images with the RGB-D camera. Due to the effect of changing lighting conditions on image quality, a constant lighting angle was maintained throughout the experimental procedures to facilitate image analysis and comparison.

DATA PRE-PROCESSING

Adhesive manufacturing has many advantages, but one of the most important is the ability to develop complex physical objects from a digital CAD model and a 3D printer. The machine commands that represent the print head trajectories are generated using software known as a slicer, which is obtained first from a digital 3D model. For CNC milling, turning, as well as 3D printing processes, Gcode is used as a machine programming language. The code instructs the Printer's sensors, build platform, & print head to extrude the semi-molten feedstock material. There is, however, no guarantee that the 3D Printer will accurately execute the Gcode commands, resulting in manufacturing errors. Diffraction of light FDM is a technique used to detect dimensional differences between printed objects and 3D CAD models. A 3D printer can malfunction during printing, or for the user to use incompatible printing conditions to cause these errors. To identify dimensional variance errors during printing, the physical print piece is checked against the model. If the slicer software doesn't know how to make an FDM model out of a CAD model, it needs the CAD model to be in Standard Triangulation Language STL format Gibson, 2014. It's important to note that STL files only contain information about the part's external surface geometry and do not include any information about its internal structure such as the lattice structure or supporting facilities. As a result of the high cost of feedstock material and the lengthy process, some AM processes, like FDM, do not print solid objects. The internal structure of an FDM part is defined by the lattice structure filling pattern. The layer-by-layer construction of FDM 3D printing models necessitates constructing a previous layer on such. Depending on how complicated the 3D model is and how far the new layer is overhanging, you may need support structures.

A slicer program is used to make a digital 3D model. It shows how the print head moves to make a point cloud of a digital model. In the first step, you must procure the 3D CAD model into CuraEngine, a powerful engine for 3D printing commands. Once this step is complete, the build platform and print head are given Gcode instructions for how to move throughout order to build the 3D model. Following that, a Gcode simulation of the 3D printing process is performed. Following that, the generated Gcode was further processed using a modified version of gcode2vtk. They simulated the 3D Printer's functionality by parsing the Gcode and extracting the planar-layered trajectories of the print head. The following phase involves the creation of a structure that contains all of the paths used by the 3D Printer. It is important to note that only Gcode commands are looked at because they give information about the material that is being extruded and the movements of a print head which ultimately creates the physical model. Finally, these trajectories are uniformly sampled, resulting in a theoretical point cloud of the three-dimensional model.

In the last step of a pre-processing data module, a lot of extra information is sent to the 3D Printer. A precise image of a printed part can only be captured when the Printer's print head is at the Printer's zero X=0 and Y=0. To do this, the Gcode was changed so that after every "n" layer, the print head was sent back to a zero position of a 3D printer. A sensor was also added. Print heads are instructed to move toward zero in the AM and then pause for two Seconds. With this method, the printing head does not block the RGB-D sensor's view of a building plate during two because the accelerometer sends a signal to a monitoring system. After this, the printing head moves back to where it was before, and the printing process is restarted. In this way, the FDM process captures the printed part's real-time point cloud and compares it to the theoretical point cloud.

Experiments have been conducted in the same manner as previously described; the print head of a 3D printer was reset to zero every 'N' layer to allow the camera used to acquire a straightforward point cloud of a printed part during the FDM procedure. To remove the background and define the crop area, a fiducial marker has been used to filter the point cloud acquired with the Asus Xtion sensor's wide field of view. In virtual reality, robotic systems, and camera calibration, the AprilTag visual fiducial system is widely used to measure distances and coordinates Olson, 2011. Any size target can be printed on the standard sheet, making it easy to incorporate the marker into an experiment at a reasonable cost. When a small AprilTag is attached to a build platform of a 3D printer, it can be used to figure out the orthogonal area of interest and remove any outliers from the point cloud. The orthogonal region includes the build platform, AprilTag, and 3D printed model. This method allows for quick noise filtering and printing of a model in a small area. To accurately evaluate the model's quality, it's necessary to separate the original point cloud from the printed layer. This is a plane detection and segmentation problem.

3D point cloud information can be segmented into planes using a variety of methods in the scientific literature. In this study, the Ransac algorithm was used because it is easy to accomplish in PCL and provides accurate results. Although Ransac is slow with large data sets, cropping the original contention cloud reduced its size and therefore did not affect the want to introduce an optical monitoring system. When the algorithm was used on a cropped point cloud, it came up with reasonably accurate segmentation of a build platform, but some areas still had a lot of outlier points, such as where the AprilTag was placed, as shown in the left part. To eliminate any outliers from the point cloud representation, a method has been devised only to include points associated with the model's inliers. We were able to divide away points from the printed model by increasing Ransac's distance threshold. To resolve the problem, a distance-based algorithm was used to cluster the remaining point cloud and then segment the largest cluster, resulting in a point cloud entirely composed of points that belonged to the printed layer of a part. We used a 3D Euclidean distance clustering technique to cluster the remaining issues. Using a distance between two points as just a metric to press the point cloud resulted in the model being satisfactorily filtered because of its sparseness.

The final stage was to recreate the printed model and compare it to the 3D model's conceptual point cloud. Concatenating the point clouds from each individual deposited layer would be sufficient in instances where its 3D Printer's build platform stayed stationary. However, just on 3D Printer used, the print head stays permanently in the vertical direction while a build platform moves. As a result, pre-processing the acquired point clouds before concatenating them was essential. The first step computed the transition between two consecutive layers by trying to align their build platforms. We used the Iterative Closest Point ICP algorithm to minimize the difference between two point clouds. As a result, the ICP method was employed between two consecutive point clouds before segregating the build platform. Given the high point density of the construction platform, this was possible to match the same individual construction platforms of two consecutive point clouds, resulting in the transformation of both point clouds as a result of this matchup. As a result of segmenting the build platform and transforming the extracted layers, the physical printed object can be represented. As a final step, all the printed layers must be in one point cloud to create the 3D model.

In addition, the relationship between the conceptual and real point clouds was also looked at. Compared to the theoretical model, the reconstructed model shows a reasonable degree of agreement in terms of quality. To make a comparison of the reconstructed point cloud to its academic counterpart, we were using an octree-based point cloud spatial change detection method. Each internal node of an octree has exactly eight children, resulting in a tree data structure with eight internal nodes. It is commonly used to partition three-dimensional space into octants recursively. By calculating as well point cloud octrees, this was possible to recover points stored in one octree framework but not in the other. The above method used generated point clouds with voxel sizes of 0.5 mm to calculate the spatial difference between theoretical and reconstructed models. It is 9.6%. When two point clouds were examined, their resolution was a big difference. This is because the optical sensor used had a shallow resolution. Although the suggested visual monitoring system is adequate for monitoring FDM processes, it can also detect real-time errors in AM procedures if a more precise sensor is used.

CASE STUDIES

Cylinder

It is used in mechanical applications like shafts and pivot pins. Because these parts are often subject to tolerances, it is essential to look into the dimensional accuracy of this geometrical feature. Models that use closed volumes must include at least two additional surfaces. A technique that compares the boundary surfaces to the actual boundaries of the model's variables u and v while also altering the scaling variable. It is based on a specimen fabricated from the original model. The manufacturing support structures have overhanging surfaces to deposition angles of 135° to 225°. Each specimen's section was acquired using a flatbed scanner. A morphological operator has been applied to the binary image to remove the noise. As a result, the Canny algorithm produces excellent results when used to process the cleaned data. Following that, data can be analyzed by converting the position of each white pixel to two-dimensional coordinates. The procedure was then replicated with the specimen modified according to the proposed methodology. Aligning the data concerning their nominal geometry and one another is necessary before running a comparison. Least squares and the Levenberg Marquardt technique were used to nonlinearly the data. After the regression, the center coordinates were used to locate the experimental data in a circle. A horizontal wall is more favorable to the methodology than other surfaces, and a vertical wall is treated as if the method used an isotropic offset equal to zero. The data were shown to be aligned with the nominal geometry. Some interesting ideas can be gleaned by zooming in on the areas where evidence has been found. The filaments create an anomalous profile with variable spacing when near horizontal walls are magnified. As expected, there is a 0.2 mm deviation from the nominal circle. The proposed procedure offsets the new peaks by 0.2 mm in just this zone, ensuring that they coincide with the negligible ring: the result is entirely satisfactory. The same offset was used for horizontal walls inside the bottom half of the section, which had deposition angles of about 180°. So because model has been deposited more than a come up with strict with a convex profile, it doesn't form a free shape but rather copies the convex profile in negative. The procedure's application shows that the modified profile, as well as the original one, are in good alignment again.

Spherical Joint

In the second case study, we'll look at a spherical joint. It has two parts: a sphere-bearing stud and a socket enclosed in a casing. The fabricated original spherical joint is Assembled manually revealing a few issues during the coupling process. This is almost certainly a result of an incomplete clearance. To describe this joint, this same sphere and socket were measured with a micrometer and a digital image analysis program. The map made with the micrometer has 187 diameter measurements that are evenly spread across the surface of the sphere. The data were processed to determine deviations from the nominal sphere diameter value. This is how it worked: The original specimen's diameter was only in this range when it was deposited at an angle close to 90°. Wherever the deviations exceed 0.3 mm, they increase to 0.4 mm. DFM results in a very consistent map: deviance is relative to nominal values with a scattering of about 0.05 millimeters after applying the DFM. During the methodology used, data were plotted in box-and-whiskers plots, indexed by. The FDM component's typical behavior: there are two distinct zones, one adjacent to vertical surfaces and the other located elsewhere. The term is defined by specifications-conforming measurements, while the latter exhibits deviances. A scattering pattern that is both broad and variable can be seen in the first zone, while only a 0.1-mm difference is seen in the second zone in some cases until 0.3 mm. Following the DFM, the angle of deposition has a negligible impact. All measurements are all within 0.05 mm of a nominal value, with data pairs scattering similarly. Because measurements have been made on curved surfaces instead of flat surfaces, a slight overestimation results in the small mean shift. By conducting a thorough analysis, it is evident that a mechanical coupling improves the performance of the modified component significantly. To complete the combined analysis, the socket's interior was characterized using digital image processing. The comparison of the socket measurement results to the sphere. Due to the presence of support structures in these inner zones, certain defects and profile changes occur, leading to a more significant deviation than occurs in external zones. Thus, while using DFM slightly enhances the scattering of experiments, the median is correctly centered on the wanted nominal value. The resulting coupling is significantly improved: prototypes produced with original shapes have interference and clearance issues, likely to result in an unsure coupling; after the DFM, the desired clearance is achieved. Additionally, the volumetric reducing error is significantly reduced in this case, from 0.055 to 0.01 millimeters for sphere and from 0.099 to 0.001 millimeters again for a socket, a reduction of 82 percent and 99.2 percent, respectively.

Air Fan Blade

With the help of digital image analysis, we could dig deeper into the airfoil section. Observers could see the discrepancy between the expected and actual profile shapes. The initial section demonstrates a distinct staircase effect, with peaks exceeding the nominal profile. The fabricated profile can be more closely compared to its nominal section using DFM. It can be seen that the modified upper profile is more precise than the original in the zones adjacent to the leading edge, while the bottom profile displays some local issues. A regional assessment of the top and bottom profiles was performed to elucidate these points further. The shapes were divided into 4 mm-long segments before and following the DFM application, and the deviance from the nominal contour for every component was determined by calculating and expressed using a box-and-whiskers chart. DFM's upper profile clearly reduces x-axis deviations between 30 and 10 mm. An unfortunate problem occurs between 10 and 0 mm: because the profile's height cannot assume fractional values of layer thickness all along the stratification direction, the DFM application removes an extra layer, reducing dimensional accuracy. The remaining sections of the profile could be enhanced. The bottom profile has few deviations, except for the sharp end between 26 and 18 mm. There is a lot of damage to the prototype's surface in this area because of support structures. It is possible to conclude that the scientific method results in an overall improvement in dimensional accuracy, despite the aforementioned local issues. The box-and-whiskers plots demonstrate this for the upper and bottom profiles, respectively, which show an improvement in dimensional accuracy. The average reduction in volumetric error in this case study is 36%, or from 0.115 mm to 0.073 mm. This is comparable to the declines observed in Tong's test case

CONCLUSION

This chapter examines the applications of image processing techniques to the modelling of fused deposition. A study of the fused deposition modelling procedure demonstrates the importance of incorporating image processing techniques throughout the entire procedure. Image processing has multiple applications, including process design, planning, and optimization, as well as quality control and improvement of FDM products. The associated issues have been described in detail to include the benefits and drawbacks of the FDM process, as well as other obstacles to the technology's use. In addition, the intersection of FDM and image processing as well as the existing barriers have been reviewed. The majority of research when employing FDM is devoted to optimizing process parameters. These are effective for enhancing production process parameters for at least one quality indicator. Image processing is improving the prospects for AM adoption, and it is expected that this trend will continue in the foreseeable future. The potential for expanding AM adoption and strengthening its value proposition have both benefited from image processing. These tools should be further developed for use in the real world, and case studies from the industry should be published to demonstrate their efficacy.

REFERENCES

Armillotta, A. (2006). Assessment of surface quality on textured FDM prototypes. *Rapid Prototyping Journal, 12*(1), 35–41. doi:10.1108/13552540610637255

Bellaire, G. & Talmi, K., Oezguer, E., & Koschan, A. (1998). *Object recognition: Obtaining 2-D reconstructions from color edges.* . doi:10.1109/IAI.1998.666884

Campbell, I., Bourell, D., & Gibson, I. (2012). Additive manufacturing: Rapid prototyping comes of age. *Rapid Prototyping Journal, 18*(4), 255–258. doi:10.1108/13552541211231563

Charalampous, P., Kostavelis, I., Kontodina, T., & Tzovaras, D. (2021). Learning-based error modeling in FDM 3D printing process. *Rapid Prototyping Journal.* . doi:10.1108/RPJ-03-2020-0046

Charalampous & Kostavelis. (2020). In Situ Visual Quality Control in 3D Printing. *IVAPP 2020 - 11th International Conference on Information Visualization Theory and Application, 3*, 317-324.

Devesse, W., De Baere, D., Hinderdael, M., & Guillaume, P. (2017). Modelbased temperature feedback control of laser cladding using highresolution hyperspectral imaging. *IEEE/ASME Transactions on Mechatronics, 22*(6), 2714–2722. doi:10.1109/TMECH.2017.2754550

Espalin, D. (2014). A review of melt extrusion additive manufacturing processes: I. Process design and modeling. *Rapid Prototyping Journal, 20*(3), 192–204. doi:10.1108/RPJ-01-2013-0012

Gibson, Rosen, & Stucker. (2014). *Additive Manufacturing Technologies: 3D Printing, Rapid Prototyping, and Direct Digital Manufacturing*. Springer.

Godoy & Carmo. (2020). *Study of Defects and Geometric Anomalies on Monolayer Parts Obtained by Fused Deposition Modeling Process*. Proceedings.

Gonzalez, R. C., & Woods, R. E. (2002). Digital image processing. Prentice-Hall.

Jiang, J., Xu, X., & Stringer, J. (2019). Optimization of Process Planning for Reducing Material Waste in Extrusion Based Additive Manufacturing. *Robotics and Computer-integrated Manufacturing*, *59*, 317–325. doi:10.1016/j.rcim.2019.05.007

Jiang, J., Xu, X., & Stringer, J. (2019). Optimisation of multi-part production in additive manufacturing for reducing support waste. *Virtual and Physical Prototyping*, *14*(3), 219–228. doi:10.1080/17452759.2019.1585555

Kousiatza, C., & Karalekas, D. (2016). In-situ monitoring of strain and temperature distributions during fused deposition modeling process. *Materials & Design*, *97*, 400–406. doi:10.1016/j.matdes.2016.02.099

Lin, W., Tong, T., Gao, Q., Guo, D., Du, X., Yang, Y., Guo, G., Xiao, M., Du, M., & Qu, X. (2018). Convolutional neural networks-based MRI image analysis for the Alzheimer's disease prediction from mild cognitive impairment. *Frontiers in Neuroscience*, *12*, 777. doi:10.3389/fnins.2018.00777 PMID:30455622

Lu, L., Zheng, J., & Mishra, S. (2015). A layer-to-layer model and feedback control of ink-jet 3-D printing. *IEEE/ASME Transactions on Mechatronics*, *20*(3), 1056–1068. doi:10.1109/TMECH.2014.2366123

Lu, Q. Y., & Wong, C. H. (2018). Additive manufacturing process monitoring and control by nondestructive testing techniques: Challenges and in-process monitoring. *Virtual and Physical Prototyping*, *13*(2), 39–48. doi:10.1080/17452759.2017.1351201

Neumann, L., & Matas, J. (2011). A Method for Text Localization and Recognition in Real-World Images. In R. Kimmel, R. Klette, & A. Sugimoto (Eds.), Lecture Notes in Computer Science: Vol. 6494. *Computer Vision – ACCV 2010. ACCV 2010*. Springer. doi:10.1007/978-3-642-19318-7_60

Olson, E. (2011). AprilTag: A robust and flexible visual fiducial system. *2011 IEEE International Conference on Robotics and Automation*, 3400-3407. 10.1109/ICRA.2011.5979561

Ozsoy, Aksoy, & Salman. (2021). Investigation of the dimensional accuracy using image processing techniques in powder bed fusion. *Proc IMechE Part E: J Process Mechanical Engineering*.

He, Zhang, & Hong. (2018). Profile monitoring based quality control method for fused deposition modeling process. *Journal of Intelligent Manufacturing*, (May). Advance online publication. doi:10.100710845-018-1424-9 21

Rengier, F., Mehndiratta, A., von Tengg-Kobligk, H., Zechmann, C. M., Unterhinninghofen, R., Kauczor, H. U., & Giesel, F. L. (2010, July). 3D printing based on imaging data: Review of medical applications. *International Journal of Computer Assisted Radiology and Surgery*, *5*(4), 335–341. doi:10.100711548-010-0476-x PMID:20467825

Russ. (2016). *Feature-Specific Measurements*. doi:10.1201/b10720-14

Samtas & Gulesin. (2011). *Digital image processing and its applications in different fields*. Academic Press.

Shen, H., Du, W., Sun, W., Xu, Y., & Fu, J. (2020). Visual Detection of Surface Defects Based on Self-Feature Comparison in Robot 3-D Printing. *Applied Sciences (Basel, Switzerland)*, *10*(1), 235. doi:10.3390/app10010235

Tlegenov, Y., Hong, G. S., & Lu, W. F. (2018). Nozzle condition monitoring in 3D printing. Robotics and Computer-Integrated Manufacturing, 54, 45–55. doi:10.1016/j.rcim.2018.05.010

Von Tengg-Kobligk, W. T., Rengier, F., Kotelis, D., Geisbusch, P., Bockler, D., Schumacher, H., & Ley, S. (2008). Imaging modalities for the thoracic aorta. *The Journal of Cardiovascular Surgery*, *49*(4), 429–447. PMID:18665106

Wang, Y., Lin, Y., Zhong, R. Y., & Xu, X. (2019). IoT-enabled cloud-based additive manufacturing platform to support rapid product development. *International Journal of Production Research*, *57*(12), 3975–3991. doi:10.1080/00207543.2018.1516905

Wang, Y., Zheng, P., Peng, T., Yang, H., & Zou, J. (2020). *Smart additive manufacturing: current artificial intelligence-enabled methods and future perspectives*. Science China Technological Sciences., doi:10.100711431-020-1581-2

Wu, He, Hu, & Zhao. (2018). *Process Monitoring of Fused Deposition Modeling through Profile Control*. doi:10.1109/CBS.2018.8612192

Chapter 10
The Importance of Polymers in Medicine and Their FTIR and Raman Spectroscopic Investigations

Sefa Celik
https://orcid.org/0000-0001-6216-1297
Istanbul University, Turkey

Sevim Akyuz
Istanbul Kultur University, Turkey

Aysen E. Ozel
Istanbul University, Turkey

ABSTRACT

Polymers are macromolecules with a very high molecular weight that are formed as a result of one or more small molecules bonding to each other, typically by covalent bonds. The concept of macromolecules consisting of many repeating units are published in the article titled "Uber Polymerization," which was first published in 1920 by Hermann Staudinger. The idea of polymers, which have been in our lives for a long time, has thus become an important field of study. Polymers are commonly used in tissue engineering because of their properties such as biocompatibility and biodegradability. In addition, polymers have a wide range of applications in tissue regeneration, drug delivery, and wound healing. Polymers have varying structural properties based on their backbone structure, molecular weight, and crystallinity. For this cause, a great deal of study in the medicine has been based on this category of materials. This work focuses on the importance of polymers in the medicine, 3D printed polymer composites and Fourier transform infrared, and Raman spectroscopic investigations of polymeric structures.

DOI: 10.4018/978-1-6684-6009-2.ch010

THE EMERGENCE OF THE POLYMER CONCEPT

The concept of macromolecule consisting of large number of repeating units was first introduced by Hermann Staudinger in an article entitled "On Polymerization" in 1920 (Schubert et al., 2020). In this article, several reactions that produced high molecular weight molecules from repeating covalently bonded units, and called polymerization, were shown. At the end of 1920, Staudinger confirmed that molecular weights were unchanged during the polymerization process, by giving additional evidence based on viscosity measurements. However, despite his substantial empirical evidence, Staudinger had to face criticism from leading organic chemists for nearly twenty years. One of them was Heinrich Wieland, the winner of the 1927 Nobel Prize in Chemistry, wrote that Staudinger should give up the idea of large molecules because organic molecules that have molecular weights higher than 5000 could not exist. The first real breakthrough in industrial production came with the phenolic resin Bakelite, the first synthetic polymer discovered by Leo Hendrik Baekeland (1863–1944) in 1908. No other material is as versatile as polymeric materials, for example, they can be hard, soft or elastic, permeable or impermeable, transparent or opaque. Due to their versatile properties, polymers have large applications in packaging, toys, furniture and fabrics industry, circuit boards, composites for spaceships. Moreover, synthetic polymers have also medical applications in absorbable sutures and implant materials (Mulhaupt, 1999).

Polymer chemistry and physics have become an important research area, and polymers used in many areas have made great progress economically. (Schubert et al., 2020).

NATURAL AND SYNTHETIC BIOMEDICAL POLYMERS

Biomedical polymers are either naturally occurring polymers or synthetic polymers.

Proteins, polysaccharides and nucleic acids are natural polymers (Donnaloja et al., 2020).

Collagen is found in the bones, muscles, skin, and tendons. It is the single most abundant protein in animals. Although, in molecular level, twenty-eight collagen types have been characterized, over 90% of the collagen in the human body is type I collagen (Lodish et al., 2000; Ricard-Blum, 2011).

Another remarkable natural fibrous protein is silk fibroin, obtained from cocoons of silkworm's larvae. It has a unique hierarchical structure and due to this structural property, it has strong mechanical strength and notable biocompatibility. Moreover, silk fibroin is frequently employed as biomaterials for medical purposes due to its tunable biodegradation and slightly aqueous processing capabilities (Nguyen et al., 2019). Silk fibroin is composed of glycine, alanine and serine amino acids that found in other natural biomaterials (Zhang et al., 2019).

Chitosan is a linear polysaccharide obtained from chitin and found in abundance after cellulose. Chitosan offers significant opportunities for the development of biomedical applications (Ibrahim et al., 2015).

Alginate is a natural unbranched anionic polymer, composed of two uronic acid blocks (guluronic acid and mannuronic acid block units). It is extensively used for many biomedical applications for biocompatibility, low toxicity properties, and being relatively low cost (Lee and Mooney, 2012). It shows natural biological affinity for a variety of enzymes (Jain et al., 2006).

Hyaluronic acid (or hyaluronan) is the simplest glycosaminoglycan (a class of negatively charged polysaccharides) found in various tissues and fluids throughout the body and binds to specific cell surface receptors (Hench et al., 2005).

Cellulose is a polysaccharide (Sun et al., 2016; Gupta et al., 2019).

In tissue engineering, synthetic polymers are extensively used. Polycaprolactone (PCL), polyethylene glycol (PEG) and polyurethanes are some examples of synthetic biomedical polymers.

Polycaprolactone (PCL) has a low melting point (60° C) (**Figure 1**). It is made by ring-opening polymerization of ε-caprolactone and easily be degraded by enzymatic hydrolysis of microbial esterases (Nair et al., 2017).

Figure 1. Molecular structure of Polycaprolactone

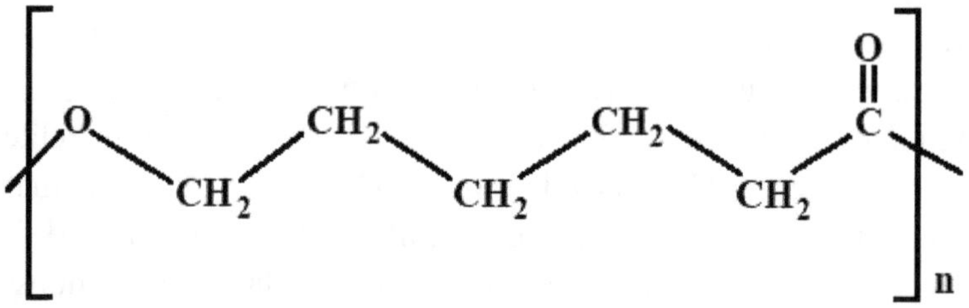

Polyethylene glycol (PEG) (polyethylene oxide; PEO) is a flexible polyether compound that has many applications, especially in medicine (**Figure 2**) (Zarrintaj et al., 2020).

Figure 2. Molecular structure of Polyethylene glycol

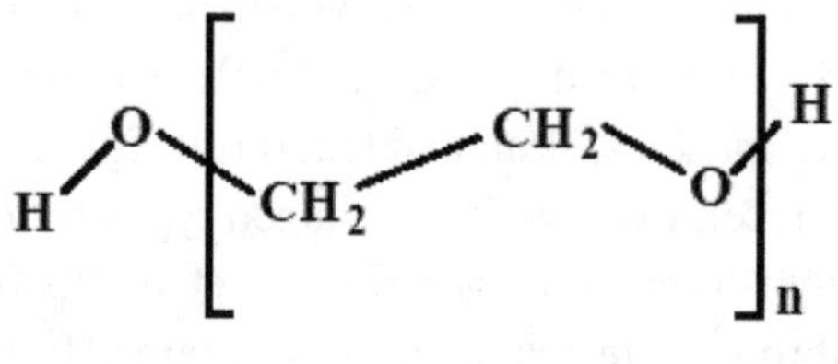

The synthetic copolymers known as polyurethanes are composed of organic molecules connected by urethane groups (**Figure 3**). They make up of three monomers: a diisocyanate, a polyol, and a chain extender (Pivec et al., 2017).

Figure 3. Molecular structure of Polyurethane

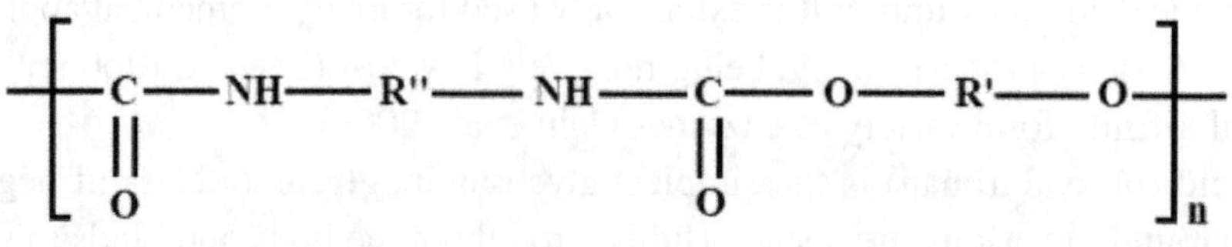

IMPORTANCE AND USES OF POLYMERS IN MEDICINE

The use of synthetic or natural materials in the body to repair damaged tissue dates back to the sixteenth century. The use of a gold plate to repair a cleft palate in 1588 is the first recorded use. In the late 1800s, there were numerous reports of the use of metal plates to repair broken bones. In addition, with the developments in the polymer industry and the discovery of materials (polymers) compatible with the body, there has been a noticeable growth in the use of materials in surgery after the 1950s (Lyman, 1974).

Firstly the synthetic polymer poly (methyl methacrylate) (PMMA), had been used by Sir Nicholas Harold, a British ophthalmologist in 1949, for making an intraocular lens (Damodaran et al., 2012). The two most crucial factors taken into account when choosing a polymer for biomedical usage are biostability and biodegradability (Carmali et al., 2014). Biostability is the ability of a material to maintain its properties after implantation into a living tissue (Wexler, 2014). On the other hand, the capacity of an organic substance to be biodegraded by living organisms to basic substances, is its biodegrability property (Goswami et al., 2016).

Polymers are important that are used widely in biotechnology and medicine, due to hydrolytically degradable property, and are called as smart polymers (Carmali et al., 2014; Joseph Francis et al., 2018).

For more than a quarter of a century, the use of polymeric compounds as important aids in drug delivery in chemotherapy has importance (Neuse, 2008).

In the healthcare field, degradable polymers started to be used as in vivo resorbing sutures in the middle of the 20th century. Since then, the potential applications of these polymers have grown rapidly. Poly (glycolic acid) (PGA), the stereoisomer forms of poly (lactic acid) (PLA), and their copolymer poly(lactic-coglycolide) (PLGA), which are biodegradable synthetic aliphatic polymers, are the most widely used polymers for bone tissue engineering. However, it was determined that the degradation products of lactic acid-based polymers reduce the PH, as a result, the degradation rate of polyesters may increase and inflammatory reaction is induced due to acidosis. Nowadays, important information has been obtained on the biocompatibility of biodegradable polymers, which can greatly reduce the occurrence of undesirable reactions. (Kroeze et al., 2009).

Biopolymers produced by peptide bonding of many α-amino acids have significant advantages when used as carriers for drug delivery. These polymers are biodegradable and has importance because of their (a) biocompatibility, (b) bio-absorbability, and (c) mechanical properties. These polymers have hydrolysable ester groups (COO) in the main chain and form a family of biodegradable materials that have good thermal and mechanical properties with strong intermolecular hydrogen bonding interactions between amide groups. (Khan et al., 2011).

Polymers are also used in Orthopedic Surgery as biomaterials (Ambrose et al., 2015).

Antimicrobial polymers are the materials that have ability to kill microbes in their environment. They are found to be more usable than small molecules in terms of improved efficacy, reduced toxic effect, minimal environmental problems, durability (Jain et al., 2014).

Biodegradable polymeric biomaterials provide important advantages such as being degraded and removed after they fulfill their functions. Degradable polymers have a wide variety of applications, as are surgical sutures and implants used clinically (Ulery et al., 2011).

Since colon targeting has primary importance in the treatment of colon-related diseases, colon-targeted drug delivery has gained great importance. Various natural products are used for the treatment of colon diseases. These biodegradable polymers are getting more attention day by day due to being non-toxic, economical, and chemically compatible (Kumar et al., 2016).

The application of polymeric materials to use in medicine is a highly specialized field due to having variety of specific requirements so the total polymer volume used in this area is very small compared to the annual polyethylene production. However, the amount of money spent annually on prosthetics and biomedical devices exceeds $ 16 billion in the United States alone. These applications include prostheses, dental fillings, contact lenses, prosthetic joints, plastic surgical surgeries, more than 25,000 heart valves, and 60,000 pacemakers. More than 40,000 patients regularly need hemodialysis units, and every year, more than 90,000 cardiac bypass surgeries are carried out; frequently, synthetic polymers are used. (Gebelein, 1985).

Drug forms based on polymer carriers with long-lasting action have been made to reduce the toxicological effect of drugs despite long-term disease treatment. Various synthetic and natural polymers have been investigated as carriers of antituberculosis drugs for controlled delivery application. Drugs such as isoniazid and ethionamide are incorporated into the polymeric matrix and chemically linked to the polymer chain by covalent or electrostatic forces. Biodegradation of polymeric systems and release of drugs have been studied by various physicochemical methods and it has been found that drug release depends on the method of immobilization, the type of drug / polymer bond, and drug loading. (Batyrbekov et al., 1998).

Table 1. Some applications of polymers

Polymers	Applications	References
Polydimethyl siloxane	Catheters, heart valves, drug delivery vehicles and blood-contacting biomaterials,	(Robinson et al., 2001; Subramaniam et al., 2014)
Polyurethane	Ventricular assist devices, drug delivery systems, artificial heart systems, catheters, mammary implants, semiocclusive dressings	(Robinson et al., 2001; Cherng et al., 2013; Szycher et al., 1991)
Polyethylene	Catheters, hip prostheses, medical implants, medical devices	(Robinson et al., 2001; Trusler et al., 1958; Paxton et al., 2019; Silva, 2012; Steiner et al., 1965)
Polymethylmethacrylate	Fracture fixation, bone cement, lenses, bone substitutes, and drug delivery systems	(Robinson et al., 2001; Manoukian et al., 2019)
Cellophane	Dialysis membranes,	(Robinson et al., 2001; Kazmi et al., 1973)
Poly (D, L-lactide-co-glycolide)	Drug delivery devices, drug carrier	(Kerimoglu et al., 2012; Lee et al., 2004)
Polylactic acid	Drug delivery systems, bone fixation devices	(Davachi et al., 2015)
Poly-ε-coprolactone	Implant delivery device	(Cipurkovic et al., 2018)
Polybutylene succinate	Drug delivery system	(Mohanraj et al., 2013)
Polystyrene	Drug delivery	(Canal et al., 2012)
Polyvinylchloride	Medical Device	(Hong, 1996)
Polyorthoesters	Drug delivery system, implanted biomaterials and drug delivery vehicles	(Einmahl et al., 2001; Kwon et al., 2007)
Polyphosphazenes	Drug delivery and skeletal tissue regeneration	(Mateos-Timoneda, 2009; Allcock, 1990; Laurencin et al., 1993)
Polyanhydrides	Drug delivery materials, tissue engineering	(Arakawa et al., 2017; Burkoth et al., 2000; McHale et al., 2013)

For ophthalmological applications, both synthetic and natural polymeric biomaterials are used. Especially in recent years, natural biocompatible materials such as gelatin, hyaluronan, chitosan, gums etc. are of great interest in ophthalmic applications, and studies focus on these issues (Calles et al., 2015).

Natural and synthetic polyanions, such as heparin and pyran, have many biological activities. Among these biological activities, Rauscher leukemia, Friend leukemia, vesicular stomatitis virus, and foot and mouth diseases take place (Ottenbrite et al., 1983).

Various researches are carried out on biodegradable polymers. Materials that can be decomposed and disappeared in the body are desired to be used for short-term applications in orthopedics, tissue engineering, and other fields. These biomaterials are found in many different medical devices (Robinson et al., 2001).

3D PRINTING OF POLYMER COMPOSITES

3D printing is a new technology that uses computer-aided design and works similarly to typical printers (Zhu et al., 2020). Hideo Kodama is regarded as a pioneer of additive manufacturing, having invented methods for creating 3D items in the 1980s. To make 3D sculptures, he employed polymers that could be hardened or polymerized under UV radiation. Kodama, the first 3D printing patent holder, was working at the Nagoya Municipal Industrial Research Institute in Japan. To make 3D sculptures, he employed polymers that could be hardened or polymerized under UV radiation. Kodama, the first 3D printing patent holder, was working at the Nagoya Municipal Industrial Research Institute in Japan. Chuck Hull created 3D objects using stereolithography technology and additive manufacturing. Hull is known as the "Father of 3D Printing" (Ravi, 2015). Different materials can be manufactured using the 3D printing method and they are customized items with complex designs that use a wide range of materials such as polymers and metals (Zhu et al., 2020). Polymer Matrix Composite materials with different mechanical, thermal, electrical and tribological properties are used according to the properties of the desired product (Soyler, 2019) in 3D printing. With their ability to produce 3D printers, their use in the industry has become widespread today. 3D printing can be used to process acrylonitrile butadiene styrene (Tymrak et al. 2014; Sun et al., 2008; Tran et al., 2017), polylactic acid(Tymrak et al. 2014; Tran et al., 2017; Melnikova et al. 2014), polyamide (Caulfield et al., 2007), thermoplastic materials like polycarbonate (Garcia et al., 2012), and thermosetting polymer materials such as epoxy resins (Wang et al., 2017). Thermal or UV curing is required to finish the polymerization of epoxy resins. These resins have a low viscosity that rises as they cure (Gu et al., 2016[a]; 2016[b]; Dou et al., 2012). Epoxy resins are used in the UV assisted printing process. Depending on material choice, the aerospace industry can use 3D printing of polymers to make lightweight structures (Kroll et al., 2011), architectural industries for structural models (Wong et al., 2012), arts fields for education (Short, 2015) and it has various applications such as medical fields (Murphy et al., 2014) for printing (Wang et al., 2017) tissues-organs.

Polylactic acid (PLA), an environmentally friendly bio-plastic, is employed for the production process of safety and fused filament fabrication (FFF) and is widely used in many different industries to create low-cost, adaptable material extraction systems (Zhang, 2019). As it uses a series of heat treatments for continuous polymer filaments, therefore, the work focuses on improving thermal behavior of PLA polymer filaments. (Wang, et al. 2019; Zhang, 2019).

Fourier-transform infrared spectroscopy (FT-IR) is the most effective method for identifying material molecular-level changes. Based on the FT-IR technology, a classifier method using artificial neural networks (ANN) has been developed by Zhang (2019). The most tsensitive spectral region for investigation of the thermal degradation of 3D-Printed PLA specimens was determined as 650-4000 cm^{-1} region. (Zhang, 2019). It was demonstrated that thermal degradation degrees of PLA 3D printing may be determined busing ANNs methodology based on FTIR spectral results.

3D printable hydrogel based on cellulose nanocrystals (CNCs)-for tissue engineering applications was fabricated and characterized by Patel et al. (2021). The interaction between the polymer matrix and the added CNCs (Cellulose nanocrystals) was investigated using FT-IR spectroscopy and it was discovered that polymer chains and CNCs interacted more closely compared to controls. (hBMSCs) cell survival was found to be higher in the presence of composite scaffolds, which had a rougher surface characteristics than pure polymer scaffolds. This roughness was thought to help to improve the cellular activity. Moreover, it was reported that the CNC content in the produced hydrogels had a mayor effect on cellular activity.

In media containing composite scaffolds as compared to control, the cells found to have an outstretched and flattened shape, as well as a greater intensity of fluorescence. Additionally, greater composite mineralization potential was seen when treated medium with composite scaffolds were present compared to control. It was also reported that in the scaffold-treated medium, osteogenic-related genes were expressed more frequently than in the control, and this increased osteogenic potential was evident. These findings led them to the conclusion that the manufactured composite scaffold is a desirable biomaterial for applications in tissue engineering particularly those involving bone tissue, based on their findings. (Patel et al., 2021).

The growing need and production of strong and lightweight plastic components has attracted increasing attention from both science and industry to the recycling of polymeric materials. Such materials are recycled primarily by thermomechanical processes, which have a substantial impact on both the mechanical and overall physicochemical characteristics of polymers. The study by Vidakis et al. in 2021 focused on the recyclability of 3D-printed Polypropylene (PP) using fused filament fabrication (FFF) and reported how the mechanical characteristics of those parts how affected for a particular number of recycling courses In particular, the mechanical properties of the 3D printed samples in terms of stretching, bending, impact, and microhardness were examined by thorough Differantial Scanning Calorimetry (DSC), thermogravimetric analysis (TGA), Raman spectroscopy, and Scanning Electron Microscopy (SEM). While the DSC characterization suggested an increase in polymer crystallinity throughout recycling, the overall results showed a minor overall increase in the material's mechanical properties, both in strain and bend mode (Vidakis et al., 2021).

FT-IR AND RAMAN SPECTROSCOPIC APPLICATIONS ON POLYMERS

Infrared spectroscopy (IR) technique is regarded as a useful tool for data collecting and as a fingerprinting technique due to its ability to concurrently provide information about numerous properties of the samples. Fourier transform infrared bands that coincide to the molecule's fundamental vibrations can be connected to functional groups (Colthup et al., 1975; Griffith et al., 1986). There are two types of vibrations that can be seen in the mid-infrared range (4000-1000 cm^{-1}): bond stretching vibrations, which result in changes in bond length, and bending vibrations, which involve changes in bond angles (Berthomieu et al., 2009).

Raman spectroscopy: When a molecule is exposed to monochromatic (Laser) light, elastic and inelastic light scattering always occurs. In the elastic scattering, the frequency, wavelength or energy of the photon is not affected, while in inelastic scattering, the frequency of the photon changes as a result of the interaction of laser light with molecular system and the photon either loses or gains energy (Kaur, 2006). The energy shift reveals details about the vibrational modes of the molecular system. (Bugay et al., 2008; Das et al., 2011). IR and Raman spectroscopy techniques are complementary techniques to each other. Both gives information about the molecular systems, from a different perspective.

Vibrational spectroscopy includes Fourier transform infrared (FT-IR) and Raman spectroscopy and both techniques have an effective potential for molecular analysis of polymers. Vibrational spectroscopy gives information about molecular bonds, and intra- and inter-molecular interactions. Any molecular structural changes can be detected. Thus, FT-IR and Raman spectroscopy techniques are suitable for analysis of polymers, for determination of their crystallinity degree and additives. Both techniques are simple, fast, and non-destructive. Moreover, these methods are simple as no excessive sample pretreatment is required, so they are also called "green analytical methods". (Rohman et al., 2019. Beneath favorable conditions, time-consuming chemical separation procedures for multi-component systems can be successfully performed with the spectroscopic absorbance subtraction technique. The difference spectrum between the standard pure polymer and modified polymer gives information about the additives (Siesler, 1980).

The identification of end groups of the polymer is important for the characterization of chain length and branching degree (Siesler, 1980).

In the FT-IR and Raman spectroscopy, for characterization of polymeric structure the fingerprint region is important. Determination of the crystallinity and order of a polymeric material are important and usually determined by using X-ray diffraction (XRD) and nuclear magnetic resonance (NMR) techniques, respectively. Both techniques require a high consumption of electrical energy and expensive reagents compared to FT-IR spectral analysis, thus more costly. Therefore, it is quite useful to obtain the crystallinity and structural arrangement of the polymeric materials from fingerprint region analysis of the IR spectra. (Ramirez-Hernandez et al., 2019). On the other hand, polymer powders, pellets, molded articles, fibers, fabrics, and films can be analyzed by Raman spectrometry technique without any sample preparation, by simply sending a laser beam onto sample and analyzing the backscattered radiation (Boerio et al., 1971, Paradkar et al., 2009). Both IR and Raman spectroscopy techniques were used to investigate the morphological and structural changes of modified polymers and their crystallinity.

Although vibrational spectroscopy cannot effectively detect crystal order, it is sensitive to the molecular environment that surrounds the polymer chain. Vibrational spectroscopy can also recognize the difference between the crystalline and amorphous states of the material. Compared to an amorphous material spectrum, the presence of crystallinity can also be identified by the changes in the spectrum, including a decrease in the full width at half maximum of certain bands, the emergence of new peaks, and peak splitting. (Snively et al., 2017).

FT-IR and Raman spectroscopic applications of polymeric structures are given below:

The vibration spectrum of a new type of $[NP(O_2C_{12}H_8)]_n$ polymer was investigated in 1998 by Carriedo et al. Although the vibration modes of the main chain were observed in high intensity in the IR, the vibration bands of the peripheral biphenyl groups were observed as high intense in Raman spectrum. It was also reported a decrease in the full width at half maximum value of the CC stretching modes of biphenyl noticed at 1609 cm^{-1} in the Raman spectrum, as the polymer sample's degree of crystallinity increases (Carriedo et al., 1998).

Papke et al. studied different alkali metal salt complexes of poly (ethylene oxide) polymer using IR and Raman spectroscopy, and determined the polymer-metal salt interactions (Papke et al., 1981).

In a study conducted by Davies et al in 1990, the advantages of FT-Raman spectroscopy in the analysis of some Drug-Polymer mixtures were shown and FT-Raman spectra of some polymeric biomaterials were given. In this study, it was also demonstrated that FT- Raman spectroscopy could have a promising importance for quantitative analysis of changes in drug concentration in these systems during the drug release period. (Davies et al., 1990).

In the study performed by Tudor et al. in 1990, it has been shown that FT-Raman spectroscopy is an important tool for the molecular analysis of biomedical polymers (Tudor et al., 1990).

In the FT-IR and Raman spectroscopic study conducted by Sahoo et al. in 2012, Norfloxacin/Carbopol934 Polymeric Suspension was investigated. In order to maintain a safe and effective blood level of norfloxacin for a long time, the existence of several formulations in which the drug is absorbed homogeneously has been demonstrated. As a result of vibrational spectral analysis, the carboxylic groups of norfloxacin and the hydroxyl groups of carbopol934 interacted, resulting in micelle formation due to esterification and hydrogen bonding, leading to more drug retention and a stable formulation. Therefore, it was concluded that norfloxacin formulation could provide a more controlled release and mucoadhesive effect in the gastrointestinal tract and carbopol934 could be considered an effective carrier for norfloxacin (Sahoo et al., 2012).

Surface corrosion of microwave digestion vessels based on polytetrafluoroethylene (PTFE) was investigated by reflection FT-IR and FT-Raman techniques (Mihaly et al. 2006). After 4 different digestion cycles of the milk powder, samples were taken from the digestion vessels and the reflection IR spectral research was examined. It has been shown that after different cycles, new CF2 and CF3 groups are formed, describing the surface degradation of the polymer chain. Additionally, the PTFE samples' Raman spectra also showed the formation of new CF3 groups, showing that the polymer chain changed after different cycles

Raman and FT-IR Poly (4-methyl-1-pentene) polymer spectra were recorded in the 4000-400 and 4000-100 cm-1 ranges (Samuel et al., 2004), and the reported basic bands of poly(4-methyl-1-pentene) were compared to the normal coordinate analysis. The potential energy distribution contribution corresponding to each of the observed frequencies demonstrated the spectral analysis's reliability and validity.

In the study conducted by Todica et al. in 2015, the modification of the polymeric structure of poly (acrylic) acid (PAA) was investigated by IR and Raman spectroscopy. It was reported that neutralization with triethanolamine affected carboxylic groups and modified the conformation of the polymer. These changes were linked to shifts and modifications in the IR and Raman spectra (Todica et al., 2015).

In a 2016 study, the effect of polydimethylsiloxane (PDMS) silicone oil, which is used in vitreoretinal surgery, on the internal structures of polymer intraocular lenses was investigated. It was demonstrated by FT-IR spectroscopy technique that poly (2-hydroxyethyl methacrylate) (PHEMA) -based lenses absorb silicone oil, whereas polymethyl methacrylate (PMMA) lenses do not absorb. The results obtained using the Raman spectroscopy method were inconclusive, possibly as a result of the lower intensity of the characteristic PDMS bands (Chamerski et al., 2016).

Macocinschi et al. In 2015, polyurethane-extracellular matrix bio-nano composites containing antibacterial biocompatible polymers developed for medical devices were obtained by solvent method. The hydrolytic stability of silver bio-nano composites in PBS solutions was investigated by FT-IR spectroscopy analysis. Hydrolytic degradation of ester and urethane bonds has been found to occur in poly (ester urethanes). Added extracellular matrix components have been found to increase hydrophilicity (Macocinschi et al., 2015).

In the study conducted in 2007, it was stated that although the Poly (3-hydroxybutyrate) (PHB) was a significant component of medicine, due to being brittle and being a slightly hydrophobic polymer, it has some deficiencies, which limited its applications. Gonzalez et al. (2007) showed that by using radiation the chemical properties of PHB could be improved. The PHB was modified by radiation-induced graft copolymerization with vinyl acetate (VAc), and PHB-graft-PVAc was obtained. FTIR and Raman spectroscopy was used to confirm vinyl acetate grafting to the PHB.

In another study, it was stated that the majority of polymer-based drug delivery films were created through the solution or dispersion states and that estimating the physicochemical characteristics of drugs was important for designing new formulations. According to studies, ATR-FTIR imaging provided crucial details regarding the entry of water into the films as well as the presence, distribution, and physicochemical status of the drug. It was concluded that ATR-FTIR imaging is an important method to look into physicochemical operations for films made of medicinal polymers (Hifumi et al., 2016).

In a 2001 study, it was demonstrated how to use laser-Raman spectroscopy to determine the monomer content in poly(methylmethacrylate) under experimental conditions by using the intensities of the bands at 1640 and 1725 cm^{-1} in the poly(methylmethacrylate) spectrum (Pallikari et al., 2001). Infrared and Raman spectra of poly (trimethylene terephthalate, poly (tetramethylene terephthalate) and poly(ethylene terephthalate) samples were compared in a 1977 study. Spectrum changes during crystallization were examined. Stress was applied to the samples, and the observed spectral differences in the spectrum have been proposed as changes in the conformational of molecular chains (Ward et al., 1977).

The Raman spectra of polyacrylamide in the range of 200-3600 cm^{-1} were investigated in a paper published in 1981. The spectra of various polymers were compared to determine the vibration bands for polyacrylamide. Polyacrylamide exhibits identical spectral characteristics in solid phase and aqueous solution (Gupta et al., 1981).

The emulsion polymerization of vinylacetate was examined using FT-Raman in a 1997 study. The Raman technique has been found to be reliable in monitoring the emulsion polymerization process of vinylacetate as a consequence of the research. In addition, Özpozan et al. (1997) discovered that this technique may be used to investigate the kinetics of vinylacetate polymerization in real time.

Raman spectra of poly (ethylene glycol) were recorded in solid and molten phases, as well as in aqueous and chloroform solutions, in Koenig's research. Several new lines were detected in the Raman spectrum of the solid state using argon ion laser as the source. Additional Raman lines were detected in the molten polymer and chloroform solution Raman spectra, indicating a disordered structure (Koenig et al., 1970).

Nielsen devised a method depends on the intensities of the Raman bands at 808 and 841 cm^{-1}, for estimating crystallinity in isotactic polypropylene, and revealed three distinct conformational states. The 808 cm^{-1} Raman band of isotactic polypropylene was allocated to the helical chains within the crystals. The relationship between Raman intensity and crystallinity was investigated in detail, using samples with various cooling rates and crystallization temperatures (Nielsen et al., 2002).

CONCLUSION

Polymers have varying structural properties based on their backbone structure, molecular weight, and crystallinity so the use of polymers in the biomedical industry is more significant and effectual. Polymers are used numerous medical applications. In general, some additives were used, and modifications are made to obtain polymer with desired properties. By using vibrational spectroscopy (IR, Raman), the additives and polymeric modifications can be easily detected. This chapter focuses on the applications of FT-IR and Raman spectroscopy on polymers as well as the areas in which polymers can be employed in health research based on their molecular structures.

REFERENCES

Allcock, H. R. (1990). Polyphosphazenes as new biomedical and bioactive materials. *Drugs and the Pharmaceutical Sciences, 45*, 163-193.

Ambrose, C. G., Hartline, B. E., Clanton, T. O., Lowe, W. R., & McGarvey, W. C. (2015). Polymers in orthopaedic surgery. In *Advanced Polymers in Medicine* (pp. 129–145). Springer.

Arakawa, C. K., & DeForest, C. A. (2017). Polymer Design and Development. In Biology and Engineering of Stem Cell Niches (pp. 295-314). Academic Press. doi:10.1016/B978-0-12-802734-9.00019-6

Batyrbekov, E. O., Iskakov, R., & Zhubanov, B. A. (1998, February). Synthetic and natural polymers as drug carriers for tuberculosis treatment. *Macromolecular Symposia, 127*(1), 251–256. doi:10.1002/masy.19981270132

Berthomieu, C., & Hienerwadel, R. (2009). Fourier transform infrared (FTIR) spectroscopy. *Photosynthesis Research, 101*(2), 157–170. doi:10.100711120-009-9439-x PMID:19513810

Boerio, F. J., & Koenig, J. L. (1971). Raman spectroscopy of polymers. In *Polymer Characterization Interdisciplinary Approaches* (pp. 1–13). Springer. doi:10.1007/978-1-4684-1929-0_1

Bugay, D. E., & Martogolio Smith, P. A. (2008). In S. Jickells & A. Negrusz (Eds.), *Clarke's Analytical Forensic Toxicology* (3rd ed., pp. 455–468). Pharmaceutical Press.

Burkoth, A. K., & Anseth, K. S. (2000). A review of photocrosslinked polyanhydrides: In situ forming degradable networks. *Biomaterials, 21*(23), 2395–2404. doi:10.1016/S0142-9612(00)00107-1 PMID:11055287

Calles, J. A., Bermúdez, J., Vallés, E., Allemandi, D., & Palma, S. (2015). Polymers in ophthalmology. In *Advanced Polymers in Medicine* (pp. 147–176). Springer.

Canal, C., Aparicio, R. M., Vilchez, A., Esquena, J., & García-Celma, M. J. (2012). Drug delivery properties of macroporous polystyrene solid foams. *Journal of Pharmacy & Pharmaceutical Sciences, 15*(1), 197–207. doi:10.18433/J3X884 PMID:22365097

Carmali, S., & Brocchini, S. (2014). Polyacetals. In Natural and Synthetic Biomedical Polymers (pp. 219-233). Elsevier.

Carriedo, G. A., García Alonso, F. J., González, P. A., & Menéndez, J. R. (1998). Infrared and Raman spectra of the phosphazene high polymer $[NP(O_2C_{12}H_8)]_n$. *Journal of Raman Spectroscopy: JRS*, *29*(4), 327–330. doi:10.1002/(SICI)1097-4555(199804)29:4<327::AID-JRS245>3.0.CO;2-5

Caulfield, B., McHugh, P. E., & Lohfeld, S. (2007). Dependence of mechanical properties of polyamide components on build parameters in the SLS process. *Journal of Materials Processing Technology*, *182*(1-3), 477–488. doi:10.1016/j.jmatprotec.2006.09.007

Chamerski, K., Lesniak, M., Sitarz, M., Stopa, M., & Filipecki, J. (2016). An investigation of the effect of silicone oil on polymer intraocular lenses by means of PALS, FT-IR and Raman spectroscopies. *Spectrochimica Acta. Part A: Molecular and Biomolecular Spectroscopy*, *167*, 96–100. doi:10.1016/j.saa.2016.05.036 PMID:27261889

Cherng, J. Y., Hou, T. Y., Shih, M. F., Talsma, H., & Hennink, W. E. (2013). Polyurethane-based drug delivery systems. *International Journal of Pharmaceutics*, *450*(1-2), 145–162. doi:10.1016/j.ijpharm.2013.04.063 PMID:23632262

Cipurkovic, A., Horozic, E., Donlagic, N., Maric, S., Saletovic, M., & Ademovic, Z. (2018). *Biodegradable Polymers: Production, properties and application in medicine*. Academic Press.

Colthup, N. B., Daly, L. H., & Wiberly, S. E. (1975). *Introduction to infrared and Raman spectroscopy*. Academic Press.

Damodaran, V. B., Joslin, J., & Reynolds, M. M. (2012). Preparing biocompatible materials for non-permanent medical devices. *European Pharmaceutical Review*, *17*(5), 71–77.

Das, R. S., & Agrawal, Y. K. (2011). Raman spectroscopy: Recent advancements, techniques and applications. *Vibrational Spectroscopy*, *57*(2), 163–176. doi:10.1016/j.vibspec.2011.08.003

Davachi, S. M., & Kaffashi, B. (2015). Polylactic acid in medicine. *Polymer-Plastics Technology and Engineering*, *54*(9), 944–967. doi:10.1080/03602559.2014.979507

Davies, M. C., Binns, J. S., Melia, C. D., Hendra, P. J., Bourgeois, D., Church, S. P., & Stephenson, P. J. (1990). FT Raman spectroscopy of drugs in polymers. *International Journal of Pharmaceutics*, *66*(1-3), 223–232. doi:10.1016/0378-5173(90)90403-Q

Donnaloja, F., Jacchetti, E., Soncini, M., & Raimondi, M. T. (2020). Natural and Synthetic Polymers for Bone Scaffolds Optimization. *Polymers*, *12*(4), 905. doi:10.3390/polym12040905 PMID:32295115

Dou, J., Zhang, Q., Ma, M., & Gu, J. (2012). Fast fabrication of epoxy-functionalized magnetic polymer core-shell microspheres using glycidyl methacrylate as monomer via photo-initiated miniemulsion polymerization. *Journal of Magnetism and Magnetic Materials*, *324*(19), 3078–3082. doi:10.1016/j.jmmm.2012.05.005

Einmahl, S., Behar–Cohen, F., D'Hermies, F., Rudaz, S., Tabatabay, C., Renard, G., & Gurny, R. (2001). A new poly (ortho ester)-based drug delivery system as an adjunct treatment in filtering surgery. *Investigative Ophthalmology & Visual Science*, *42*(3), 695–700. PMID:11222529

Garcia, C. R., Correa, J., Espalin, D., Barton, J. H., Rumpf, R. C., Wicker, R., & Gonzalez, V. (2012). 3D printing of anisotropic metamaterials. *Progress in Electromagnetics Research Letters*, *34*, 75–82. doi:10.2528/PIERL12070311

Gebelein, C. G. (1985). Medical applications of polymers. In *Applied polymer science* (vol. 2, pp. 535–556). American Chemical Society. doi:10.1021/bk-1985-0285.ch023

Gonzalez, M., Galego, F., Ortiz, D. T., & Rapado, M. (2007). *Raman spectroscopy of poly (3-hydroxy-butyrate) modified with poly (vinyl acetate) by radiation-induced copolymerization*. Nucleus.

Goswami, P., & O'Haire, T. (2016). Developments in the use of green (biodegradable), recycled and biopolymer materials in technical nonwovens. In *Advances in Technical Nonwovens* (pp. 97–114). Woodhead Publishing. doi:10.1016/B978-0-08-100575-0.00003-6

Griffith, P. R., & de Haseth, J. A. (1986). *Fourier transform infrared spectroscopy*. Wiley.

Gu, H., Ma, C., Gu, J., Guo, J., Yan, X., Huang, J., Zhang, Q., & Guo, Z. (2016). An overview of multifunctional epoxy nanocomposites. *Journal of Materials Chemistry. C, Materials for Optical and Electronic Devices*, *4*(25), 5890–5906. doi:10.1039/C6TC01210H

Gu, J., Yang, X., Lv, Z., Li, N., Liang, C., & Zhang, Q. (2016). Functionalized graphite nanoplatelets/ epoxy resin nanocomposites with high thermal conductivity. *International Journal of Heat and Mass Transfer*, *92*, 15–22. doi:10.1016/j.ijheatmasstransfer.2015.08.081

Gupta, M. K., & Bansil, R. (1981). Laser Raman spectroscopy of polyacrylamide. *Journal of Polymer Science. Polymer Physics Edition*, *19*(2), 353–360. doi:10.1002/pol.1981.180190214

Gupta, P. K., Raghunath, S. S., Prasanna, D. V., Venkat, P., Shree, V., Chithananthan, C., ... Geetha, K. (2019). An update on overview of cellulose, its structure and applications. *Cellulose (London, England)*, 846–1297.

Hench, L., & Jones, J. (Eds.). (2005). *Biomaterials, artificial organs and tissue engineering*. Elsevier.

Hifumi, H., Ewing, A. V., & Kazarian, S. G. (2016). ATR-FTIR spectroscopic imaging to study the drying and dissolution of pharmaceutical polymer-based films. *International Journal of Pharmaceutics*, *515*(1-2), 57–68. doi:10.1016/j.ijpharm.2016.09.085 PMID:27720876

Hong, K. Z. (1996). Poly (vinyl chloride) in medical device and packaging applications. *Journal of Vinyl and Additive Technology*, *2*(3), 193–197. doi:10.1002/vnl.10123

Ibrahim, H. M., & El-Zairy, E. M. R. (2015). Chitosan as a biomaterial—structure, properties, and electrospun nanofibers. *Concepts, compounds and the alternatives of antibacterials*, 81-101.

Jain, A., Duvvuri, L. S., Farah, S., Beyth, N., Domb, A. J., & Khan, W. (2014). Antimicrobial polymers. *Advanced Healthcare Materials*, *3*(12), 1969–1985. doi:10.1002/adhm.201400418 PMID:25408272

Jain, S., Mondal, K., & Gupta, M. N. (2006). Applications of alginate in bioseparation of proteins. *Artificial Cells, Blood Substitutes, and Biotechnology, 34*(2), 127-144.

Joseph Francis, P. J., Arun, K. J., Navas, A. A., & Joseph, I. (2018). Biomedical Applications of Polymers-An Overview. *Current Trends in Biomedical Engineering & Biosciences*, *15*(2), 555909.

Kaur, H. S. (2006). *Instrumental Methods of Chemical Analysis* (3rd ed.). Pragati Prakashan.

Kazmi, S. J. A., & Mitchell, A. G. (1973). Cellophane as a dialysis membrane for studying nonionic surfactants. *Journal of Pharmaceutical Sciences*, *62*(8), 1299–1302. doi:10.1002/jps.2600620814 PMID:4725174

Kerimoglu, O., & Alarçin, E. (2012). Poly (lactic-co-glycolic acid) based drug delivery devices for tissue engineering and regenerative medicine. *Ankem Derg*, *26*(2), 86–98. doi:10.5222/ankem.2012.086

Khan, W., Muthupandian, S., Farah, S., Kumar, N., & Domb, A. J. (2011). Biodegradable polymers derived from amino acids. *Macromolecular Bioscience*, *11*(12), 1625–1636. doi:10.1002/mabi.201100324 PMID:22052719

Koenig, J. L., & Angood, A. C. (1970). Raman spectra of poly (ethylene glycols) in solution. *Journal of Polymer Science. Part A-2, Polymer Physics*, *8*(10), 1787–1796. doi:10.1002/pol.1970.160081013

Kroeze, R. J., Helder, M. N., Govaert, L. E., & Smit, T. H. (2009). Biodegradable polymers in bone tissue engineering. *Materials (Basel)*, *2*(3), 833–856. doi:10.3390/ma2030833

Kroll, E., & Artzi, D. (2011). Enhancing aerospace engineering students' learning with 3D printing wind-tunnel models. *Rapid Prototyping Journal*, *17*(5), 393–402. doi:10.1108/13552541111156522

Kumar, D., Sharma, M., Verma, S., & Saroha, K. (2016). Natural polymers and herbal medicine based therapy for colonic diseases. *International Journal of Herbal Medicine*, *4*(3), 49–56.

Kwon, G. S., & Furgeson, D. Y. (2007). Biodegradable polymers for drug delivery systems. In *Biomedical polymers* (pp. 83–110). Woodhead Publishing. doi:10.1533/9781845693640.83

Laurencin, C. T., Norman, M. E., Elgendy, H. M., El-Amin, S. F., Allcock, H. R., Pucher, S. R., & Ambrosio, A. A. (1993). Use of polyphosphazenes for skeletal tissue regeneration. *Journal of Biomedical Materials Research*, *27*(7), 963–973. doi:10.1002/jbm.820270716 PMID:8360223

Lee, J. S., Chae, G. S., Kim, M. S., Cho, S. H., Lee, H. B., & Khang, G. (2004). Degradation behaviour in vitro for poly (D, L-lactide-co-glycolide) as drug carrier. *Bio-Medical Materials and Engineering*, *14*(2), 185–192. PMID:15156109

Lee, K. Y., & Mooney, D. J. (2012). Alginate: Properties and biomedical applications. *Progress in Polymer Science*, *37*(1), 106–126. doi:10.1016/j.progpolymsci.2011.06.003 PMID:22125349

Lodish, H., Berk, A., Zipursky, S. L., Matsudaira, P., Baltimore, D., & Darnell, J. (2000). Collagen: The fibrous proteins of the matrix. *Molecular and Cellular Biology*, 4.

Lyman, D. J. (1974). Polymers in medicine. *Angewandte Chemie International Edition in English*, *13*(2), 108–112. doi:10.1002/anie.197401081 PMID:4207434

Macocinschi, D., Filip, D., Zaltariov, M. F., & Varganici, C. D. (2015). Thermal and hydrolytic stability of silver nanoparticle polyurethane biocomposites for medical applications. *Polymer Degradation & Stability*, *121*, 238–246. doi:10.1016/j.polymdegradstab.2015.09.017

Manoukian, O. S., Sardashti, N., Stedman, T., Gailiunas, K., Ojha, A., Penalosa, A., ... Kumbar, S. G. (2019). *Biomaterials for tissue engineering and regenerative medicine*. Academic Press.

Mateos-Timoneda, M. A. (2009). Polymers for bone repair. In *Bone repair biomaterials* (pp. 231–251). Woodhead Publishing. doi:10.1533/9781845696610.2.231

McHale, M. K., Bergmann, N. M., & West, J. L. (2013). *Histogenesis in Three-Dimensional Scaffolds. Handbook of Stem Cells* (2nd ed.). Academic Press.

Melnikova, R., Ehrmann, A., & Finsterbusch, K. (2014, August). 3D printing of textile-based structures by Fused Deposition Modelling (FDM) with different polymer materials. *IOP Conference Series. Materials Science and Engineering*, *62*(1), 012018. doi:10.1088/1757-899X/62/1/012018

Mihaly, J., Sterkel, S., Ortner, H. M., Kocsis, L., Hajba, L., Furdyga, E., & Mink, J. (2006). FTIR and FT-Raman spectroscopic study on polymer based high pressure digestion vessels. *Croatica Chemica Acta*, *79*(3), 497–501.

Mohanraj, K., Sethuraman, S., & Krishnan, U. M. (2013). Development of poly (butylene succinate) microspheres for delivery of levodopa in the treatment of parkinson's disease. *Journal of Biomedical Materials Research. Part B, Applied Biomaterials*, *101*(5), 840–847. doi:10.1002/jbm.b.32888 PMID:23401377

Morris, B. A. (2017). *4-Commonly Used Resins and Substrates in Flexible Packaging. The Science and Technology of Flexible Packaging*. William Andrew Publishing.

Mulhaupt, R. (1999). Essay: International Historic Chemical Landmark: The Foundation of Polymer Science by Hermann Staudinger (1881-1965). *Macromolecular Chemistry and Physics*, *200*(8).

Murphy, S. V., & Atala, A. (2014). 3D bioprinting of tissues and organs. *Nature Biotechnology*, *32*(8), 773–785. doi:10.1038/nbt.2958 PMID:25093879

Nair, N. R., Sekhar, V. C., Nampoothiri, K. M., & Pandey, A. (2017). Biodegradation of biopolymers. In *Current Developments in Biotechnology and Bioengineering* (pp. 739–755). Elsevier. doi:10.1016/B978-0-444-63662-1.00032-4

Neuse, E. W. (2008). Synthetic polymers as drug-delivery vehicles in medicine. *Metal-Based Drugs*, *2008*, 2008. doi:10.1155/2008/469531 PMID:18497867

Nguyen, T. P., Nguyen, Q. V., Nguyen, V.-H., Le, T.-H., Huynh, V. Q. N., Vo, D.-V. N., Trinh, Q. T., Kim, S. Y., & Le, Q. V. (2019). Silk Fibroin-Based Biomaterials for Biomedical Applications: A Review. *Polymers*, *11*(12), 1933. doi:10.3390/polym11121933 PMID:31771251

Nielsen, A. S., Batchelder, D. N., & Pyrz, R. (2002). Estimation of crystallinity of isotactic polypropylene using Raman spectroscopy. *Polymer*, *43*(9), 2671–2676. doi:10.1016/S0032-3861(02)00053-8

Ottenbrite, R. M., Kuus, K., & Kaplan, A. M. (1983). Characteristic biological effects of anionic polymers. In *Polymers in Medicine* (pp. 3–22). Springer. doi:10.1007/978-1-4615-7643-3_1

Özpozan, T., Schrader, B., & Keller, S. (1997). Monitoring of the polymerization of vinylacetate by near IR FT Raman spectroscopy. *Spectrochimica Acta. Part A: Molecular and Biomolecular Spectroscopy*, *53*(1), 1–7. doi:10.1016/S1386-1425(97)83003-X

Pallikari, F., Chondrokoukis, G., Rebelakis, M., & Kotsalas, Y. (2001). Raman spectroscopy: A technique for estimating extent of polymerization in PMMA. *Materials Research Innovations*, *4*(2-3), 89–92. doi:10.1007100190000076

Papke, B. L., Ratner, M. A., & Shriver, D. F. (1981). Vibrational spectroscopy and structure of polymer electrolytes, poly (ethylene oxide) complexes of alkali metal salts. *Journal of Physics and Chemistry of Solids*, *42*(6), 493–500. doi:10.1016/0022-3697(81)90030-5

Paradkar, R. P., Leugers, A., Patel, R., Chen, X., & Meyers, G. (2009). Raman Spectroscopy of Polymers. In Encyclopedia of Analytical Chemistry: Applications, Theory and Instrumentation. John Wiley & Sons, Ltd. doi:10.1002/9780470027318.a9032

Patel, D. K., Dutta, S. D., Shin, W. C., Ganguly, K., & Lim, K. T. (2021). Fabrication and characterization of 3D printable nanocellulose-based hydrogels for tissue engineering. *RSC Advances*, *11*(13), 7466–7478. doi:10.1039/D0RA09620B PMID:35423276

Paxton, N. C., Allenby, M. C., Lewis, P. M., & Woodruff, M. A. (2019). Biomedical applications of polyethylene. *European Polymer Journal*, *118*, 412–428. doi:10.1016/j.eurpolymj.2019.05.037

Pivec, T., Smole, M. S., Gasparic, P., & Kleinschek, K. S. (2017). Polyurethanes for Medical Us. *Tekstilec, 60*(4).

Ramirez-Hernandez, A., Aguilar-Flores, C., & Aparicio-Saguilan, A. (2019). Fingerprint analysis of FTIR spectra of polymers containing vinyl acetate. *Dyna*, *86*(209), 198–205. doi:10.15446/dyna.v86n209.77513

Ravi, M. (2015). *3D Printing - A Current Opinion*. Academic Press.

Ricard-Blum, S. (2011). The Collagen Family. *Cold Spring Harbor Perspectives in Biology*, *3*(1), a004978. doi:10.1101/cshperspect.a004978 PMID:21421911

Robinson, J., Saint Louis, P. M., & Padmaraju, A. (2001). *Polymer in medicine*. Introduciton to Polymers.

Rohman, A., Windarsih, A., Lukitaningsih, E., Rafi, M., Betania, K., & Fadzillah, N. A. (2019). The use of FTIR and Raman spectroscopy in combination with chemometrics for analysis of biomolecules in biomedical fluids: A review. *Biomedical Spectroscopy and Imaging*, *8*(3-4), 55–71. doi:10.3233/BSI-200189

Sahoo, S., Chakraborti, C. K., Behera, P. K., & Mishra, S. C. (2012). FTIR and Raman spectroscopic investigations of a Norfloxacin/Carbopol934 Polymerie suspension. *Journal of Young Pharmacists*, *4*(3), 138–145. doi:10.4103/0975-1483.100017 PMID:23112532

Samuel, E. J. J., & Mohan, S. (2004). FTIR and FT Raman spectra and analysis of poly (4-methyl-1-pentene). *Spectrochimica Acta. Part A: Molecular and Biomolecular Spectroscopy*, *60*(1-2), 19–24. doi:10.1016/S1386-1425(03)00212-9 PMID:14670457

Schubert, U. S., & Zechel, S. (2020). The Year of Polymers–100 Years of Macromolecular Chemistry. *Macromolecular Chemistry and Physics*, *221*(1), 1900530. doi:10.1002/macp.201900530

Short, D. B. (2015). Use of 3D printing by museums: educational exhibits, artifact education, and artifact restoration. *3D Printing and Additive Manufacturing, 2*(4), 209-215.

Siesler, H. W. (1980). Fourier transform infrared (ftir) spectroscopy in polymer research. *Journal of Molecular Structure*, *59*, 15–37. doi:10.1016/0022-2860(80)85063-0

Silva, M.A. (2012). *Coating of polyethylene medical devices*. Academic Press.

Snively, C. M., & Koenig, J. L. (2017). IR and Raman Spectroscopies, Polymer Applications. In Encyclopedia of Spectroscopy and Spectrometry (3rd ed., pp. 365-371). Elsevier.

Soyler, M. (2019). *3B yazicilarda kullanilan mumlarin üretilmesi ve geri kazanilmasi* [Master's thesis]. Kahramanmaras Sutcu Imam University.

Steiner, M. L., Bartley, T. D., Byers, F. M., & Krovetz, L. J. (1965). Polyethylene catheter in the heart: Report of a case with successful removal. *Journal of the American Medical Association*, *193*(12), 1054–1056. doi:10.1001/jama.1965.03090120062022 PMID:14338809

Subramaniam, A., & Sethuraman, S. (2014). Biomedical applications of nondegradable polymers. In *Natural and synthetic biomedical polymers* (pp. 301–308). Elsevier. doi:10.1016/B978-0-12-396983-5.00019-3

Sun, Q., Rizvi, G. M., Bellehumeur, C. T., & Gu, P. (2008). Effect of processing conditions on the bonding quality of FDM polymer filaments. *Rapid Prototyping Journal*, *14*(2), 72–80. doi:10.1108/13552540810862028

Sun, S., Sun, S., Cao, X., & Sun, R. (2016). The role of pretreatment in improving the enzymatic hydrolysis of lignocellulosic materials. *Bioresource Technology*, *199*, 49–58. doi:10.1016/j.biortech.2015.08.061 PMID:26321216

Szycher, M., Siciliano, A. A., & Reed, A. M. (1991). Polyurethanes in medical devices. *Medical Design and Material*, *1*(2), 18–25. PMID:10171592

Todica, M., Stefan, R., Pop, C. V., & Olar, L. (2015). IR and Raman investigation of some poly (acrylic) acid gels in aqueous and neutralized state. *Acta Physica Polonica A*, *128*(1), 128–135. doi:10.12693/APhysPolA.128.128

Tran, P., Ngo, T. D., Ghazlan, A., & Hui, D. (2017). Bimaterial 3D printing and numerical analysis of bio-inspired composite structures under in-plane and transverse loadings. *Composites. Part B, Engineering*, *108*, 210–223. doi:10.1016/j.compositesb.2016.09.083

Trusler, G. A., & Mustard, W. T. (1958). Intravenous polyethylene catheter successfully removed from the heart. *Canadian Medical Association Journal*, *79*(7), 558. PMID:13585269

Tudor, A. M., Melia, C. D., Binns, J. S., Hendra, P. J., Church, S., & Davies, M. C. (1990). The application of Fourier-transform Raman spectroscopy to the analysis of pharmaceuticals and biomaterials. *Journal of Pharmaceutical and Biomedical Analysis*, *8*(8-12), 717–720. doi:10.1016/0731-7085(90)80109-3 PMID:2100613

Tymrak, B. M., Kreiger, M., & Pearce, J. M. (2014). Mechanical properties of components fabricated with open-source 3-D printers under realistic environmental conditions. *Materials & Design*, *58*, 242–246. doi:10.1016/j.matdes.2014.02.038

Ulery, B. D., Nair, L. S., & Laurencin, C. T. (2011). Biomedical applications of biodegradable polymers. *Journal of Polymer Science. Part B, Polymer Physics*, *49*(12), 832–864. doi:10.1002/polb.22259 PMID:21769165

Vidakis, N., Petousis, M., Tzounis, L., Maniadi, A., Velidakis, E., Mountakis, N., Papageorgiou, D., Liebscher, M., & Mechtcherine, V. (2020). Sustainable additive manufacturing: Mechanical response of polypropylene over multiple recycling processes. *Sustainability*, *13*(1), 159. doi:10.3390u13010159

Wang, Q., Ji, C., Sun, J., Yao, Q., Liu, J., Saeed, R. M. Y., & Zhu, Q. (2019). Kinetic thermal behavior of nanocellulose filled polylactic acid filament for fused filament fabrication 3D printing. *Journal of Applied Polymer Science*. Advance online publication. doi:10.1002/app.48374

Wang, X., Jiang, M., Zhou, Z., Gou, J., & Hui, D. (2017). 3D printing of polymer matrix composites: A review and prospective. *Composites. Part B, Engineering*, *110*, 442–458. doi:10.1016/j.compositesb.2016.11.034

Ward, I. M., & Wilding, M. A. (1977). Infra-red and Raman spectra of poly (m-methylene terephthalate) polymers. *Polymer*, *18*(4), 327–335. doi:10.1016/0032-3861(77)90077-5

Wexler, P. (2014). *Encyclopedia of Toxicology*. Academic Press.

Wong, K. V., & Hernandez, A. (2012). A review of additive manufacturing. *International Scholarly Research Notices*.

Zarrintaj, P., Saeb, M. R., Jafari, S. H., & Mozafari, M. (2020). Application of compatibilized polymer blends in biomedical fields. In *Compatibilization of Polymer Blends* (pp. 511–537). Elsevier. doi:10.1016/B978-0-12-816006-0.00018-9

Zhang, S. U. (2019). Classifying Degraded Three-Dimensionally Printed Polylactic Acid Specimens Using Artificial Neural Networks based on Fourier Transform Infrared Spectroscopy. *Applied Sciences (Basel, Switzerland)*, *9*(13), 2772. doi:10.3390/app9132772

Zhang, Y., & Atala, A. (2019). Regenerative medicine of the bladder. In *Principles of Regenerative Medicine* (pp. 1263–1279). Academic Press. doi:10.1016/B978-0-12-809880-6.00072-2

Zhu, Z. H., Zhang, N., Wang, T., & Hao, M. Y. (2020, February). Short review of polymer composites for 3D Printing. *IOP Conference Series. Materials Science and Engineering*, *758*(1), 012046. doi:10.1088/1757-899X/758/1/012046

Chapter 11
Industrial Robot-Integrated Fused Deposition Modelling for the 3D Printing Process

T. S. Senthil
Panimalar Engineering College, India

R. Ohmsakthi vel
Chennai Institute of Technology, India

M. Puviyarasan
Panimalar Engineering College, India

S. Ramesh Babu
Sri Venkateswara College of Engineering, India

Raviteja Surakasi
https://orcid.org/0000-0002-0786-0105
Lendi institute of Engineering and Technology, India

B. Sampath
https://orcid.org/0000-0002-2065-6539
Muthayammal Engineering College, India

ABSTRACT

In this chapter, six axis robots-integrated fused deposition modelling (FDM) processes have been used to fabricate the polymer three-dimensional (3D) printing objects. A unique method of printing 3D things uses an industrial robot with an FDM extruder as its end-effector. This robot is controlled by sophisticated controller technology and Robot Ware control software. The robot's end effector will travel along the designed toolpath as it is modelled by Robot Studio software. Printing a cube serves as a demonstration of the methods used to combine the FDM process with a six-axis industrial robot. This demonstrates how components can be produced using additive manufacturing and robotics. By applying different process parameters of the innovative system, the tensile and flexural strengths of printed specimens have been optimised using the Taguchi method.

DOI: 10.4018/978-1-6684-6009-2.ch011

INTRODUCTION

A three-dimensional object is created using the additive manufacturing technique of fused deposition modelling (FDM), in which material is deposited in horizontal layers on top of one another. A huge spool of thermoplastic material is used to feed a moving, heated extruder head with filament. An industrial robot is a mechanical device that can be programmed to undertake repetitive or risky activities with high levels of accuracy in place of a human. The welding, painting, assembly, pick-and-place work, packaging, palletizing, product inspection, and testing are typical uses for these robots. Extruder orientation changes and printing on inclination planes are not possible with traditional FDM machines. An industrial robot with an extruder as its end effector can be used to get around this. The goal of the current effort is to investigate a novel industrial robot application to get over the constraints of traditional machines. The current work's goal is to create an FDM method that is coupled with an industrial robot to print 3D items. Additionally, this project seeks to print items on curved surfaces, numerous planes, and inclined planes. The six-axis industrial robot has a maximum payload capacity of 6 kg. The extruder in this work is moved using quick programming (meant for ABB robot) commands. The STL file created from the 3D model is utilised as the input file. Using a MATLAB tool, it is divided into layers based on layer thickness. The influence of various operating parameters such as raster angle, layer thickness, deposition speed and deposition angle on tensile strength and flexural strength is evaluated. By extruding the same build material from both nozzles of the extruder attached to an industrial robot, a new method has been created to reduce printing time. The findings showed that the tensile and flexural strength are considerably influenced by the deposition feed rate, scanline width, and raster angle parameters. Contrary to the traditional FDM technique, which slices items parallel to the horizontal plane, an industrial robot has been created that can slice objects in a variety of planes. The outcomes showed that with the aid of the robot, homogeneous material deposition was seen at the intersection of the planes. Using the methodology, a NACA0015 airfoil-shaped section of an aeroplane wing is printed. The wing is cut in a series of concentric curves, and each layer's curved surface is also printed along. By using the industrial robot to deposit material over nonplanar layers, the staircase effect is minimised.

OVERVIEW OF FDM PROCESS USING INDUSTRIAL ROBOT

The suggested system improves the CNC procedure. The proposed system only yielded a small number of prototypes, but the outcomes showed that the requisite dimensional accuracy was attained. In this study, a trajectory generating algorithm that creates trajectories for various paint thicknesses has been constructed. In this study, the deviation angle of the spray cone is adjusted via patch formation. The created trajectory generation algorithms have demonstrated a reliable level of performance, according to the paint thickness verification method (G. Q. Jin et al., 2013). A brand-new toolpath planning method for robotic manipulator applications like spray painting that is based on the CAD model of the work part. By segmenting the part's surface into smaller sections or patches, the system creates efficient toolpaths automatically. According to experimental findings, the technology that has been designed is useful for robotic applications. Aims to use a novel hybrid toolpath generating method to increase accuracy and decrease building time for additive manufacturing operations. To shorten the building time, algorithms were created to produce the toolpath for each slice based on its geometry. Applying the suggested tool route generating technique to five biomedical models produced successful results (Keating & Oxman,

2013). Additive, subtractive, and formative processes are all included in the multi-functional fabricating process known as robotic manufacturing. A robotic arm is created to do jobs like 3D printing, carving, and milling operations in the suggested fabricating process. It suggested that the robotic arm fabrication platform offers to work effectively and in a broader workspace (Y. an Jin et al., 2014). A 5-axis machine tool has a spindle and an FDM extruder placed on the opposite side of the rotary axis. After being printed, the FDM items can be machined to improve their dimensional accuracy (Ding et al., 2014). To create direction parallel toolpaths for the FDM process, a new method has been created. In order to reduce acceleration or deceleration when moving the tool from one location to another, parametric curves are used. This technology was used to fabricate various parts, and the results showed that the requisite efficiency and precision were attained (Allen & Trask, 2015). Toolpath generation algorithm for wire and arc additive manufacturing. Reduced starting and stopping positions are a goal of the algorithm. The industrial robot was used to create two pieces, one using the current hybrid technique and the other using the created algorithm (Song et al., 2015). Direct Energy Deposition (DED) techniques have an overhang/undercut problem, and it has been suggested to use tool mechanisms with more than three axes. To generate overhang features, a complex slicing technique was suggested for the control of tool paths on a 5-axis base table. The authors created and applied an auto-partitioning approach to produce 3D layer data for the multi-axis tool path (Lee & Jee, 2015). RevoMaker, a novel 3D printing method, can create items using less build material and less support material for overhangs. This "RevoMaker" was used to print several prototypes, and the findings showed that there was less material waste and the printing time was shorter (B. Huang & Singamneni, 2015). The foundation of the system is the use of six linear actuators to control the mobility of the platform. The faults that are produced during the printing process are corrected using a laser and camera system. Results showed that multidirectional printing allows for the printing of items with fewer support structures and without the staircase effect. A problem with the FDM process and created a fix for it. According to the scientists, the cited flaw in the existing printing technique can be fixed by deposition of material in curved layers utilising a parallel robot. According to authors, the staircase effect will cause the component printed using this method to have weak strength and bonding along the z-axis. According to the findings, printing durations for components with better surface finishes were also significantly shorter than those for components made using the traditional FDM process(Gao et al., 2015).A slicing algorithm for robotic material deposition systems to print structures with overhangs in many directions. The created algorithm makes various cuts in the STL model of the object. It has been shown that the suggested approach provides a straightforward method for slicing objects with sharp edges. To create parts with superior mechanical qualities, a research team has proposed curved layer adaptive slicing (CLAS), a revolutionary slicing technique for the FDM process. The object is adaptively divided into flat and curved layers using the CLAS approach. It said that applying CLAS to a test part revealed that its mechanical behaviour is comparable to the part that is printed traditionally(Y. Huang et al., 2016). In addition to preserving fibre continuity, CLAS is good at capturing sharply variable surface profiles and other finer component characteristics. Thicker curved layers produce better mechanical qualities, as demonstrated by three-point bending tests on light curved objects built of curved layers of different thicknesses.

A robotic frame manufacturing algorithm that leads to a logical fabrication process. Based on the suggested methodology, a prototype was created and tested using a 6-axis robotic arm. It suggested that a variety of items that can be constructed using lightweight frames can be printed using the provided algorithm(Lim et al., 2016). Researchers have improved 3D printing using the spider's natural construction methods. Robot is programmed to print ABS-based 3D geometries in an open volume. The findings showed that ABS material has insufficient strength and cannot be used for large-scale architectural printing(Yuan et al., 2016). Researchers at Aberystwyth University in the UK have created an AM fabrication device that uses a robotic arm to deposit the material in various planes. The staircase effect is eliminated by a 3D printer that can print parts in various planes, according to a study team at the University of British Columbia in Canada. According to the results, the suggested system may be used to print components with improved mechanical properties, and the surface quality of printed components has also improved(Y. Jin et al., 2017). A method of fabricating curved layers in concrete for large-scale building. Grasshopper script is used to build a zig-zag toolpath along the curved layers. A sizable three-axes, gantry-style printing device is used to carry out the printing(Ertay et al., 2018). To reduce retraction during printing, a novel route filling pattern for the FDM process has been created. In this approach, discontinuous deposition is also prevented. The entire region is divided into a number of smaller sub-regions and sub-paths in this work. A method has been created that uses a number of interconnected sub-paths to create a continuous path filling pattern. By avoiding retraction motions, the Fast Fouquémier Descent Manufacturing (FDM) method becomes more effective(Li et al., 2018). Assisted support structures are not needed when using the incline layer printing (ILP) technique to print overhang features in 3D. In this method, the object's overhanging feature is cut at an angle to prevent the nozzle from interfering with the previously deposited layers. In contrast to the base of a standard gable, more material is deposited at the point where two planes with different orientations meet (gable roof). Controlling the deposition of many planar layers at the same location in a geometrical structure necessitates the use of a different tactic(Zhao et al., 2018). Utilising a 6DOF industrial robotic arm, a revolutionary hybrid technology. The robotic arm is equipped with various interchangeable heads that can do both subtractive and additive operations. A milling tool is used for subtractive operations and an FDM extruder for additive activities. For this hybrid system, which consists of a heat bed and fixture configuration, an integrated platform is created. Multiple directions can be printed by a robotic arm, cutting down on building time and expense. This hybrid process can provide high-quality items. To create high-quality sculptures, material is removed with the milling tool as it is deposited in layers by the FDM printer(Kubalak et al., 2018). A method for the FDM process that uses a six-axis industrial robot to deposit material on an item to enhance its mechanical qualities. It said that the object's interlayer bonding would be improved by this skin. In this work, two tensile test specimens were printed in accordance with ASTM D638 specifications. An innovative method of producing printed items using the conventional FDM procedure and a skinned surface has been put forth by a research team. The mechanical qualities of the skinned specimen are superior to those of the conventionally printed specimen, according to the findings of the tensile test. [(IEEE Robotics and Automation Society et al., 2019) A technique for the FDM process that synchronises the material's extrusion rate with the tangential speed and deposition temperature when the material is deposited along curved routes By carefully managing the current supplied to the extruder's heater, the temperature is kept at the desired consistent level. The tangential velocity is tuned in relation to the rate of material deposition while taking the machine jerk into account. After applying this technology to various pieces, it became clear that the printed parts' dimensional correctness was better than it was using the traditional FDM procedure(Jensen et al., 2019). A 3D printing method that employs multiple mobile robots to manufacture massive, single-piece constructions. Using Open RAVE software, careful planning is done in this effort to prevent mutual encounters. Concrete is delivered to the printing nozzles through a pump mechanism. A group of researchers has created a 3D printing device that they believe is capable of printing well on-site. The researchers claim that it is more scalable and time-effective than existing 3D concrete printing processes since it uses movable robotic arms to carry out printing. To increase the dimensional precision of the FDM procedure, a team from the University of Bristol created a robotic arm that can cooperate with a unique FDM extrusion head. In this work, a dedicated computer is regarded as the controller. Throughout the printing process, 1900 is used to regulate the extrusion temperature. According to the findings, the weight of the extruder setup and the robot's acceleration or deacceleration both have an impact on the printed part's dimensional correctness (Ishak & Larochelle, 2019).

An integrated method for multi-plane printing and 3D printing of lattice structures has been created by a research team. A 6-DOF robot arm platform for FDM and an interface programme to control the robot arm make up the system. The Microsoft Foundation Class (MFC) library was used in its development. Additionally, a simulation tool is created in MatLab to enable visual verification of the printing process. The 3D lattice structure with an overhang design is printed without the support structure using this integrated system. (Jiang et al., 2019). According to a study, print path and orientation have a significant impact on how much support material is consumed with printed goods. It implies that a fresh approach to support creation could save waste by as much as 50%. Applying the suggested technique could also cut down on production time and energy usage. According to the team's findings, printing parts required less material, time, and energy than traditional printing. To print 3D objects, there are two novel deposition techniques. The first is an additive manufacturing strategy based on extrusion with five degrees of freedom. In contrast to standard extruder-based additive manufacturing procedures, rounder surfaces are printed in this work with an improved surface polish. Researchers at the Massachusetts Institute of Technology have created a 3D printer that can produce complicated pieces devoid of support structures (MIT). It creates components for 3D printing by using a fixed extruder and an articulated robot arm with a build plate as the end effector. (Bhatt et al., 2020). For the FDM process, a novel slicing algorithm is used to reduce the staircase effect and print stronger objects. This technique is able to identify an object's curved surfaces and produce nonplanar layers for each of those surfaces. The programme produces toolpaths for the extruder that are free of collisions along the planar levels. On a 3DOF FDM machine,

Figure 1. The Robot model and their working ranges

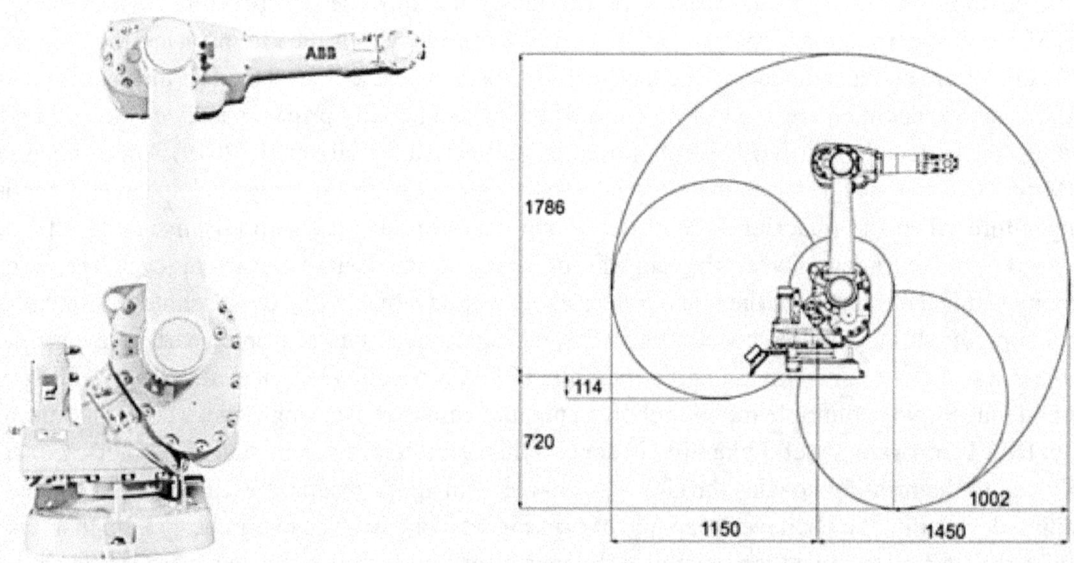

a method for printing curved objects has been devised. The outcomes showed that, contrary to other reports, the printed objects exhibited a smooth surface finish due to the material being deposited along nonplanar layers. For the robotic printer, the researchers employed an algorithm to produce a toolpath with numerous directions. It has been asserted that the tensile strength of the specimens printed utilising the created technique is comparable to that of human hands. Tensile strength and young's modulus have been proven in tests to increase with increasing printing angle and decrease with decreasing layer thickness. In this study, a slicing method is designed to print thin shell objects with a robotic extrusion system and reorienting deposition platform. The goal is to manufacture items with overhanging features without the usage of support structures at 3D printer speeds up to 10 times faster than conventional techniques. Because there is no need to print additional material for the support framework, the build time is dramatically reduced. It claimed that because the suggested technology can print nonplanar layers, the surface smoothness is also improved.

INDUSTRIAL ROBOT INTEGRATION FOR FUSED DEPOSITION MODELING PROCESS

An industrial robot with an FDM extruder as its end-effector is used in a novel approach to print 3D things. In additive manufacturing, it is ideal to combine the fused deposition modelling technique with a 6-axis industrial robotic arm. This chapter describes several elements that went into this integration. Short cycle times and good path accuracy are characteristics of this robot. Arc welding, grinding, material handling, and deburring applications are better suited for it. Industrial robot ABB IRB 1600 has 6 DOF, a 6 kg payload capacity, and a 1.45 m reach. The robot has a position repeatability of less than 0.05mm. The six-axis robot model and their working ranges are illustrated in Figure 1. The motion ranges and motion speeds of six axis robot are shown in Table1.

Each of the robot's six joints is moved by powerful servo motors. These joints all revolve around a common axis. There are two groups assigned to these axes. Because they are used to place the wrist at the proper area in the workspace, the first three axes are known as the main axes.

Table 1. Motion ranges and motion speeds of six axis robot

Joint	1	2	3	4	5	6
Range (°)	±180	+150 to - 90	+65 to – 245	±200	±115	±400
Speed (°/s)	180	160	170	320	400	460

The fifth-generation Industrial Robot Controller (IRC5) uses a system-level controller and a servo-loop controller to govern how the robot moves. Industrial robots frequently carry out machining tasks like cutting and stamping. Robot controller technology is standardised by the IRC5, according to the robotics industry. It offers support for PC tooling, flexibility, safety, modularity, application interfaces, and multi-robot control. It has been discovered that the servo controller is the primary bottleneck controlling the accuracy of the robot operations. A robot is an autonomous machine with three primary components that all function cohesively. The controller, actuator, and sensors make up the three essential components. Robot control systems regulate and direct the robot's operations to produce the intended outcome. The robot's movements is managed by the fifth-generation Industrial Robot Controller (IRC5). All the programmes required to control the robotic arm are included in here, which is the system's main brain. Controlling four drive modules that can drive 36 axes is made easier by the IRC5 controller.

The user interface between the operator and the controller is the Flex Pendant, also known as the Teach Pendant. It has both hardware and software and performs a variety of tasks to control the robot. The Flex Pendant has a joystick that may be used to jog the robotic arm and direct its movements.

Robot Coordinate System

At the apex of the hierarchy, a universal or world coordinate system stands in for the entire robotic work cell. In light of this, other coordinate systems, such as base coordinate, tool coordinate, work object co-ordinates, and user coordinate systems, are created. In the robotic work cell, each robot has its own base coordinate system. The various coordinate system of the robot is shown in Figure 2. Tool Center Point is the coordinate system's intersection point for all axes (TCP). When performing the task as specified by the operator, the robot moves the TCP along the predetermined spot. When work object coordinate systems are established with reference to the position of the work object, it is possible to easily alter the robot programmes with an offset if the workpiece's location changes. To make the programming step easier, reference points are created using a user coordinate system.

Rapid Prototyping Programming

Robots from ABB are programmed using the high-level programming language Rapid. This programming language's objective is to construct a sequential path that the robot can follow to complete the desired tasks. Each module in a robot programme has a set of routines. An instruction set makes up each routine. Target instructions, Motion instructions, and Digital Output instructions were all employed in this project. These commands specify how the robot's end-effector is to move to the desired location. End effectors are the sections of the robot that communicate with other elements or objects in the surrounding space. A robot's arm's end effectors, which are used to move and position items, are located there. These robots will be capable of picking up, carrying, and setting down objects. Robot movement can be divided into three categories: circular, joint, and linear. Servo robots are equipped with manipulators, effectors, and robotic appendages that serve as the robot's hands and arms. The components of the robot that carry out the work are called effectors. Any tool that you can mount on your robot and operate with its computer is an effector. They are frequently tailored to the jobs that you want your robot to perform. The most widely used AM technology at the moment is FDM. High-strength thermoplastic components can be manufactured using this technique. The component of the 3D printer that propels the filament, melts it, and deposits it on the bed to construct the model is called an extruder. It is utilised for a variety of hues. mechanically robust and long-lasting. Removal is supported by hands-free. most Stratasys FDM 3D printers are compatible with it. Excellent ratios of cost to quality and cost to performance.

Figure 2. Various Coordinate system of the Robot

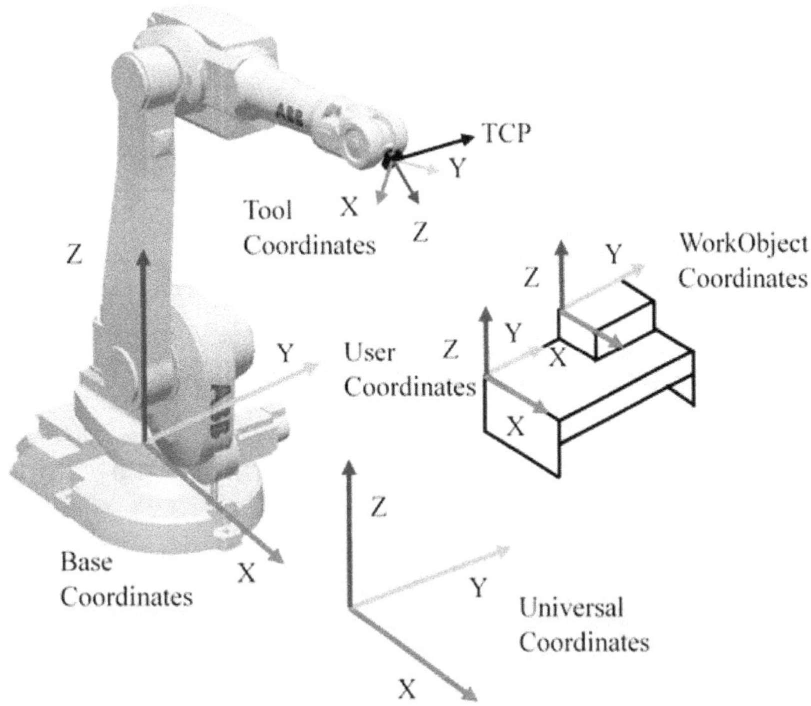

Robot Studio Software

Robot Studio is a simulation programme used to visualise and programme the ABB robots in offline mode on a computer. Both actual and virtual robots can be managed using this. This simulation environment can be used to construct virtual robots and tools. Within this environment, work objects and TCPs (Tool Center Points) can be created.

With this software, CAD models may be imported as well as designed. Based on the CAD data, the software enables the programmer to construct precise placements. Utilizing the "play" and "synchronise" options will replicate the journey. The programme will generate errors if there are any singularity or reachability problems. By highlighting problems in the early phases of a robot's development, simulation software is tremendously helpful in reducing the programming time. These mistakes can be fixed by altering the course. The generated software can also be altered. To perform the appropriate duties, this application can be uploaded to the actual robotic cell.

Instruction for Robot Motion Control

The programme is originally built using the simulation application Robot Studio. After that, a USB drive is used to upload it to the flexible Pendant. The drive module will provide electricity to the robotic arm's joints. They will be turned at a particular angle in accordance with the controller's instructions. The robot arm's movement is managed in this manner. The procedure and instructions used in the Robot Studio software are displayed in Figure 3.

Figure 3. Procedure and Instructions for Robot Studio software

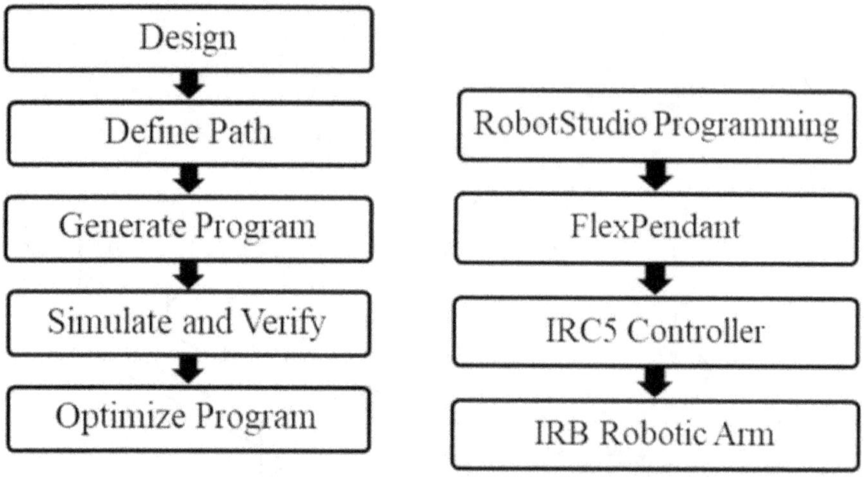

METHODOLOGY FOR INTEGRATING INDUSTRIAL ROBOT WITH FDM EXTRUDER

This section outlines the process used to print an object with an industrial robot. The sequential processes for robot integrated 3D printing is illustrated in Figure 4.

- Making a CAD model of the object is the initial stage in this process. Making it with design software like CATIA is one method to do it. The alternative method involves creating a CAD model of an existing part using a 3D scanner.
- The process of converting a 3D model of an object into the format required for 3D printing comes next. In the STL format, triangles are used to roughly represent the object's surface rather than squares and rectangles as in the preceding example.
- The next step is to slice the CAD model using a horizontal plane after it has been converted into STL format. Up until the entire object is sliced, this plane's vertical increment is used. Each layer's thickness is then considered to be equivalent to the layer thickness in this example.
- This stage involves extracting a set of boundaries that are spaced apart by the same amount as the layer thickness. The following step is extracting each layer's border. If the boundary is a convex polygon, no additional processing is necessary to construct the toolpath.
- A line segment can be used to separate a concave polygon into convex polygons. Some of the diagonals of a convex polygon are outside the polygon. It features at least one interior angle that is close to 180 degrees.
- Based on the information gathered from the layer boundary, a toolpath is constructed. The same algorithm may be applied to concave and convex polygons as well as convex and concave boundaries. The required toolpath is generated by this algorithm using the input of the polygon's vertices.
- Using a customised MATLAB application, the technique is applied to produce the toolpath. The scanline polygon filling procedure is used for each of the individual convex polygons if the boundary is shaped like a concave polygon. The MATML application will be used to create a zigzag toolpath.

Figure 4. Sequential processes for Robotic Printing

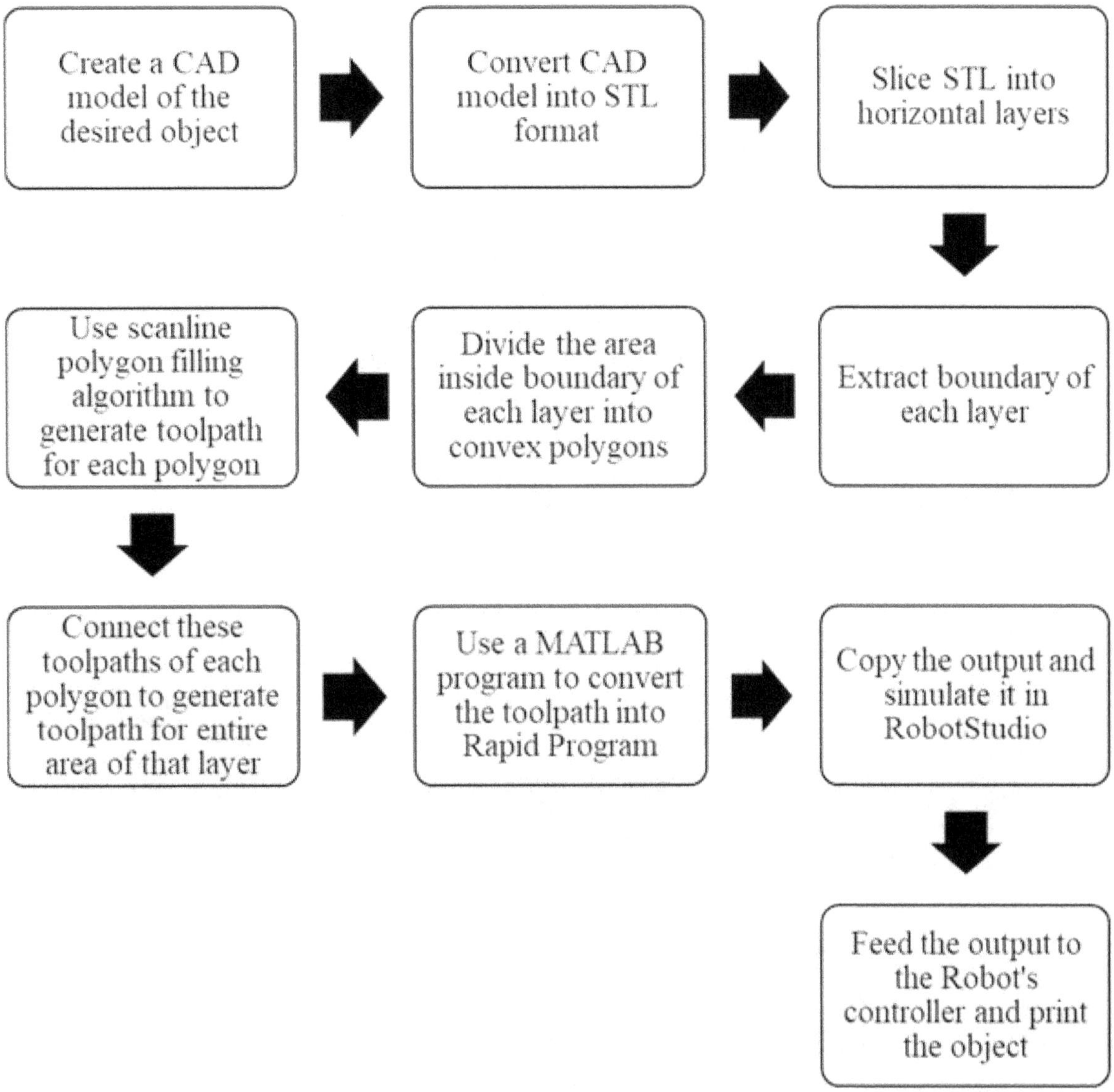

- Similar to levels on a 3D computer network, toolpaths are generated for each layer of an item. Using the extruder, the robot must deposit material at specific points along the toolpath. These segments are picked based on the sequence in which a computer sent the digital output signals to it.
- In a simulation environment, an ABB IRB 1600 robot is imported, and the flange of a two-nozzle extruder model is attached. TCP is simulated in a virtual environment that has been loaded with a quick software. The logs show any reachability or presence of singularity errors that may have occurred.

- A quick programme is performed, and then the robotic arm prints the desired thing. The programme is loaded into the actual controller after this simulation has been verified in the Robot Studio software.

FDM EXTRUDER AND HEAT BED SETUP

This is the first time that PLA (polylactic acid or polylactide) material has been deposited by a robot in an industrial setting. A dual nozzle extruder was used to deposit the material in a controlled environment while heating it to 195°C to melt PLA. Stepper motors are used to feed PLA into the heat block as 1.75 mm-diameter filament. The substance melts and is then forced into the nozzle. Due to its thermoplastic nature, the material that is extruded from the nozzle solidifies instantaneously when it is exposed to the outside atmosphere. The robot's temperature is maintained at 195°C with an Arduino board and a Ramps

Figure 5. Various components of FDM Extruder and Heater

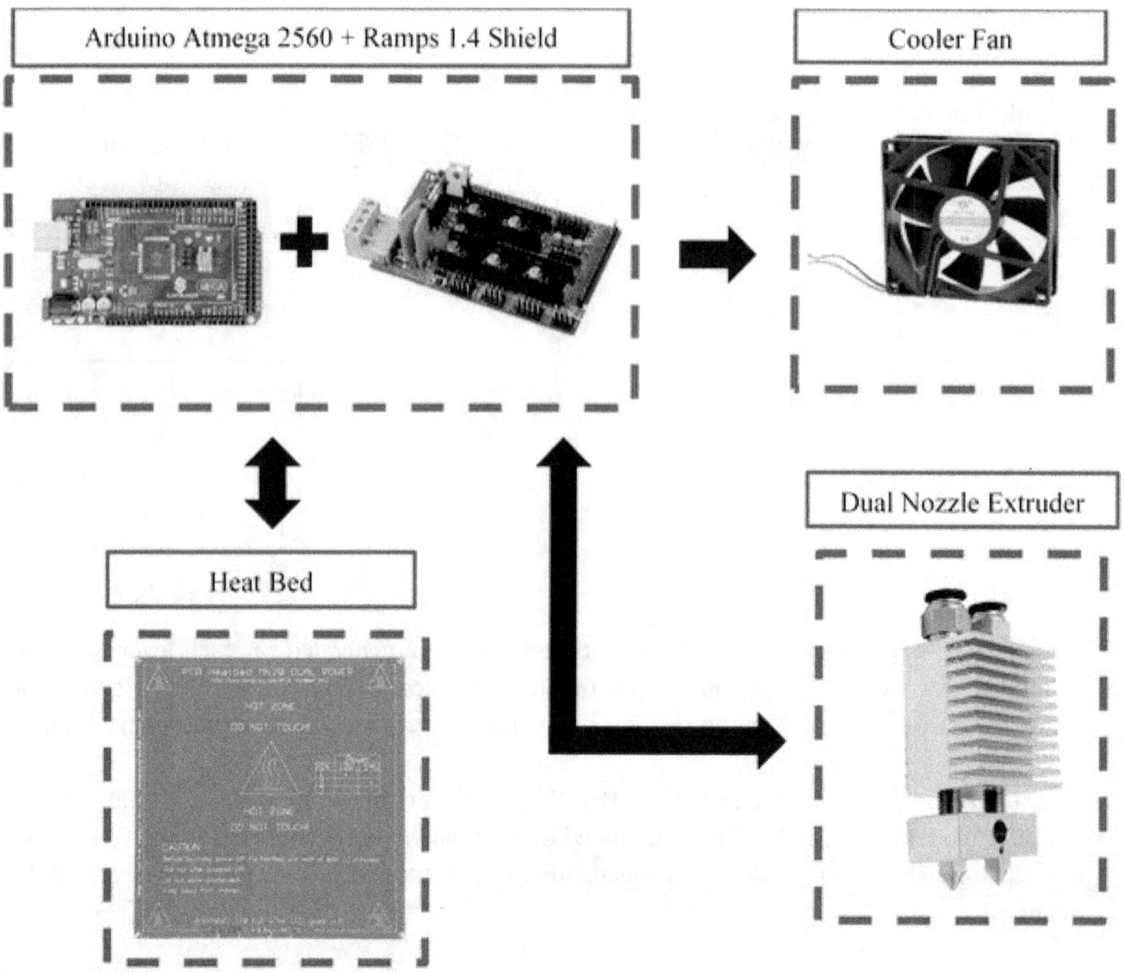

1.4 Shield with continuous feedback control. On a thin glass plate that is set in front of the robot, the desired object is allowed to construct itself. To regulate the cooling of the deposition of material, this plate is positioned on a heat bed. The same Arduino board control with constant feedback is used to keep the heat bed's temperature at 70°C. The likelihood of thermal stress buildup and warping in the printed object owing to abrupt cooling is reduced by the use of a heat bed. The components' temperatures (heat block and heat bed) are attained in less than 5 minutes after the Arduino board is powered up.

COMPONENTS AND SETUP FOR FDM PROCESS AND INDUSTRIAL ROBOT

Prior to being executed by the robot's primary computer, programmes are uploaded to the FlexPendant. These instructions include both digital output signals and motion instructions. The main computer in the controller processes these commands and activates the robot's joints appropriately to move toward the intended targets in the intended direction. The various components of FDM extruder and heater are illustrated in Figure 5. The components and their connections for the Robot-3D printer setup is illustrated in Figure 6.

L293D Motor Driver are available on every Arduino Uno board. Each motor feeds filament-shaped material into the nozzle's heat block. The stepper motor keeps its position and material deposition ceases if the output signal is set to "LOW." The I/O board processes the digital output signals.

In this study, two distinct stepper motors in a PLA extruder are controlled by two digital output signals. Material is deposited through the nozzle when the stepper motor turns and the Arduino board's input signal is "HIGH." If it is set to "LOW," the motor maintains its position and no material is deposited.

An exclusive module for the Arduino UNO Board is the L293D. A full-featured Arduino shield that can operate DC motors, stepper motors, and servo motors is included in this motor driver shield. Material is fed via the extruder's nozzles by each stepper motor.

The stepper motor's four-terminal wires are linked to the L293D motor shield's corresponding pins. The stepper motor will be powered on by the L293D motor shield if the digital input signal for the Arduino board is "HIGH." In this work, stepper motors of the Nema17 type are utilised. They have a 12V DC rated voltage. These motors have a holding torque of 0.52 Nm (Figure 7).

A glass plate's surface serves as the definition of a work object. The material is deposited between the desired targets while the first nozzle's tip (TCP) travels along the predetermined path. This relies on the digital signals the controller sends, as well as how near the cooler fan the nozzle is. The real time robot setup for 3D printing of models are photographed in Figure 8.

Figure 6. Components and their connections for the Robot-3D printer setup

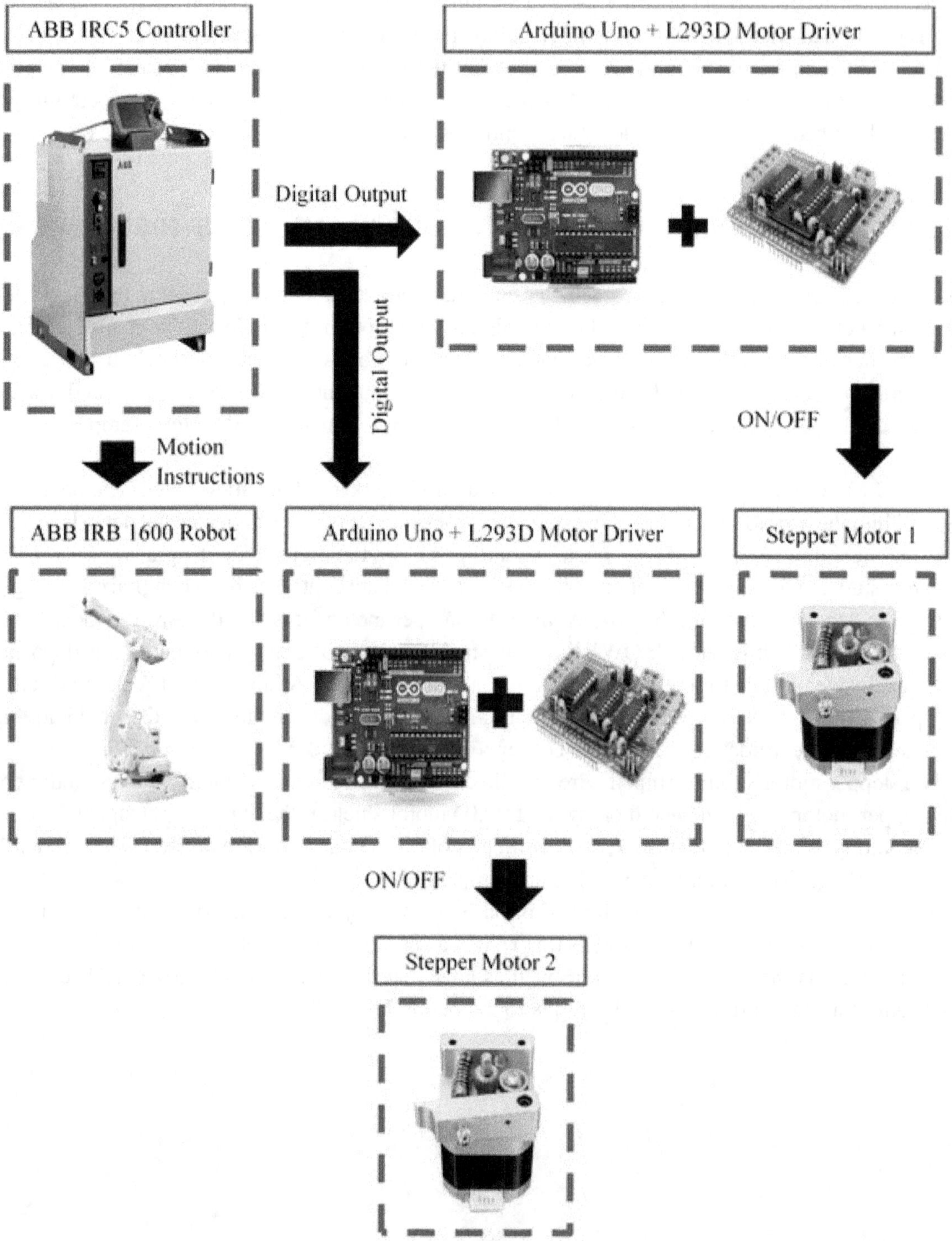

Figure 7. Stepper motor controller

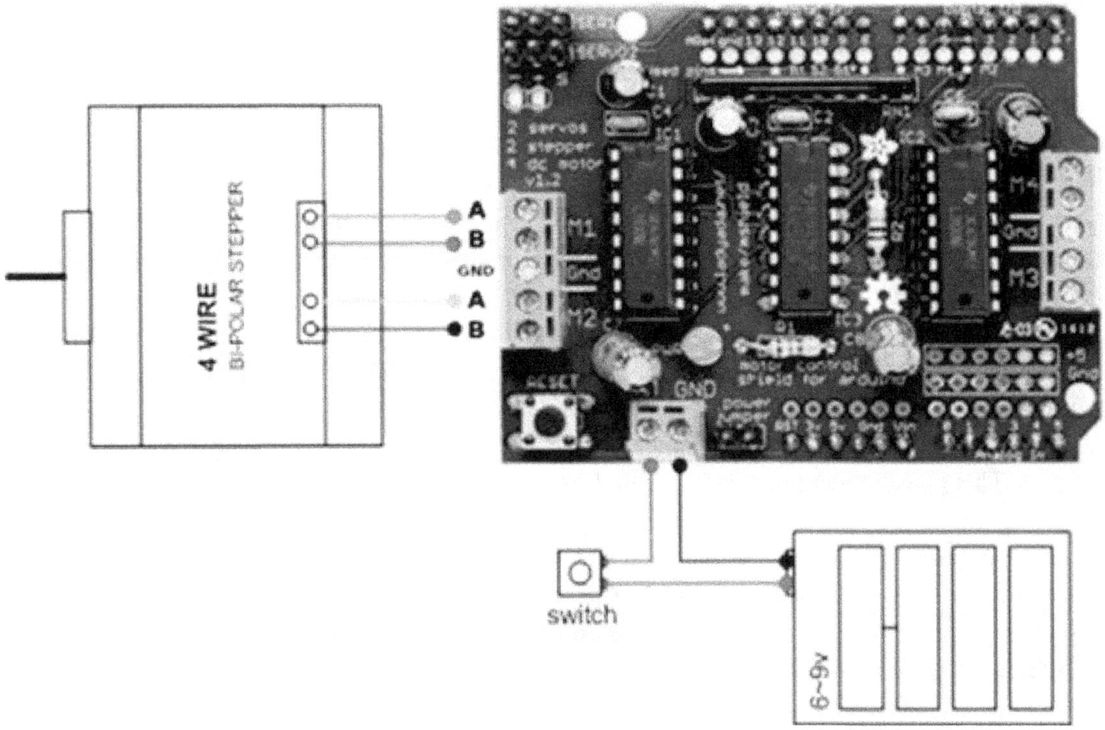

Figure 8. Robot for 3D Printing Process

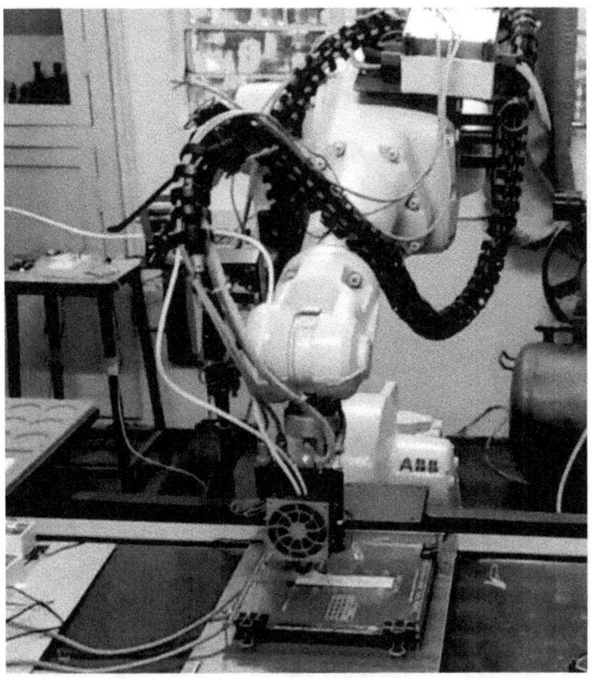

PROCEDURE FOR INTEGRATING FUSED DEPOSITION MODELING PROCESS AND INDUSTRIAL ROBOT

How to simulate and print a cube using an industrial robot and FDM technology. The MATLAB programme is used to create a toolpath for material deposition, and it is translated into commands for the robot. Stepper motors are controlled to deposit material along the required route by using digital output signals. ABB Robot Studio simulation software is used to validate the generated toolpath. The influence of various operating parameters on the mechanical properties of the printed object is also presented in this chapter. The actual printing of the test part is done using the industrial robot after validating the generated toypath.

The methodology developed for printing 3D objects using the Industrial robot integrated with the FDM process is discussed in the previous chapter. The developed methodology was evaluated by printing a cube having an edge of length 30 mm. A horizontal plane that is parallel to the XY-plane is considered to slice the STL format of the cube to generate layers. The layer thickness considered to print the cube is 0.5 mm. This plane is incremented in the multiples of 1.5 vertically starting from the height of 0 mm till 30 mm. The horizontal plane is intersecting the cube at a height of 10 mm from the base, and it intersects the four faces of the cube. For each intersection, a set of intersection points are obtained. The horizontal plane is used to slice the cube along with the height in steps of 0.5 mm. After obtaining the intersection points, by joining all boundary points, the boundary of the layer at 10mm height is obtained. The different steps of FDM using industrial robot programming is illustrated in Figure 9.

Figure 9. Fusion Deposition Modelling using Industrial Robot

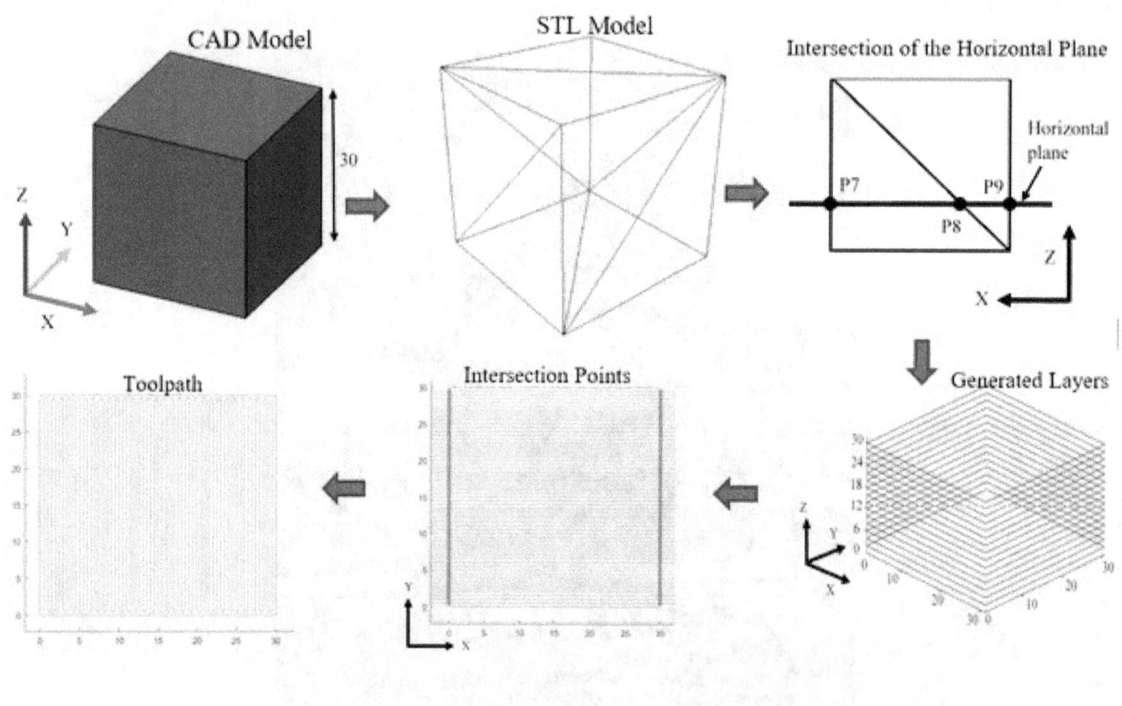

Generation of Toolpath

The layers are laid out in a bottom-to-top approach, with the bottom-most layer selected to generate the toolpath. As the boundary of each layer is in a convex shape (square), the scanline polygon filling algorithm is directly applied. A scanline is used to intersect the sides of the square from bottom to top (towards positive Y-axis) with an increment of 0.5 mm to obtain a set of intersection points. This is done by a program that is developed in MATLAB software.

The direction of the infill's toolpath is set to perpendicular for any two consecutive layers. In this way, all the layers are scanned and a set of intersection points is obtained for each layer. A MATLAB program is used to generate the toolpath for the base layer of a digital image. As the considered boundary of the layer is in a convex shape, every horizontal scanline will intersect the sides of the sides at two points i.e. left interaction point and right intersection point. Let's consider the left and right intersection points on a 2x2 grid. The left intersection point is H1 (H1X, H1Y, 0) and the right intersection point as H2 (H2X, H2Y)/. The coordinates of H2 are modified to (H2X-0.5, H2Y, 0.5) using the same MATLAB program used to generate the toolpath for infill. The newly obtained set of modified intersection points are connected in a jig-jag fashion using the Jig-Jag program. For the next consecutive top layer (2nd) the direction of the toolpath for infill is perpendicular to the previous one. As it is scanned from left to right (towards positive X-axis), it intersects the sides of the square at two points i.e. bottom intersection point and top intersection point. The "y" coordinate of the bottom intersection point is subtracted by 0.5 mm from the top intersection point. This means that there is a gap between the toolpath on the top side and the infill on the bottom side. All the "y" coordinates of V2 are modified to (V2X, V2Y-0.5, 0.5) and all

Figure 10. Simulation of the robot integrated 3D Printing Process

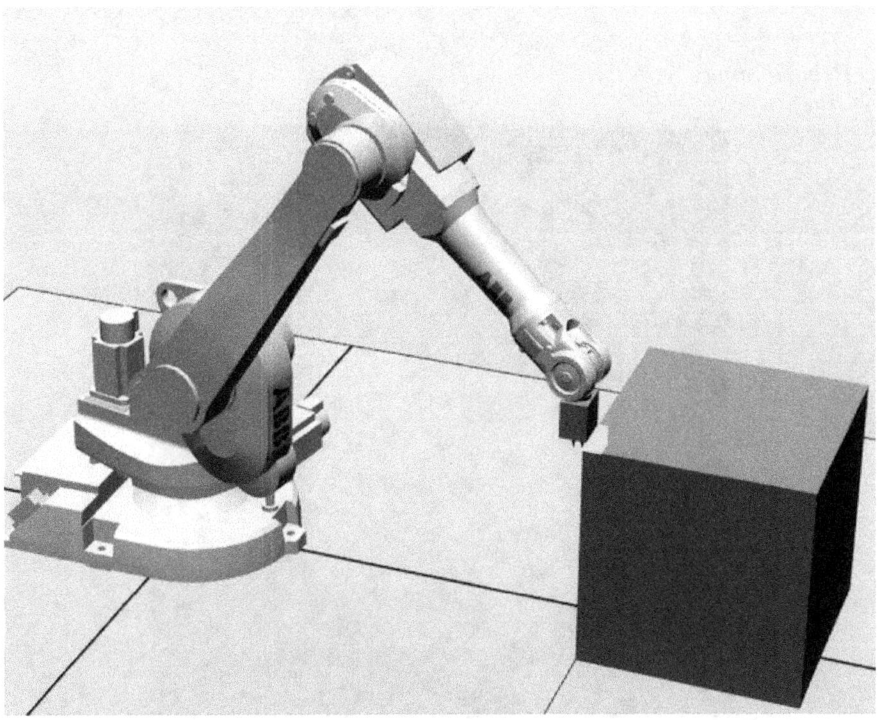

the intersection points are modified for the 2nd layer. This gives a gap between the toolpath of the shell and the infill on the top side. Toolpaths for all the even-numbered layers in the cube are generated in this way. The infill's direction of each layer is perpendicular to the direction of the next consecutive layer.

Simulation of the Printing Process

The simulation of the robot integrated 3D Printing Process is illustrated in Figure 10. toolpath is generated by a command-line program that makes the extruder move linearly between two points at a velocity of 50 mm/s. The commands are then converted into digital output signals using the MATLAB program. All the motion commands are in form of MoveL statements that make the extrusion occur at a rate of 50mm/s or more. A rapid program is uploaded to RobotStudio software for validating the generated path. ABB IRB 1600 robot is imported into the simulation environment. Digital output signals are used to control the material deposition along the path. The velocity is considered based upon the viscosity of the PLA material and deposition feed rate just like a conventional FDM process. A dual nozzle extruder model is designed in CATIA software by considering the dimensions of the actual extruder. It has its nozzles separated by a distance of 20 mm, and is attached to the robot's flange via an international link-up system. A table model is created that resembles the printing bed. Workobject is defined on the surface of the table in such a way that the extruder model's orientation stays perpendicular to the table's surface. The virtual controller is created and the generated rapid program is uploaded to it.

Printing of Cube using the Industrial Robot

The rapid code is uploaded to the IRC5 controller. After validating the simulation of TCP movement along the toolpath, the actual printing of the cube is carried out using the industrial robot. Objects with

Figure 11. 3D Printed object

overhanging structures can also be printed by depositing the support material where ever needed. An industrial robot has been developed which can be used as a material depositing machine for printing in the same way as a conventional FDM machine. The robot can print on inclined planes as well as on curved surfaces, according to researchers at the University of Bristol. The final product of the 3D printed object is shown in Figure 11.

TESTING OF MECHANICAL PROPERTIES OF 3D PRINTED PARTS

The various process parameters to test the mechanical properties of 3D printed object is shown in Table 2. Engineering's understanding and analysis of any process or system's performance depend heavily on the experimental design, or planning of tests. The conventional method involves maintaining all the parameters constant while changing one parameter at a time. It takes a lot of time, necessitates running more experiments, and costs money. By using orthogonal arrays, Taguchi's resilient design gives each factor the same weight(Boopathi, 2021a; Boopathi et al., 2021b; Boopathi, Sureshkumar, et al., 2022; Myilsamy & Sampath, 2021; Sampath & Myilsamy, 2021). The chosen process parameters have a significant impact on the quality of the parts produced by the industrial robot. The mechanical characteristics of printed items are influenced by a number of process variables, including raster angle, layer thickness, deposition speed, and deposition angle.

Table 2. Factors and their levels

Factor	Raster angle (°)	Scanline width (mm)	Layer thickness (mm)	Deposition speed (mm/s)	Deposition feed rate (mm/min)	Deposition Angle (°)
Level	90-135	0.45, 0.5, 0.55	0.4, 0.5, 0.6	45, 50, 55	200, 220, 240	700, 900

The 6 DOF industrial robot that was employed in this project as the equipment for depositing the material provides the option to deposit the material at an angle. In order to examine its influence alongside the other components, this work has considered the deposition angle parameter. The angle at which a thin strip of thermoplastic material is extruded through the extruder's nozzle is known as the deposition angle. The thickness of the deposited layer is referred to as the layer's thickness. Layer width is referred to as raster width, while the space between two neighbouring roadways inside a layer is known as air-gap. To create an L18 mixed-level design with 6 components, the Taguchi method is utilised (Boopathi, 2022b, 2022a, 2022d, 2022c; Boopathi, Haribalaji, et al., 2022; Boopathi et al., 2023; Kannan et al., 2022; Kumar et al., 2023; Saravanan et al., 2022; Yupapin et al., 2022a). The extruder moves at a certain pace as it deposits material in the XY plane, which is known as the printing speed. Printing time is the total amount of time needed to print the object under consideration.

For tensile and flexural tests, specimens are printed in accordance with the L18 design matrix(Boopathi, Lewise, et al., 2022; Boopathi, Thillaivanan, et al., 2022; Gunasekaran et al., 2022; Sampath & N. Naveenkumar P. Sampathkumar, 2022; Trojovsky et al., 2023). These samples were examined using the Universal Testing Device INSTRON 3969. This machine was created specifically to perform tensile, compression, flexure, shear, and torsion testing. The procedure is inserting the test specimen into an apparatus and exerting tension on it until it breaks. The gauge section's length is measured against the applied force as tension is applied. The industrial robot is used to print flexural specimens in accordance with ASTM D790 standard(Boopathi, 2021b; Boopathi et al., 2021a; Boopathi, Haribalaji, et al., 2022; Haribalaji et al., 2022; Jeyakymar et al., 2022; Kavitha et al., 2022; Yupapin et al., 2022b). The most used flexural test for thermoplastic specimens is the 3-point flexural bending test. The test specimen is put into the testing machine, and force is applied until the specimen breaks. Tensile specimen 11 achieves its maximum tensile strength. With a raster angle of 60°/150°, a scanline width of 0.45 mm, a layer thickness of 0.5 mm, a deposition speed of 45 mm/s, a deposition feed rate of 240 mm/min, and a deposition angle of 110°, this sample was produced. Deposition feed rate (which contributes 40.87 percent) has the greatest impact on tensile strength, while deposition speed has the least impact (0.94 percent). The deposition angle (36.79%) and scanline width (36.79%) have the greatest effects on flexural strength (1.24 percent).

SUMMARY

An industrial robot with an FDM extruder as its end-effector is used in a novel approach to print 3D things. Robot Ware control software and a sophisticated IRC5 controller are used to control this robot. ABB Robot Studio software simulates how the robot's end effector will move along the created toolpath. Printing a cube serves as a demonstration of the methods used in this chapter to combine the FDM process with a 6-axis industrial robot. This shows that robotics and additive manufacturing can work together to produce components. On the tensile strength and flexural strength of printed specimens, the effects of different operation conditions are investigated.

REFERENCES

Allen, R. J. A., & Trask, R. S. (2015). An experimental demonstration of effective Curved Layer Fused Filament Fabrication utilising a parallel deposition robot. *Additive Manufacturing*, *8*, 78–87. doi:10.1016/j.addma.2015.09.001

Bhatt, P. M., Malhan, R. K., Rajendran, P., & Gupta, S. K. (2020). Building free-form thin shell parts using supportless extrusion-based additive manufacturing. *Additive Manufacturing*, *32*, 101003. doi:10.1016/j.addma.2019.101003

Boopathi, S. (2021a). An investigation on gas emission concentration and relative emission rate of the near-dry wire-cut electrical discharge machining process. *Environmental Science and Pollution Research International*, *29*(57), 86237–86246. doi:10.100711356-021-17658-1 PMID:34837614

Boopathi, S. (2021b). Improving of Green Sand-Mould Quality using Taguchi Technique. *Journal of Engineering Research*. doi:10.36909/jer.14079

Boopathi, S. (2022a). An experimental investigation of Quench Polish Quench (QPQ) coating on AISI 4150 steel. *Engineering Research Express*, *4*(4), 045009. doi:10.1088/2631-8695/ac9ddd

Boopathi, S. (2022b). An Extensive Review on Sustainable Developments of Dry and Near-Dry Electrical Discharge Machining Processes. *Journal of Manufacturing Science and Engineering*, *144*(5), 1–37. doi:10.1115/1.4052527

Boopathi, S. (2022c). Cryogenically treated and untreated stainless steel grade 317 in sustainable wire electrical discharge machining process: A comparative study. *Environmental Science and Pollution Research International*, 1–10. doi:10.100711356-022-22843-x PMID:36057706

Boopathi, S. (2022d). Performance Improvement of Eco-Friendly Near-Dry Wire-Cut Electrical Discharge Machining Process Using Coconut Oil-Mist Dielectric Fluid. *Journal of Advanced Manufacturing Systems*, 1–20. Advance online publication. doi:10.1142/S0219686723500178

Boopathi, S., Balasubramani, V., Kumar, R. S., & Singh, G. R. (2021a). The influence of human hair on kenaf and Grewia fiber-based hybrid natural composite material: an experimental study. In Functional Composites and Structures (Vol. 3, Issue 4). doi:10.1088/2631-6331/ac3afc

Boopathi, S., Balasubramani, V., Kumar, R. S., & Singh, G. R. (2021b). The influence of human hair on kenaf and Grewia fiber-based hybrid natural composite material: An experimental study. *Functional Composites and Structures*, *3*(4), 045011. doi:10.1088/2631-6331/ac3afc

Boopathi, S., Haribalaji, V., Mageswari, M., & Asif, M. M. (2022). Influences of Boron Carbide Particles on the Wear Rate and Tensile Strength of Aa2014 Surface Composite Fabricated By Friction-Stir Processing. *Materiali in Tehnologije*, *56*(3), 263–270. doi:10.17222/mit.2022.409

Boopathi, S., Lewise, K. A. S., Subbiah, R., & Sivaraman, G. (2022). Near-dry wire-cut electrical discharge machining process using water—air-mist dielectric fluid: An experimental study. *Elsevier: Materials Today: Proceedings*, *50*(5), 1885–1890. doi:10.1016/j.matpr.2021.08.077

Boopathi, S., Sureshkumar, M., Jeyakumar, M., Kumar, R. S., & Subbiah, R. (2022). Influences of Fabrication Parameters on Natural Fiber Reinforced Polymer Composite (NFRPC) Material: A Review. *Materials Science Forum*, *1075*, 115–124. doi:10.4028/p-095f0t

Boopathi, S., Thillaivanan, A., Pandian, M., Subbiah, R., & Shanmugam, P. (2022). Friction stir processing of boron carbide reinforced aluminium surface (Al-B4C) composite: Mechanical characteristics analysis. *Materials Today: Proceedings*, *50*(5), 2430–2435. doi:10.1016/j.matpr.2021.10.261

Boopathi, S., Venkatesan, G., & Anton Savio Lewise, K. (2023). Mechanical Properties Analysis of Kenaf–Grewia–Hair Fiber-Reinforced Composite. In *Lecture Notes in Mechanical Engineering* (pp. 101–110). Springer. doi:10.1007/978-981-16-9057-0_11

Ding, D., Pan, Z., Cuiuri, D., & Li, H. (2014). A tool-path generation strategy for wire and arc additive manufacturing. *International Journal of Advanced Manufacturing Technology*, *73*(1–4), 173–183. doi:10.100700170-014-5808-5

Ertay, D. S., Yuen, A., & Altintas, Y. (2018). Synchronized material deposition rate control with path velocity on fused filament fabrication machines. *Additive Manufacturing*, *19*, 205–213. doi:10.1016/j.addma.2017.05.011

Gao, W., Zhang, Y., Nazzetta, D. C., Ramani, K., & Cipra, R. J. (2015). RevoMaker: Enabling multi-directional and functionally-embedded 3D printing using a rotational cuboidal platform. *UIST 2015 - Proceedings of the 28th Annual ACM Symposium on User Interface Software and Technology*, 437–446. 10.1145/2807442.2807476

Gunasekaran, K., Boopathi, S., & Sureshkumar, M. (2022). Analysis of a Cryogenically Cooled Near-Dry Wedm Process Using Different Dielectrics. *Materiali in Tehnologije*, *56*(2), 179–186. doi:10.17222/mit.2022.397

Haribalaji, V., Venkatesan, G., Asif, M. M., Pandian, M., Subbiah, R., & Boopathi, S. (2022). Investigation on corrosion and tensile Characteristics: Friction stir welding of AA7075 and AA2014. *Materials Today: Proceedings*, *66*, 743–748. doi:10.1016/j.matpr.2022.04.037

Huang, B., & Singamneni, S. B. (2015). Curved layer adaptive slicing (CLAS) for fused deposition modelling. *Rapid Prototyping Journal*, *21*(4), 354–367. doi:10.1108/RPJ-06-2013-0059

Huang, Y., Zhang, J., Hu, X., Song, G., Liu, Z., Yu, L., & Liu, L. (2016). FrameFab: Robotic Fabrication of Frame Shapes. *ACM Trans. Graph. Article*, *3516*(1112), 978–1. doi:10.1145/2980179.2982401

Ishak, I., & Larochelle, P. (2019). MotoMaker: A robot FDM platform for multi-plane and 3D lattice structure printing. *Mechanics Based Design of Structures and Machines*, *47*(6), 703–720. doi:10.1080/15397734.2019.1615943

Jensen, M. L., Mahshid, R., D'Angelo, G., Walther, J. U., Kiewning, M. K., Spangenberg, J., Hansen, H. N., & Pedersen, D. B. (2019). Toolpath strategies for 5DOF and 6DOF extrusion-based additive manufacturing. *Applied Sciences (Switzerland)*, *9*(19), 4168. Advance online publication. doi:10.3390/app9194168

Jeyakymar, M., Munusami, V., Shanmugam, P., Boopalan, N., Boopathi, S., & Sureshkumar, M. (2022). An investigation on wear loss and hardness of nitro-carburizing coated stainless-steel grade-316. *Materials Today: Proceedings*, *66*, 1398–1404. doi:10.1016/j.matpr.2022.05.223

Jiang, J., Xu, X., & Stringer, J. (2019). Optimization of process planning for reducing material waste in extrusion based additive manufacturing. *Robotics and Computer-integrated Manufacturing*, *59*(January), 317–325. doi:10.1016/j.rcim.2019.05.007

Jin, G. Q., Li, W. D., & Gao, L. (2013). An adaptive process planning approach of rapid prototyping and manufacturing. *Robotics and Computer-integrated Manufacturing*, *29*(1), 23–38. doi:10.1016/j.rcim.2012.07.001

Jin, Y., an, He, Y., & Fu, J. (2014). Optimization of tool-path generation for material extrusion-based additive manufacturing technology. *Additive Manufacturing*, *1*, 32–47. doi:10.1016/j.addma.2014.08.004

Jin, Y., He, Y., Fu, G., Zhang, A., & Du, J. (2017). A non-retraction path planning approach for extrusion-based additive manufacturing. *Robotics and Computer-Integrated Manufacturing, 48*(August), 132–144. doi:10.1016/j.rcim.2017.03.008

Kannan, E., Trabelsi, Y., Boopathi, S., & Alagesan, S. (2022). Influences of cryogenically treated work material on near-dry wire-cut electrical discharge machining process. *Surface Topography: Metrology and Properties*, *10*(1), 015027. Advance online publication. doi:10.1088/2051-672X/ac53e1

Kavitha, C., Geetha Malini, P. S., Charan Kantumuchu, V., Manoj Kumar, N., Verma, A., & Boopathi, S. (2022). An experimental study on the hardness and wear rate of carbonitride coated stainless steel. *Materials Today: Proceedings*. Advance online publication. doi:10.1016/j.matpr.2022.09.524

Keating, S., & Oxman, N. (2013). Compound fabrication: A multi-functional robotic platform for digital design and fabrication. *Robotics and Computer-integrated Manufacturing*, *29*(6), 439–448. doi:10.1016/j.rcim.2013.05.001

Kubalak, J. R., Wicks, A. L., & Williams, C. B. (2018). Using multi-axis material extrusion to improve mechanical properties through surface reinforcement. *Virtual and Physical Prototyping*, *13*(1), 32–38. doi:10.1080/17452759.2017.1392686

Kumar, P., Sampath, B., Kumar, S., Babu, B. H., & Ahalya, N. (2023). Hydroponics, Aeroponics, and Aquaponics Technologies in Modern Agricultural Cultivation. In Trends, Paradigms, and Advances in Mechatronics Engineering (pp. 223–241). IGI Global.

Lee, K., & Jee, H. (2015). Slicing algorithms for multi-axis 3-D metal printing of overhangs. *Journal of Mechanical Science and Technology*, *29*(12), 5139–5144. doi:10.100712206-015-1113-y

Li, L., Haghighi, A., & Yang, Y. (2018). A novel 6-axis hybrid additive-subtractive manufacturing process: Design and case studies. *Journal of Manufacturing Processes*, *33*(March), 150–160. doi:10.1016/j.jmapro.2018.05.008

Lim, S., Buswell, R. A., Valentine, P. J., Piker, D., Austin, S. A., & De Kestelier, X. (2016). Modelling curved-layered printing paths for fabricating large-scale construction components. *Additive Manufacturing*, *12*, 216–230. doi:10.1016/j.addma.2016.06.004

Myilsamy, S., & Sampath, B. (2021). Experimental comparison of near-dry and cryogenically cooled near-dry machining in wire-cut electrical discharge machining processes. *Surface Topography: Metrology and Properties*, *9*(3), 035015. doi:10.1088/2051-672X/ac15e0

IEEE Robotics and Automation Society, National Science Foundation (U.S.), & Institute of Electrical and Electronics Engineers. (2019). *2019 IEEE 15th International Conference on Automation Science and Engineering (CASE)*. Author.

Sampath, B., Naveenkumar, P., & Sampathkumar, P. S. A. V. (. (2022). Experimental comparative study of banana fiber composite with glass fiber composite material using Taguchi method. *Elsevier: Materials Today: Proceedings*, *49*(5), 1475–1480. doi:10.1016/j.matpr.2021.07.232

Sampath, B., & Myilsamy, S. (2021). Experimental investigation of a cryogenically cooled oxygen-mist near-dry wire-cut electrical discharge machining process. *Strojniski Vestnik. Jixie Gongcheng Xuebao*, *67*(6), 322–330. doi:10.5545v-jme.2021.7161

Saravanan, M., Vasanth, M., Boopathi, S., Sureshkumar, M., & Haribalaji, V. (2022). Optimization of Quench Polish Quench (QPQ) Coating Process Using Taguchi Method. *Key Engineering Materials*, *935*, 83–91. doi:10.4028/p-z569vy

Song, X., Pan, Y., & Chen, Y. (2015). Development of a low-cost parallel kinematic machine for multi-directional additive manufacturing. *Journal of Manufacturing Science and Engineering*, *137*(2), 021005. Advance online publication. doi:10.1115/1.4028897

Trojovsky, P., Dhasarathan, V., & Boopathi, S. (2023). Experimental investigations on cryogenic friction-stir welding of similar ZE42 magnesium alloys. *Alexandria Engineering Journal*, *66*, 1–14. doi:10.1016/j.aej.2022.12.007

Yuan, P. F., Meng, H., Yu, L., & Zhang, L. (2016). Robotic Fabrication in Architecture, Art and Design 2016. *Robotic Fabrication in Architecture, Art and Design*. Advance online publication. doi:10.1007/978-3-319-26378-6

Yupapin, P., Trabelsi, Y., Nattappan, A., & Boopathi, S. (2022a). Performance Improvement of Wire-Cut Electrical Discharge Machining Process Using Cryogenically Treated Super-Conductive State of Monel-K500 Alloy. *Iranian Journal of Science and Technology. Transaction of Mechanical Engineering*. Advance online publication. doi:10.100740997-022-00513-0

Yupapin, P., Trabelsi, Y., Nattappan, A., & Boopathi, S. (2022b). Performance Improvement of Wire-Cut Electrical Discharge Machining Process Using Cryogenically Treated Super-Conductive State of Monel-K500 Alloy. *Iranian Journal of Science and Technology - Transactions of Mechanical Engineering*, 1–17. doi:10.1007/s40997-022-00513-0

Zhao, H., He, Y., Fu, J., & Qiu, J. (2017, November). Inclined layer printing for fused deposition modeling without assisted supporting structure. *Robotics and Computer-integrated Manufacturing*, *51*, 1–13. doi:10.1016/j.rcim.2017.11.011

Chapter 12
Applications of Fused Deposition Modeling in Dentistry

Rayappa Shrinivas Mahale
https://orcid.org/0000-0002-0508-9871
REVA University, India & Jain College of Engineering and Research, India

Gangadhar M. Kanaginahal
Jain College of Engineering and Research, India

Shamanth Vasanth
REVA University, India

Vivek Kumar Tiwary
KLS Gogte Institute of Technology, India

Rajendrachari Shashanka
Bartin University, Turkey

Sharath P. C.
Jain University, India

Adarsh Patil
https://orcid.org/0000-0002-6917-6245
KLE Technological University, India

ABSTRACT

Fused deposition modelling (FDM) is a popular additive manufacturing (AM) technique for modelling, prototyping, and production. FDM is a technology that creates three-dimensional things directly from three-dimensional CAD data. Layer by layer, thermoplastic material is extruded by a temperature-controlled head. FDM, also known as fused filament fabrication (FFF), is a simple and low-cost method of additive manufacturing that was first introduced in 1989. A thermoplastic filament is fed to a heated

DOI: 10.4018/978-1-6684-6009-2.ch012

nozzle in the FFF process. The material is melted here, and the material is deposited as the nozzle travels layer by layer in the x and y axes along the geometry. FDM has proved beneficial in the medical field to produce more naturalistic models for educational, training, and research reasons, as well as treatment and surgical planning.

INTRODUCTION

In recent years, the introduction of digital technologies in dentistry has represented a significant breakthrough. In the last years, digital technologies were imbibed in therapeutic and diagnostic applications, thus bringing a revolution in medicine and dentistry (Cousley, 2020). An integrated combination of computer-aided design and computer-aided manufacturing (CAD/CAM) is AM, where the available digital data replaces the existing materials (Scribante et al., 2022). AM follows three steps, first a scanner is used to convert the part into a digital data, second the acquired data has to be edited utilizing a software and finally the transformation of the available data into a final product (Brown et al., 2018). AM is such a technology that fabricates 3D objects by depositing the fused material in form of films or layers, one over the other in a successive manner, hence the name additive, whether the material is metal, concrete, exclusive materials such as ceramics, carbon fiber, thermoplastic or even human tissues or parts in dentistry (Kenett et al., 2019) (Rayappa Shrinivas Mahale, V. Shamanth, K. Hemanth et al., 2022).

One of the hottest topics in current trend is 3D printing technique nowadays it is known as FDM/FFF technology (Chia & Wu, 2015). Thermoplastic materials are used in technology to create solid items. Thermoplastic materials are heated till melting temperature is reached, at which point the printer deposits the melted material in a predetermined pattern using extrusion heads to build an object layer by layer. Layers of the substance are put one after the other such that they fuse together and later it solidifies. The main benefit of FDM is that it allows for printing with a minimum layer thickness of 127 μm and post-processing of fabricated parts is not necessary (Liu et al., 2006).

One of the major drawbacks of FDM is the shrinking of material as it solidifies and hardens, thus the material utilized for fabrication must have thermal qualities and viscoplastic. Its appeal as opposed to other 3D printing techniques can be attributed to its instant usability (Isobe et al., 2018) and lack of additional equipment (such as moulds, kiln, or tools), which results in minimal costs for equipment and procedures (Sood et al., 2010) (Sodeifian et al., 2019) (Krajangsawasdi et al., 2021).

FUSED DEPOSITION MODELLING

FDM, is the process of creating solid things out of a thermoplastic substance (Bartkowiak & Walkowiak-Śliziuk, 2018). In this process the material is deposited in a defined thickness layer, from the point of melting the material followed by passing it through a extrusion head in a predetermined manner (Chia & Wu, 2015). This causes the material to be discharged layer by layer, and as the structure hardens, these strata are bonded together. There are typically no further processes needed on the substance itself to enhance mechanical qualities apart from support removal (Liu et al., 2006). Among the most often used processes for printing three-dimensional (3D) objects) is fused deposition modelling (FDM) (Zhu et al., 2017). This technology makes it possible to produce moulds, prototypes on demand, and parts with intricate shapes (Shofner et al., 2002). Due to its minimal production time, negligible waste and developing intricate shapes, FDM is recognizing as one of the quickly developing manufacturing technologies (J. Wang et al., 2018). FDM products' subpar surface smoothness and questionable dimensional accuracy call into question their suitability for a variety of applications. The surface texturing of FDM parts has been improved by several researchers (Singh et al., 2017). The group's researchers concluded that layering thickness is indeed the factor most crucial to the exterior roughness of 3D printed parts (Lužanin et al., 2013).

S. Scott Crump, a co-founder of Stratasys, invented FDM for the first time in 1988. Making a 3D CAD design and converting in form of standard geometrical file (usually an STL file) are the first steps in the FDM process (Magdum et al., 2008). These machine directives (position, temperature, etc.) are then encoded in a known format G-code, which is decoded by the FDM machine and "sliced" into a sequence of layers utilizing one of the numerous open-source or specialized slicing programs (Ferreira et al., 2017) (Cantrell et al., 2017). A thermoplastic feed is introduced into the machine and heated in a melt pool until it melts above the transition temperature for glass (T_g), at which point it fine-tips extruded out via a fine-tipped nozzle (Ivey et al., 2017) (Geng et al., 2019) (Torrado Perez et al., 2014).

To create a layer, the deposited filaments—also referred to as rasters, road lines, or beads—are arranged adjacent to one another in the horizontal plane. The materials are layered on top of the prior grid in a thorough direction while the first level is constructed on the printed bed. Computer numeric controls (CNC) determine how the ejection head and printing bed will move about one another (Ngo et al., 2018). After cooling, the adjoining placed rasters fuse again and solidify (Onwubolu & Rayegani, 2014). The shrinkage or expansion of the printing materials during curing, could account for the growing and lowering tendencies of the created models (Suparyanto dan Rosad (2015, 2020) (Brown et al., 2018). There are few studies comparing the precision of 3D dentistry models generated using the FDM technology to printing methods that are more expensive (Jaber et al., 2021). In comparison to traditional model-shaping techniques, Fused Deposition Modeling (FDM) technology sought to increase success rates, decrease labour and material costs, and verifying the precision of process comparing to conventional

Figure 1. The FDM 3D printer [24]

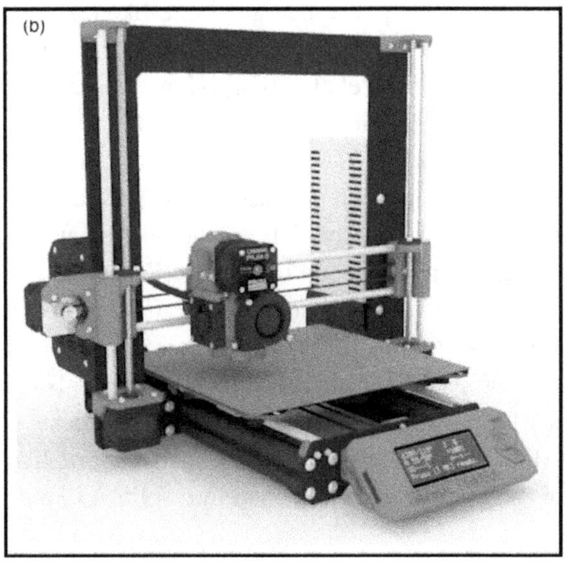

Figure 2. Original work carried out at center of excellence in additive manufacturing and reverse engineering lab at KLS Gogte Institute of Technology Belagavi

plaster models. Finally, it sought to determine whether traditional methods could be replaced by FDM technology (S. Wang et al., 2017). Fig. 1. is FDM 3D printer that will move in X, Y and Z directions to fabricate a complete material.

FDM IN DENTISTRY

There is a large selection of feedstock materials available to FFF. The most common semi-crystalline thermoplastic, polylactic acid (PLA), and the second group of amorphous polymers, such as (1) polycarbonates (PC), (2) acrylonitrile-butadiene styrene copolymers (ABS), (3) water-soluble materials for support structures, and (4) polyethylene terephthalate glycol (PETG), such as polyvinyl alcohol, are the standard materials (PVA). Polyphenylsulfone (PPSU) filaments, which are approved for brief contact with the skin and bodily fluids, are accessible for medical application. Fig. 2 shows the fabrication of typical teeth set with cavities that will processed further. The images have been chosen for a particular patient and were developed including the cavities (Tigmeanu et al., 2022) (Chen et al., 2016). A biocompatible filament produced from a mixture of polyamide-polyolefin and cellulose (PAPC) is used to make an operative guide for oral implant placement and bone-like replicas. Additionally, FFF printers are being tested to produce polyetherketon materials like PEEK for application in craniomaxillofacial implants (Honigmann et al., 2018) (Sharma et al., 2020) (Han et al., 2019).

Other filaments, such as PETG, poly (lactic-co-glycolic acid) (PLGA), and Polycaprocalactone (PCL) filaments, are approved as class 1 medical kit, which are used for minimal amount of time in body due their low degree of invasiveness. FFF is currently used most frequently in dentistry to create dental models using intraoral images (Figure 4). FFF models are thought to have poorer precision and trueness, they nonetheless satisfy the needs of many dental industries (Kim et al., 2018) (Etemad-Shahidi et al., 2020). For instance, the accuracy needed for orthodontic models is lower than that needed for models used to manufacture prosthetic restorations. FFF is economically appealing due to the pricing of the materials and equipment as well as its practically ready-to-utilize models without post-processing. Even though there was no necessity for post-processing steps and material costs were cut by up to 80%, printing times were equivalent to those of SLA systems. Additionally, alternative materials can be utilized to create dental models, which can then be thermoformed into splints, like gypsum-containing filaments (Lüchtenborg et al., 2021).

Figure 3. Original work carried out at center of excellence in additive manufacturing and reverse engineering lab at KLS Gogte Institute of Technology Belagavi

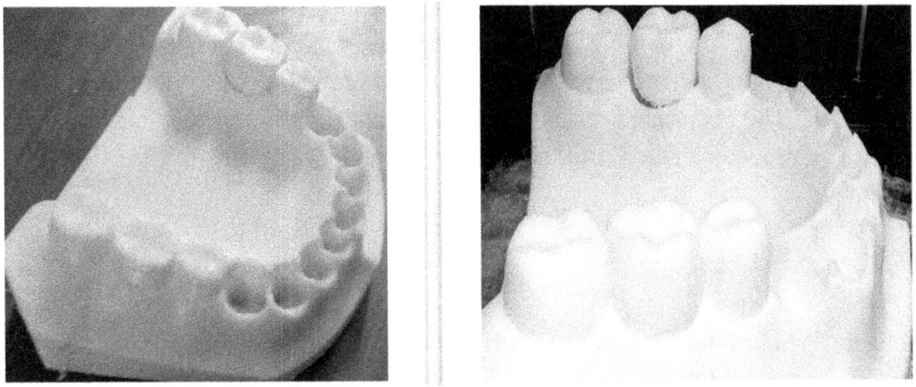

The Invisalign System employs computerized realignment to progressively shift the participant's teeth into the ideal position. Fig. 3 shows the extended work of fabrication where the tooth was fitted in the cavities created as per the patient. It will send the patient a newly developed set of aligners regularly every two weeks. This technique, which also permits the digital storing and copying of client data sets while decreasing the requirement for physical storage, will save time (Dawood et al., 2015). With the use of CAD/CAM technology, the manufacture of brackets and their placement are two separate procedures that are integrated into one. This method satisfies the desire for maximum distinctiveness with the least amount of space restrictions (Nayar et al., 2015). A comparison of machining processes and 3D printing to produce orthodontic equipment was done in the study. Six individuals with malocclusion were obligated to wear milled and 3D printed aligners. According to the study's results, there are a number of aspects to take into account while choosing the most visible and removable orthodontic aligner, including the staircase effect, system precision, layering thickness, treatment duration, and transparent of the aligners. Milling had the disadvantage of requiring more machine time when compared to 3D printing, but 3D printing is efficient with little to no human participation and creates models quickly (Martorelli et al., 2013) (A. & G., 2019).

USAGE OF PEEK AS ONE OF THE COMPOSITIONS

PEEK offers a balanced distribution in terms of tooth structure and stress transmission on abutment teeth. The cementation interface was generated as a result of the elastic modulus closer to the dentine and its deformability. The layer thickness had little effect on parameters such as print temperature, direction, and angle, although the tensile strength improved but the compressive and flexural strength were less impactful. If the filling rate is slower, the layer created will be thinner, which has an effect on mechanical properties. Smaller thicknesses do not allow for vacancies, which improves bonding between interlayers and increases contact between the surfaces of PEEK filaments, indirectly boosting mechanical qualities. An increase in nozzle diameter creates a layer of homogeneous structure, which increases compression strength and modulus. The print temperature controls filament fusion and interlayer adhesion, resulting in material plasticity. As the temperature was elevated, the rate of crystallisation accelerated, increasing the tensile strength of PEEK (Moby et al., 2022) (Tian et al., 2021) [44][45].

CASE STUDY

The 3D printing of a novel bioink Alg-Dent employs a printable alginate hydrogel with adjustable quantities of dentin matrix. The high quantities of alginate and insoluble dentin matrix improve cell viability. The inclusion of the dentin matrix increased the differentiating ability of odontogenic, resulting in an excellent biomaterial. The 3D printed resins filled the C-shaped channels that had already been created with an emphasis on root canal fills (Pillai et al., 2021).

CONCLUSION

FDM in conclusion is a layer-by-layer manufacturing, enables the quick fabrication of complicated geometries, where the component quality greatly resides on the precised configuration of material employed and print settings. The thermoplastics used in the standard FDM technique have lower mechanical performance than thermosetting materials or metals, hence tidy thermoplastic FDM items are only suitable for low-performance applications. PEEK provides a custom made dental restoration through a process of 3D printing with mechanical properties sustainable till high temperatures of 235°C also managing the surface roughness suitable for dental applications.

REFERENCES

A., J., & G., M. (2019). Applications of 3D printing in dentistry - A review. *Journal of Pharmaceutical Sciences and Research, 11*(5), 1670–1675.

Bartkowiak, T., & Walkowiak-Śliziuk, A. (2018). 3D printing technology in orthodontics – Review of current applications. *Journal of Stomatology*, *71*(4), 356–364. doi:10.5114/jos.2018.83410

Brown, G. B., Currier, G. F., Kadioglu, O., & Kierl, J. P. (2018). Accuracy of 3-dimensional printed dental models reconstructed from digital intraoral impressions. *American Journal of Orthodontics and Dentofacial Orthopedics*, *154*(5), 733–739. doi:10.1016/j.ajodo.2018.06.009 PMID:30384944

Cantrell, J. T., Rohde, S., Damiani, D., Gurnani, R., DiSandro, L., Anton, J., Young, A., Jerez, A., Steinbach, D., Kroese, C., & Ifju, P. G. (2017). Experimental characterization of the mechanical properties of 3D-printed ABS and polycarbonate parts. *Rapid Prototyping Journal*, *23*(4), 811–824. doi:10.1108/RPJ-03-2016-0042

Chen, H., Yang, X., Chen, L., Wang, Y., & Sun, Y. (2016). Application of FDM three-dimensional printing technology in the digital manufacture of custom edentulous mandible trays. *Scientific Reports*, *6*(January), 1–6. doi:10.1038rep19207 PMID:26763620

Chia, H. N., & Wu, B. M. (2015). Recent advances in 3D printing of biomaterials. *Journal of Biological Engineering*, *9*(1), 1–14. doi:10.118613036-015-0001-4 PMID:25866560

Cousley, R. R. (2020). Introducing 3D printing in your orthodontic practice. *Journal of Orthodontics*, *47*(3), 265–272. doi:10.1177/1465312520936704 PMID:32627658

Dawood, A., Marti, B. M., Sauret-Jackson, V., & Darwood, A. (2015). 3D printing in dentistry. *British Dental Journal*, *219*(11), 521–529. doi:10.1038j.bdj.2015.914 PMID:26657435

Etemad-Shahidi, Y., Qallandar, O. B., Evenden, J., Alifui-Segbaya, F., & Ahmed, K. E. (2020). Accuracy of 3-dimensionally printed full-arch dental models: A systematic review. *Journal of Clinical Medicine*, *9*(10), 1–18. doi:10.3390/jcm9103357 PMID:33092047

Ferreira, R. T. L., Amatte, I. C., Dutra, T. A., & Bürger, D. (2017). Experimental characterization and micrography of 3D printed PLA and PLA reinforced with short carbon fibers. *Composites. Part B, Engineering*, *124*, 88–100. doi:10.1016/j.compositesb.2017.05.013

Geng, P., Zhao, J., Wu, W., Ye, W., Wang, Y., Wang, S., & Zhang, S. (2019). Effects of extrusion speed and printing speed on the 3D printing stability of extruded PEEK filament. *Journal of Manufacturing Processes, 37*(December), 266–273. doi:10.1016/j.jmapro.2018.11.023

Han, X., Sharma, N., Xu, Z., Scheideler, L., Geis-Gerstorfer, J., Rupp, F., Thieringer, F. M., & Spintzyk, S. (2019). An in vitro study of osteoblast response on fused-filament fabrication 3D printed PEEK for dental and Cranio-Maxillofacial implants. *Journal of Clinical Medicine*, *8*(6), 771. Advance online publication. doi:10.3390/jcm8060771 PMID:31159171

Honigmann, P., Sharma, N., Okolo, B., Popp, U., Msallem, B., & Thieringer, F. M. (2018). Patient-specific surgical implants made of 3D printed PEEK: Material, technology, and scope of surgical application. *BioMed Research International*, *2018*, 1–8. Advance online publication. doi:10.1155/2018/4520636 PMID:29713642

Isobe, T., Tanaka, T., Nomura, T., & Yuasa, R. (2018). Comparison of strength of 3D printing objects using short fiber and continuous long fiber. *IOP Conference Series. Materials Science and Engineering*, *406*(1), 012042. Advance online publication. doi:10.1088/1757-899X/406/1/012042

Ivey, M., Melenka, G. W., Carey, J. P., & Ayranci, C. (2017). Characterizing short-fiber-reinforced composites produced using additive manufacturing. *Advanced Manufacturing: Polymer and Composites Science*, *3*(3), 81–91. doi:10.1080/20550340.2017.1341125

Jaber, S. T., Hajeer, M. Y., Khattab, T. Z., & Mahaini, L. (2021). Evaluation of the fused deposition modeling and the digital light processing techniques in terms of dimensional accuracy of printing dental models used for the fabrication of clear aligners. *Clinical and Experimental Dental Research*, *7*(4), 591–600. doi:10.1002/cre2.366 PMID:33258297

Kenett, R. S., Swarz, R. S., & Zonnenshain, A. (2019). *Systems Engineering in the Fourth Industrial Revolution: Big Data*. Novel Technologies, and Modern Systems Engineering. doi:10.1002/9781119513957

Kim, S. Y., Shin, Y. S., Jung, H. D., Hwang, C. J., Baik, H. S., & Cha, J. Y. (2018). Precision and trueness of dental models manufactured with different 3-dimensional printing techniques. *American Journal of Orthodontics and Dentofacial Orthopedics*, *153*(1), 144–153. doi:10.1016/j.ajodo.2017.05.025 PMID:29287640

Krajangsawasdi, N., Blok, L. G., Hamerton, I., & Longana, M. L. (2021). *Fused Deposition Modelling of Fibre Reinforced Polymer Composites: A Parametric Review.* Academic Press.

Liu, Q., Leu, M. C., & Schmitt, S. M. (2006). Rapid prototyping in dentistry: Technology and application. *International Journal of Advanced Manufacturing Technology*, *29*(3–4), 317–335. doi:10.100700170-005-2523-2

Lüchtenborg, J., Burkhardt, F., Nold, J., Rothlauf, S., Wesemann, C., Pieralli, S., Wemken, G., Witkowski, S., & Spies, B. C. (2021). Implementation of fused filament fabrication in dentistry. *Applied Sciences (Switzerland)*, *11*(14), 6444. Advance online publication. doi:10.3390/app11146444

Lužanin, O., Movrin, D., & Plan, M. (2013). Experimental Investigation of Extrusion Speed and Temperature Effects on Arithmetic Mean Surface Roughness in Fdm-. *Journal of Technology of Plasticity*, *38*(2), 179–190.

Magdum, Y., Pandey, D., Bankar, A., Harshe, S., Parab, V., Mahesh, M., & Kadam, S. (2008). Process Parameter Optimization for FDM 3D Printer. *International Research Journal of Engineering and Technology*, 1472.

Martorelli, M., Gerbino, S., Giudice, M., & Ausiello, P. (2013). A comparison between customized clear and removable orthodontic appliances manufactured using RP and CNC techniques. *Dental Materials*, *29*(2), e1–e10. doi:10.1016/j.dental.2012.10.011 PMID:23140842

Moby, V., Dupagne, L., Fouquet, V., Attal, J. P., François, P., & Dursun, E. (2022). Mechanical Properties of Fused Deposition Modeling of Polyetheretherketone (PEEK) and Interest for Dental Restorations: A Systematic Review. *Materials (Basel)*, *15*(19), 6801. Advance online publication. doi:10.3390/ma15196801 PMID:36234139

Nayar, S., Bhuminathan, S., & Bhat, W. (2015). Rapid prototyping and stereolithography in dentistry. *Journal of Pharmacy & Bioallied Sciences*, *7*(April), S216–S219. doi:10.4103/0975-7406.155913 PMID:26015715

Ngo, T. D., Kashani, A., Imbalzano, G., Nguyen, K. T. Q., & Hui, D. (2018). Additive manufacturing (3D printing): A review of materials, methods, applications and challenges. *Composites Part B: Engineering, 143*(December), 172–196. doi:10.1016/j.compositesb.2018.02.012

Onwubolu, G. C., & Rayegani, F. (2014). Characterization and Optimization of Mechanical Properties of ABS Parts Manufactured by the Fused Deposition Modelling Process. *International Journal of Manufacturing Engineering*, *2014*, 1–13. doi:10.1155/2014/598531

Pillai, S., Upadhyay, A., Khayambashi, P., Farooq, I., Sabri, H., Tarar, M., Lee, K. T., Harb, I., Zhou, S., Wang, Y., & Tran, S. D. (2021). Dental 3d-printing: Transferring art from the laboratories to the clinics. *Polymers*, *13*(1), 1–25. doi:10.3390/polym13010157 PMID:33406617

Rayappa Shrinivas Mahale, V., Shamanth, K., Hemanth, S. K. N., Sharath, P. C., Shashanka, R., Patil, A., & Shetty, D. (2022). Processes and applications of metal additive manufacturing. *Materials Today: Proceedings*, *54*(2), 228–233. doi:10.1016/j.matpr.2021.08.298

Scribante, A., Gallo, S., Pascadopoli, M., Canzi, P., Marconi, S., Montasser, M. A., Bressani, D., Gandini, P., & Sfondrini, M. F. (2022). Properties of CAD/CAM 3D Printing Dental Materials and Their Clinical Applications in Orthodontics: Where Are We Now? *Applied Sciences (Switzerland)*, *12*(2), 551. Advance online publication. doi:10.3390/app12020551

Sharma, N., & Honigmann, P. (2020). Dimensional characteristics of FDM 3D printed PEEK implant for craniofacial reconstructions. *Journals.Infinite-Science.De, 2*(1), 2–3. doi:10.18416/AMMM.2020.2009011

Shofner, M. L., Lozano, K., & Rodrı, F. J. (2002). Nanofiber-Reinforced Polymers Prepared by Fused Deposition Modeling. *Journal of Applied Polymer Science*, *89*(11), 3081–3090. doi:10.1002/app.12496

Singh, R., Singh, S., Singh, I. P., Fabbrocino, F., & Fraternali, F. (2017). Investigation for surface finish improvement of FDM parts by vapor smoothing process. *Composites. Part B, Engineering*, *111*, 228–234. doi:10.1016/j.compositesb.2016.11.062

Sodeifian, G., Ghaseminejad, S., & Yousefi, A. A. (2019). Preparation of polypropylene/short glass fiber composite as Fused Deposition Modeling (FDM) filament. *Results in Physics, 12*(November), 205–222. doi:10.1016/j.rinp.2018.11.065

Sood, A. K., Ohdar, R. K., & Mahapatra, S. S. (2010). Parametric appraisal of mechanical property of fused deposition modelling processed parts. *Materials & Design*, *31*(1), 287–295. doi:10.1016/j.matdes.2009.06.016

Suparyanto Dan Rosad (2015). Accuracy of Stereolithographic Models of Human Anatomy. *Suparyanto Dan Rosad*, *5*(3), 248–253.

Tian, Y., Chen, C., Xu, X., Wang, J., Hou, X., Li, K., Lu, X., Shi, H., Lee, E., & Jiang, H. B. (2021). *A Review of 3D Printing in Dentistry: Technologies, Affecting Factors, and Applications*. Academic Press.

Tigmeanu, C. V., Ardelean, L. C., Rusu, L. C., & Negrutiu, M. L. (2022). Additive Manufactured Polymers in Dentistry, Current State-of-the-Art and Future Perspectives-A Review. *Polymers*, *14*(17), 1–24. doi:10.3390/polym14173658 PMID:36080732

Torrado Perez, A. R., Roberson, D. A., & Wicker, R. B. (2014). Fracture surface analysis of 3D-printed tensile specimens of novel ABS-based materials. *Journal of Failure Analysis and Prevention*, *14*(3), 343–353. doi:10.100711668-014-9803-9

Wang, J., Li, H., Liu, R., Li, L., Lin, Y. H., & Nan, C. W. (2018). Thermoelectric and mechanical properties of PLA/Bi0·5Sb1·5Te3 composite wires used for 3D printing. *Composites Science and Technology*, *157*, 1–9. doi:10.1016/j.compscitech.2018.01.013

Wang, S., Li, M., Yang, H., Liu, Y., & Sun, Y. (2017). Application of computer aided design and fused deposition modeling technology in the digital manufacture of orthodontic study models. *Biomedical Research (India)*, *28*(10), 4425–4431.

Zhu, J., Hu, Y., Tang, Y., & Wang, B. (2017). Effects of styrene–acrylonitrile contents on the properties of ABS/SAN blends for fused deposition modeling. *Journal of Applied Polymer Science*, *134*(7), 1–5. doi:10.1002/app.44477

Chapter 13
Development of High Performance Polymer Composites by Additive Manufacturing

Pradeep Kumar G. S.
CHRIST University (Deemed), India

Sachith T. S.
Symbiosis Institute of Technology, Symbiosis International University (Deemed), India

Sasidhar Jangam
CHRIST University (Deemed), India

Shivakumar S.
CHRIST University (Deemed), India

Gurumoorthy S. Hebbar
CHRIST University (Deemed), India

ABSTRACT

Additive manufacturing, often known as 3D printing, is a process that involves the manufacture of physical things in a manner that is similar to building them up layer by layer. As a result of its utilization of automated procedures to generate complicated three-dimensional forms, which are either difficult or impossible to produce using conventional methods, fused deposition modeling is a typical approach that uses heat to assist in the extrusion process. The term "high-performance polymer" refers to a group of polymer materials that are known to maintain their desirable mechanical, thermal, and chemical properties when subjected to harsh environments such as high temperatures, high pressures, and corrosive chemicals. This chapter starts off with a brief introduction of conductive and high-performance polymer composites, followed by a rundown of how these materials are utilized in the 3D printing process.

DOI: 10.4018/978-1-6684-6009-2.ch013

INTRODUCTION

The process of producing goods is being revolutionized by the additive manufacturing (AM) technique. Inkjet printing introduced the idea of additive manufacturing processes in the 1970s, but it wasn't until the 1980s that stereo lithography, a method for printing materials rather than ink, became widely used. Charles Hull's invention of stereo lithography allowed for 3D printing, which helped firms reduce time, cost, and resources. Since then, high-performing polymer nanocomposites have been printed using novel materials and additive manufacturing technologies. Although in case of any confusion, this was first required to distinguish between the phrases "additive manufacturing," "rapid prototyping," and "3D printing" before continuing on to the rest of the paper. Background information about this industry's present state and anticipated future expansion is also given (Infographic: The History of 3D Printing, 2016).

Additive manufacturing describes the procedure of layer-by-layer addition and connecting of materials to create a component. On the contrary, subtractive manufacturing involves removing matter from bulk in order to shape the expected results. The capacity of additive manufacturing to accurately construct more complicated structures than subtractive manufacturing is a key benefit (J.Stansbury et al, 2015). Rapid prototyping, on the other side, is "a method for quickly producing systems or component representation preceding final launch or commercialization (I. Gibson et al, 2015)," according to the definition. In other terms, rapid prototyping can be viewed as a particular kind of additive manufacturing process. additive manufacturing and 3D printing are being used alternately. And although additive manufacturing is frequently seen as a much more professional term, they have the same meaning.

Additive manufacturing is expected to keep expanding in the coming decades. According to the International Data Corporation, with the compound annual growth of 26.50 percent, "the almost 11 billion dollars' sector in 2014 will inflate to 26.70 billion dollars by 2018". West Europe, Asia, and the U.S.A are the countries that are fueling this increase (Worldwide Spending on 3D Printing to Grow from Nearly $11 Billion to $26.7 Billion by 2019, 2016). In addition, this long-established method that has typically been employed for rapid prototyping can now model consisting, of finished components (A Third Industrial Revolution, 2016) in addition to the numerous purposes it now fulfills (as shown in Figure 1) (Roundup of 3D Printing Market Forecasts and Estimates, 2016). Industrial equipment, consumables, and the automobile industries, particularly passenger vehicles, are the two main sectors of additive manufacturing as of 2015 (Figure 2) (Manufacturing Our 3D Future, 2016). The National Aeronautic and Space Administration, the European Space Administration, and the China space administrations were listed as important stakeholders in the developing markets for manufacturing body tissues. Regarding their LEAP fighter jet in specifically, General Electronic Aviation has created a 3D printed fueling nozzle. Although National Aeronautic and Space Administration, in partnership with Manufactured in Space, has indeed launched a 3D printing machine to the International Space Station (Additive Manufacturing State of the Industry, 2016), they assert that this component is "approximately to twenty-five percent lighter and much more complicated than any of its counterparts and integrates within one component what was numerous combined components in the previous (GE Aviation: Additive Manufacturing, 2016). The European Space Agency (www.Ptonline.Com/Articles/Additive-Manufacturing-Materials-for-Real-Worldparts2014, 2014) experiences the very same thing. In the meantime, CASTC is working to create a metallic 3D printing machine that might function in space. But it's crucial to note that most of these high-performance 3d printing technology components, particularly those for the automobile and aviation sectors, are composed of metallic elements.

Figure 1. Purpose of 3D Printing

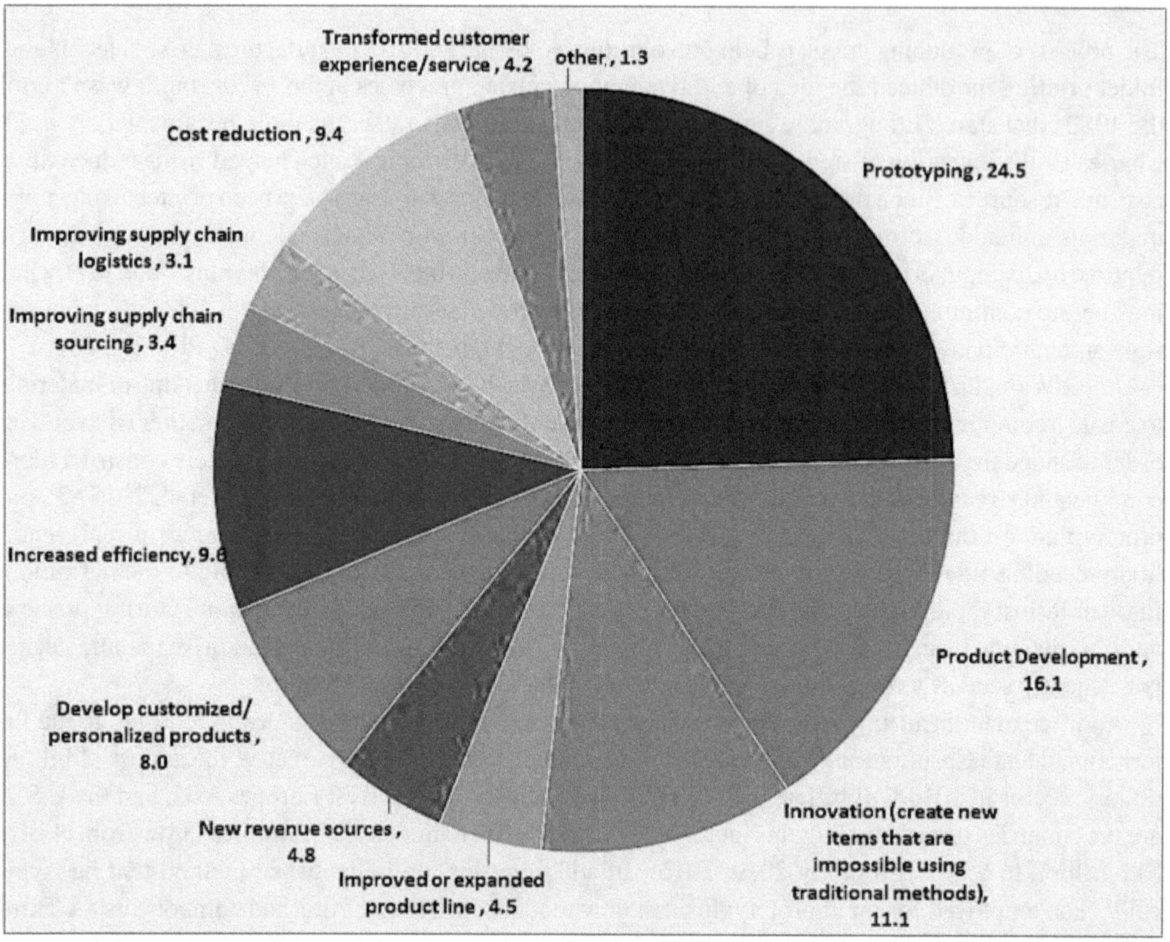

Since they often have greater strengths and melting temperatures than that polymers, metal was the material of choice (Calister et al., 2007). Metal weighs a lot and is typically costlier and difficult to manufacture, which is a drawback (Calister et al., 2007). In order to create goods with higher specific strength and appropriate qualities that can survive large loads or hazardous conditions while employing reasonably simple 3D printing procedures, were developed as a result. These days, poly-phenyl-sulfone and polyetherimide are both used in the petroleum industries (Stratasys et al., 2016). These polymers can be combined with nano-fillers to make nanocomposites for applications that require high significantly superior characteristics (Yin et al., 2013. Kalin et al., 2014. Camargo et al., 2009)

In essence, composite is multi-phase material produced up of a weak dispersion medium termed the matrix and a stronger, frequently stiff reinforcing phase. While at minimum single phase has 1, 2, or 3 of its dimension in the nanoscale range, or one nanometer of a meter, then a nanocomposite is created (Camargo et al., 2009)

The tensile strength, heat resistance, and specific stiffness of nanocomposite may all be improved by the fillers employed in them (Paul et al., 2008). Carbon nanotubes, graphene oxide, and nanoclay are several three of the most common nano-fillers being used in 3D printing applications.

Figure 2. Major Areas of Additive Manufacturing

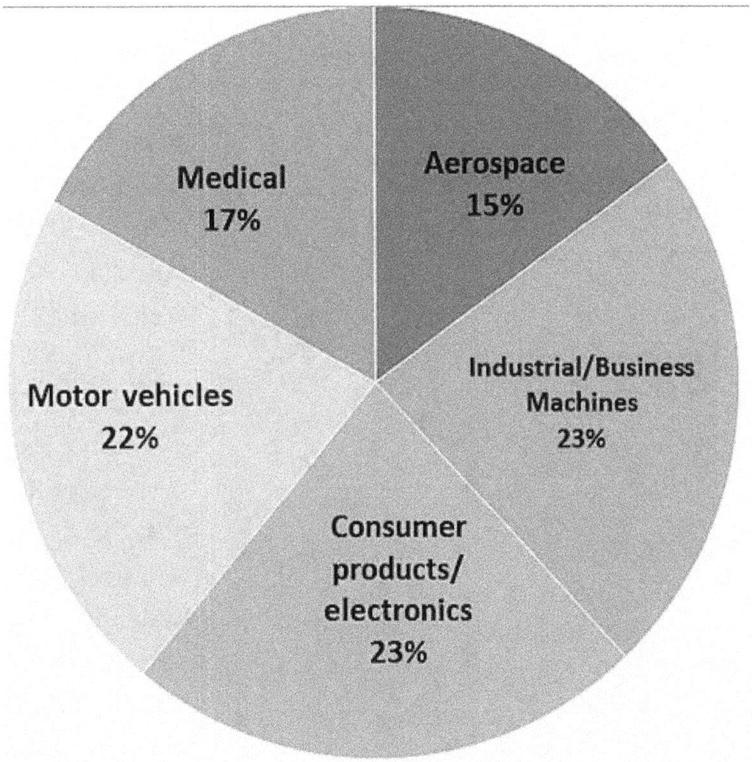

With the wrong additive manufacturing process selected, employing and nanocomposites for manufacturing is useless. Price, the variety of components equipment can manage, reliability, performance, adaptability, and precision are among the factors which have been considered throughout the selection phase. The best seven positions in the printer's distributions as of Feb-2016 (as shown in Figure 3) were taken by fused deposition modeling (FDM) printing machines, giving this the most popular additive manufacturing technology at the moment (3D Hubs and 3D Printing Trends, 2016). In the parts that follow, we'll go into more detail on fused deposition modeling and other printing methods.

The readers would learn about the present situation and the crucial part high-performance polymer nanocomposites will perform in the development of additive manufacturing in this study. This is how the review is laid out: The explanation and evaluation of several 3D printing processes that can create these nanocomposites will be the main topics of the second chapter. The characteristics of different high conductive and performance polymers and nano-fillers which are present and perhaps will be employed in additive manufacturing will be the attention of the third and fourth chapters, accordingly. For comparative purposes, the benefits and drawbacks of every methodology would be underlined. The production, print process, resultant characteristics, and uses of high-performance polymer nanocomposites will be covered in the review's last part.

Figure 3. Printer model distribution

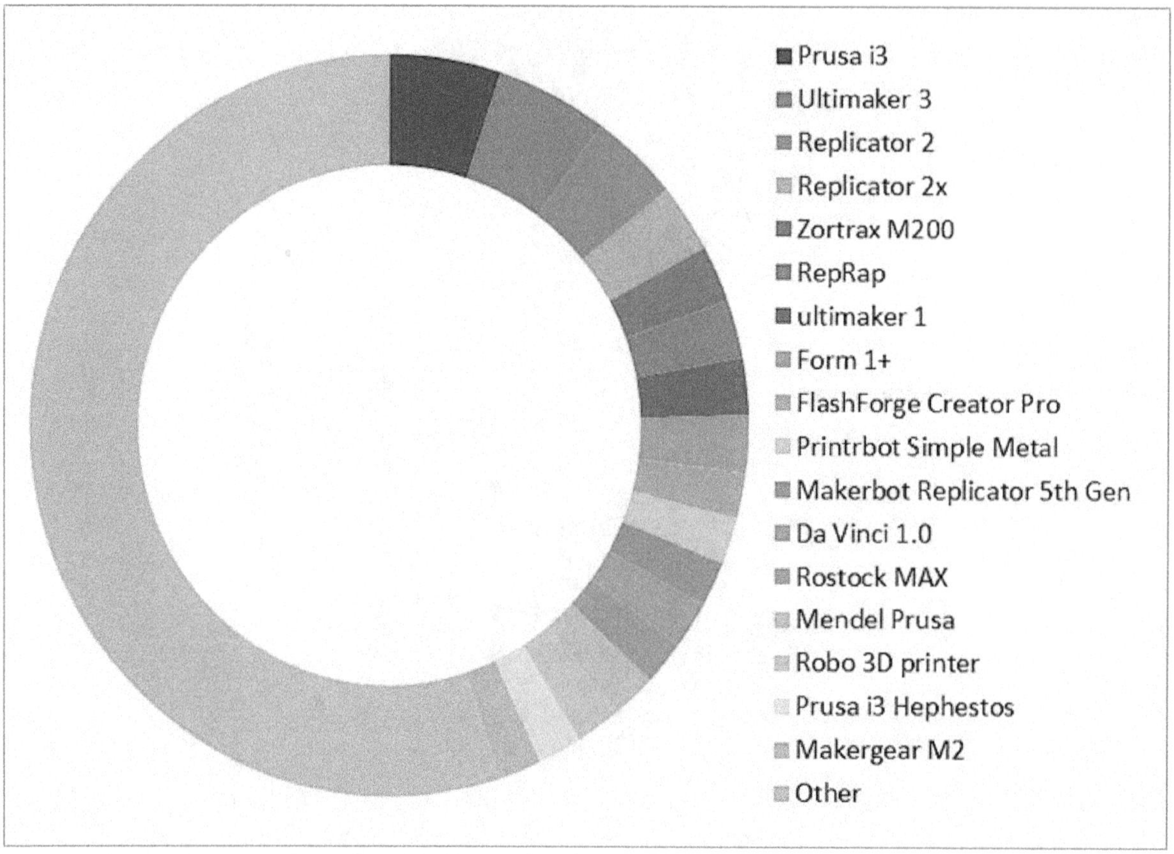

DIFFERENT AM TECHNOLOGIES

Numerous methods for additive manufacturing were created as a result of decades of studies. A typical layered deposition procedure is used by several approaches, in which a two-dimensional structure is initially printed on the x-y plane to create a single-layered structure, then further layers are deposited down the z-axis to create a three-dimensional structure. A global STL file created by a three-dimensional system is frequently utilized to construct the "multilayered" printed construction due to this new global printing method. To design the appropriate three-dimensional structure, a variety of software programs can be utilized, such as AutoCAD, Sketch-Up, and others. After analyzing the design, the programming language divides it into certain layers and converts it to STL format. The associated STL file would then be changed to a format that particular 3d printing technologies can read. The design will be built using the STL file's slice when the suitable file has been loaded into the 3D printer (Iancu et al., 2010. S S Crump et al., 1991. Noorani et al., 2006). While all approaches use this layered deposition procedure, there are significant differences in the printed mechanisms of these technologies, as will be discussed elsewhere here:

Fused Deposition Modeling

The other widely utilized methodology wherein material supply must be in filaments form is fused deposition modeling (FDM). The polymers are injected into the printer body, which is melted and ejected out, similarly to a typical extrusion. Fused deposition modeling protrudes the filament to create thin fiber with a radius of about 0.125 mm, places it across platforms in a two-dimensional pattern, refreezes under melting point, and freezes instantly rather than utilizing a mold and elevated temperatures heating. Then another layer will be added on top of that. Although fused deposition modeling's resolution is somewhat limited due to its reliance on the extrusion fiber's radius, the technology is highly cost-effective due to its straightforward print configuration and mechanisms (Elkins et al., 1997. Pham et al., 2000). Only thermoplastic polymers will be produced using fused deposition modeling because the print requires melting and extrusion. The filaments, on the other hand, are so delicate in incredibly versatile thermoplastics that they cannot withstand the compressive stresses while feeding. This frequently leads to filaments bending and feeder failures (Melchels et al., 2010).

Higher melting temperatures and big molecular mass are typical characteristics of high-performance polymers. They are renowned for their exceptional tensile strength and thermal resilience, which enables the melt of filaments challenging and requires rather more compressive and shear forces during the extruding process due to the higher viscosity of melted plastic. High-performance polymers, such as Poly-phenyl-sulfone (PPSU), Poly-ether-ether-ketone (PEEK), Poly-aryl-ether-ketone (PAEK), and others, have been successfully built with fused deposition modeling by a number of businesses.

SLA Technology

Among the most often employed additive manufacturing techniques is stereolithography (SLA), which was created by a 3D system. A photo-initiated polymerisation print mechanism serves as the foundation for this method (S S Crump et al., 1991). Reservoirs are used to manage the movement of liquid resins, which are made up of photo-curable monomers, photo-initiator, and other additive elements. Various stereolithography printing machines, as depicted in Figure 4, use bottom-up and top-down (Noorani et al., 2006) print orientations. For the top-down method, the platforms are raised to the resins bath's surface whereby the resins flow to the bottom, and just a thin layer of resin, corresponding to the layer's thickness in the "sliced" stereolithography file, is on top of the platform. A laser beam of a selected frequency would be projected from the topmost of the visible resins on the platforms during print to start the polymerisation process. The platform will be lowered after each layer has been printed, and a roller will then apply a fresh layer of resin on top of the already hardened resin for later curing.

From the other side, the bottom-up printer's construction platforms are shifted to a location near the bottom where a small space would be left seen between reservoirs and the platforms, allowing for the placement of thinner layers of resins within the space. For cure exposing resins, a laser beam would be projected directly under the reservoirs. The platform would be raised to reveal a fresh, thin coating of resins after every layer has been printed. For a bottom-up stereolithography printer, the rollers are not required because the resins may naturally take their place between both the reservoirs and the hardened resins using gravitation. Additionally, since the laser is contained within the printers, such an inkjet printer produces a finer print finish and reduces the risk of laser irradiation.

Figure 4. Schematic diagram of SLA setup

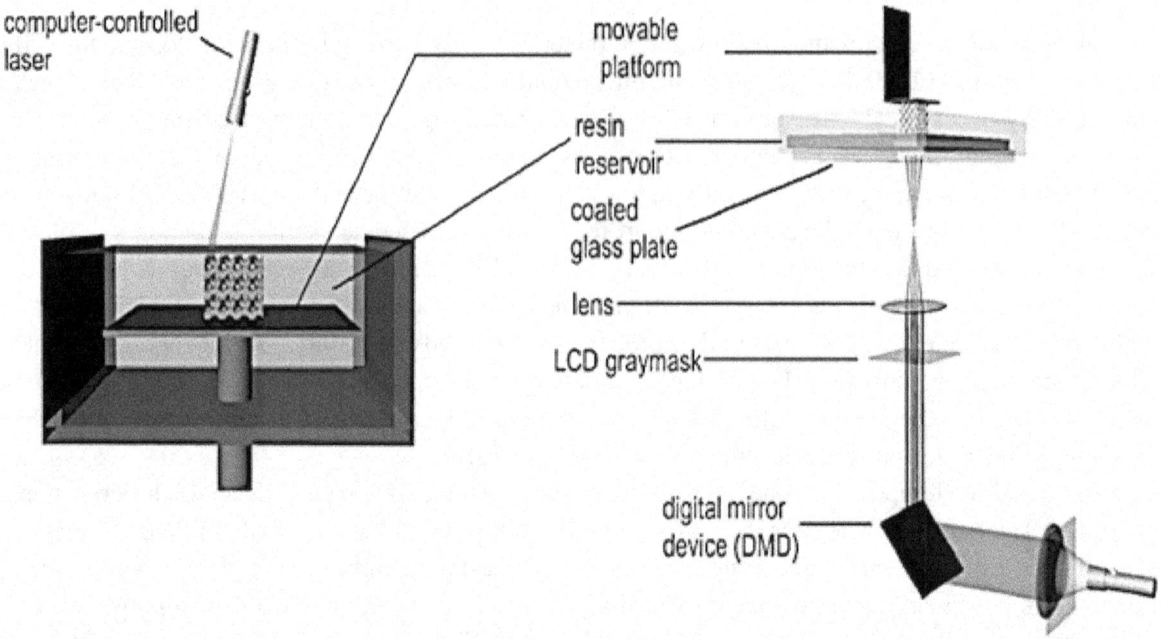

By adjusting the number of transmitted photons, the stereolithography printer of all additive manufacturing processes seems to have the best resolution (Elkins et al., 1997). 1 photon is typically used to start the polymerisation, however, 2 photons are getting popular since it has superior spatial control. Because the beginning is more confined, considerably better resolution (100nm) is feasible. Old World Laboratory had indeed made available its 100nm-resolution stereolithography printing machine, which may be used to create incredibly intricate structures. Regrettably, since the majority of high-performance polymer isn't really photo-curable, the materials that can be produced by stereolithography should be photo-curable, thus restricting its usefulness in printing high-performance polymers.

3D-Printing

MIT has a patent for 3D printing technology, that calls for powder particles and liquid binders (Halloran et al., 2011). In order to "agglomerate" the specific powdered particles and create a two-dimensional arrangement, as seen in Figure 5, water-based wet binders are injected into an inkjet printer. The following step involves adding a coating of powders over the printed layers using a roller. The adhesion force between the powdered particle is frequently strengthened with post-heating [26]. With this process, a wide range of components, including ceramics, metals, polymers, and composite components, can be manufactured. Due to the obvious quick powdered bonding technique, this is also regarded as being one of the quickest printing techniques. Poor surface smoothness, limited printing precision, and increased price are still issues with 3d printing technology, but (Lee et al., 2008). Because of the inadequate interlayer adhesiveness and weaker binding between powders in every layer, the tensile characteristics also are significantly small, which restricts its use for manufacturing high conductive and performance polymers (Sachs et al., 1990. Huang et al., 2013. Kim et al., 2008. Deckard et al., 1988).

Figure 5. Schematic diagram of 3D printing Process

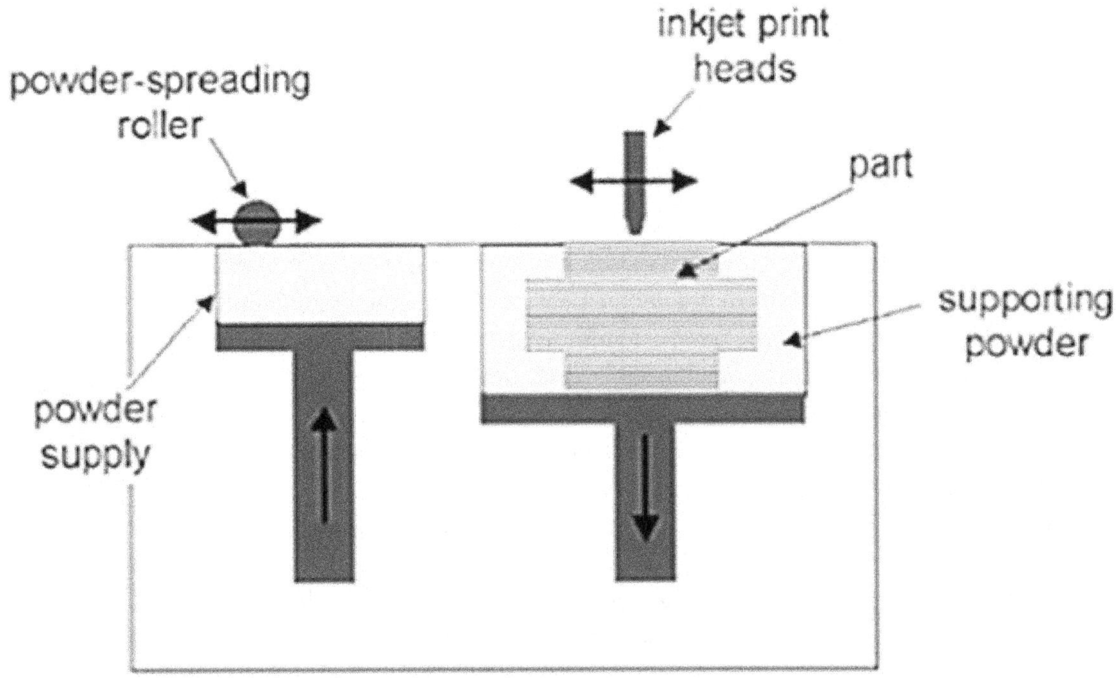

SLS Technology

The process of selective laser sintering (SLS) also uses powders. Products are received to the platforms in powdered form, similar to 3D printing technology. Selective laser sintering selectively sinters the powders with a greater power laser rather than wet binders, which allows particles to bond and create a two-dimensional design. In order to reduce thermal dispersion and distortions, the platforms are also preheated slightly underneath the melting temperature of construction particles (Beaman et al., 1996. Murr et al., 2012. Mudge et al., 2007). In contrast to selective laser sintering or fused deposition modeling which needs supportive components for the overhang layer, printed components can be maintained by the powders, making it possible to manufacture very complicated structures with ease. Only thermoplastic polymers and inorganic compounds like ceramics and metal with the introduction of organic binders can be manufactured since a sintered technique is used during printmaking (Singh et al., 2011. Le, 1998. Feygin et al., 1991. Kamrani et al., 2010). A beam of electrons is frequently employed rather than a laser to give a reached temperature for materials with high melting temperatures. Although stronger than 3D printing technology, the surface smoothness of selective laser sintering is still inferior to SLA, as well as the increased price of the print is a result of the higher price of the higher power laser source (Deckard et al., 1988. Hwa-Hasing et al., 2011. Beaman et al., 1996. Murr et al., 2012).

CONDUCTIVE AND HIGH-PERFORMANCE POLYMERS

The conductive and high-performance polymers are called polymers which can maintain their desired characteristics even when subjected to severe difficult circumstances, such as corrosion environments, extreme temperatures, and higher pressures, among others. These are also often termed as non, high-temperature polymer components, and innovative technological composites. Due to the demand for models with exceptional structural, geometrical, and biological durability at extreme temperatures and pressures, there is a significant level of interest in the use of high-performance polymers for 3D printing. The origin of high-performance polymers' power must be taken into account when creating one. The adhesion strength is perhaps the most crucial element. The adhesion dissociation potential can be used to calculate this. In general, the tougher it is to dissolve the polymer network and the more resilient the polymers would be to hostile environments, the greater the dissociation energy. Due to resonance stabilization, introducing heterocyclic compounds along the core of high-performance polymers can further increase their resistance and stability. Resonance-stabilized units are thought to increase bond strength by 40–70 kcal mol^{-1} when they are included.

Higher heterocyclic contents polymer doesn't really decompose into unstable particles at extreme temperatures since resonance-stabilized bonds are more difficult to break. When polymer degradation first starts, radical species are created at high temperatures. The polymer chain is then broken as a result of this radical species' reaction with it. However, since the radical spices are diffracted by the system when these radical species are produced from resonant stabilized subunits, these are much less responsive.

A significant drawback of getting a higher heterocyclic content is that the polymer becomes rigid and difficult to produce. The heterocyclic units are given heterocyclic like sulfur, oxygen, and nitrogen to address this issue. The polymer becomes more process able as a result, but because the adhesion strength is less, the polymer's adsorption capacity and chemical inertness are compromised. The polymers are formed to have supplementary pressures like h - binding, polar engagement, and van der Waals forces in addition to resonance stabilization to compensate for the decreased strength properties. All of these supplementary forces can significantly increase the heat consistency of high-performance polymers due to the interfacial energy between polymer chains. The polymer molecular mass and its dispersion, cross-linking density, crystalline nature, and the existence of reinforcement particles and additives are other elements that are known to increase heat stability (Therriault et al., 2003. Therriault et al., 2005. Duoss et al., 2014. Mota et al., 2013. Muth et al., 2014).

Some or all of the following requirements are met by conductive and high-performance polymers (Hergenrother, 2003): (a) The conductive and high-performance polymer must continue to function mechanically, chemically, and electrically after ten thousand hours of exposure to 178 °C, (b) At 460 °C, there must only be a five percent mass loss, (c) At high temperatures, there must be a minimum weight reduction rate, (d) There should be a minimal weight decrease rate at extreme temperatures and (e) Owing to its large volatile concentration and rigid segment, is expected to have higher glass transition temperatures and exceptional mechanical qualities.

Types of Conductive and High-Performance Polymers

Amorphous High-Performance Polymers

Amorphous polymers are defined as those containing molecules that are irregularly orientated and lack both short and long organization. The different types of amorphous high-performance polymers include:

Polysulfide

A group of amorphous high-performance known as polysulfide has the aryl-SO_2-aryl (sulfone) component as part of its foundation. One of the high-performance polymers with the greatest service temperatures is polysulfide, which has a GTT of 186 °C. Additionally, it may maintain its mechanical properties down to very cold temperatures (between 110 and 155 °C). It is resistant to oxidants, surfactants, hydrocarbons oils, electrolytes, strong acids, and alkalis. The resilience it offers to quasi to low-polar solvents including di-chloromethane, chloroform, ethanol, and aromatic compounds is, however, somewhat restricted. Due to the polar groups along its chains, Polysulfide has exceptionally strong bending strength resilience in physical and mechanical properties. Its usage in medicinal uses requiring sterilizer and steamer sterilization is made possible by Polysulfon's outstanding elevated hydrolytic stability (dimensional stability even in the presence of warm air, steam, or boiling water).

Polyetherimide

By melting polymerization of bi-sphenol A di-anhydride with a diamine, polyetherimide can be made. As a byproduct of its invention, melt-process-able, highly resistant materials are produced. Exceptional increased temperature resilience, strength properties, specific stiffness, and creep resistance are just a few of polyetherimide's exceptional mechanical and chemical qualities. Due to these factors, polyetherimide is increasingly being employed in place of metal in engineering materials. Hydrocarbon, alcohol, and halogenated solvent cannot harm it chemically. Additionally, it is believed to dim flames and produce very little ash. Despite having several exceptional qualities, it nevertheless has a lot of drawbacks, such as being notch sensitive, needing higher processing temperatures, and having reasonably high materials and process costs.

Semi-crystalline High-Performance Polymers

These are consisting of crystalline and amorphous regions and these are further divided into the following type:

Poly (phenylene sulfide)

The semi-crystalline polymer poly (phenylene sulphide), which is a member of the poly (arylene sulphides) family, has recurrent para-substituted benzene rings linked by sulphur atoms. In 1973, the Phillip Petroleum industry became the first company to commercially produce poly (phenylene sulphide) by combining sodium sulphide and di-halogenated aromatics in a polar aprotic solvent. Poly (phenylene sulphide) is renowned for its exceptional ability to withstand oxidation and can withstand steams, organic solvent, bleach, and powerful bases. Other benefits of adopting poly (phenylene sulphide) include its superior heat and structural integrity, resistance to sun radiation, dampness, aging, and other environmental factors, low coefficient of friction, electrical conductivity, and penetration resistance. However, the use of poly (phenylene sulphide) is restricted by the requirement for exceptionally high operating temperatures (melt temperatures), relatively higher cost, fracture toughness, and its sensitivity to warpage. By adding fibers and fillers, poly (phenylene sulphide) can be simply strengthened and its fracture toughness reduced.

Polyetheretherketone

A semi-crystalline heterocyclic polymer with both ether and ketone bonds is polyetheretherketone (PEEK). It is typically formed by the reaction of 1.5-benzenediol with 4.5′-di-fluorobenzenophenone in the vicinity of alkali-metal carbonic acid in highly polar solvents (such as diphenyl sulfonic acid). The polyaryletherketone family's most well-known and first completely commercialized member is polyetheretherketone. The reaction must take place in an anaerobic environment and at temperatures close to the polymer's melting temperatures, usually above 573 K. Superior thermo-mechanical resilience is a feature of polyetheretherketone. It is resistant to chemicals and fire assaults. It is hydrolytically strong, has strong abrasion resistance, and has excellent dielectric characteristics. It's been employed in a variety of sectors, including textiles, automobiles, and aviation. However, adopting polyetheretherketone has a few drawbacks, including its high cost and high extruded and molding temperature requirements.

Liquid crystalline Polymers

Partly crystallographic heterocyclic polyesters known as liquid crystalline polymers display heavily-ordered architectures both in the molten and solid phases. They often have as their foundation p-hydroxy-benzoic acid and related monomer. liquid crystalline polymers fall into one of 2 categories: thermo-tropic (molten crystallization happens whenever the polymers are heated beyond their GTT) and lyo-tropic (molten crystallization happens whenever the polymers are dissolved in a solvent). The commercially available Kevlar, a kind of aramid, is the most well-known example of lyo-tropic liquid crystalline polymers. Kevlar possesses high tensile properties and chemical resilience because of its inter-molecular hydrogen bonding between both the carbonyl groups and the nitrogen-hydrogen centers and the aromatic layering between neighboring polymer filaments. Vectra, an aromatic polyester made by poly condensing 4-hydroxybenzoic acid and 6-hydroxynaphthalene-2-carboxylic acid, is the most widely used chemotropic liquid crystalline polymer. Generally speaking, liquid crystalline polymers exhibit remarkable mechanical resilience at extreme temperatures. They are also very resistant to radioactivity, fire, and toxic assaults.

NANO-FILLERS

Graphene

Given the diversity of its binding, carbon exhibits an infinite number of physical characteristics that correlate to different structures with complex and challenging, serving as both the foundation for life and the building block of all organic synthesis. The solitary carbon atom has pairs of electrons in the 2p state because the s-orbitals are occupied. In which the attracting molecular potential outcomes intersect more effectively between nearby molecules, electron density builds up due to their bonding. The states are the name given to these bonds. The so-called states, at the other side, allow for the accumulation of charges above and below these nearby atoms. Because of this, it can create a 2-dimensional layered structure that is extremely anisotropy and resilient (Neto, 2009. Di Ventra et al., 2004. Lu et al., 2005). These structures are created utilizing p configurations that lie orthogonal to the line linking the nearby atoms.

Carbon Nanotubes

Carbon nanotubes (CNTs) are 2-dimensional graphene sheets that have been wound up with an extremely large aspect ratio to form a hollow, seamless cylinder. Changing this proportion can specify a variety of carbon nanotubes characteristics. Ceteris paribus, the diameter of the carbon nanotubes can significantly affect its electrical characteristics, whereas the lengths are known to have an influence on the mechanical behavior of the material, particularly under tensile loading (Lu et al., 2005. Abu Al-Rub et al., 2012.) The orientation in which the graphene sheets are wound up into tube could also be used to customize their characteristics, in addition to changing their sizes.

Nanocellulose

Cords of fibers, which typically have lengthy, spool compartments with strong secondary sidewalls, are the norm (Evert, 2006) These are either categorized as natural or artificially. Trees, creatures, and mineral deposits can all be used to make natural fibers (Laske, 2015). Between sixty and ninety-eight percent of them are made out of cellulose (Mohanty et al., 2005). According to Mohanty et al., cellulose is a hydrophilic glycan polymer with a degree of polymerization of about ten thousand and is made up of a linear chain of 1.5-anhydrous-glucose units that contain alcoholic hydroxyl groups that establish inter-and intramolecular hydrogen bonds. Although all-natural fibers have the same chemical structure as cellulose, the level of polymerisation differs. Additionally, the DP has a big impact on a fiber's mechanical characteristics (Frank et al., 2014).

Nanoclay

The smectite group minerals known as nanoclay materials are naturally available minerals with a plate-type anisotropy shape that is roughly 1 nanometer thick and 50 nanometers in radius (Frank et al., 2014). Averaging 760 m^2/g, every molecule does have a contact area. Montmorillonite (MMT), a 2-layer smectite clay mineral with a plate-type structure, is the primary raw material for nanoclay. The Montmorillonite's unique platelets thickness is only 1 nanometer, however, these layers' lateral sizes can vary from 200 nanometers to 2000 nanometers, giving them an exceptionally higher aspect ratio of much more than about fifty. Due to their higher aspect ratios, plate-type microstructure, complex formation capability, natural abundance, and comparatively cheap price, Montmorillonites have a major technological utility (Laske, 2015).

PREPARATION AND APPLICATIONS OF CONDUCTIVE AND HIGH-PERFORMANCE POLYMERS

Carbon Black-Filled Nylon-12 Nanocomposite

The Four weight percent carbon black-reinforced Nylon-12 (N-12) polymer nanocomposites are composed of N-12. In the investigation, N-12 powder was employed. The usual powder size diameter of N-12 powder is between 50 and 100 m, and it is a white, odorless, semi-crystalline thermoplastic polymer with a melt temperature range of 176-184 °C. The study's reinforcement is KET-JEN-BLACK EC-600 JD, a nano-sized carbon black (Akzo-Nobel, US). The carbon black pellets were crushed in a rotary tumbler using a ball mill with high wear-resistant grinding media that had a 5 mm diameter. Utilizing 139 mesh, and 176 m opening American standard sieves, the crushed carbon black was separated. The pure Nylon-12 powder and sieved carbon black were then combined for 24 h in a rotary tumbler (Athreya et al., 2010). Later, the developed composite material was used in a 3D printing system that used selective laser sintering. The electrical conductivity of the carbon black-filled N-12 polymer nanocomposite is 1.0105 S/cm, or five orders of magnitude higher than pure N-12. Furthermore, the four weight percent percolation threshold of carbon black-filled N-12 polymer nanocomposite obtained is significantly lower than that reported in the literature. The resulting nanocomposite material has a promising future in electronics and automobile manufacturing. (Athreya et al., 2010).

Epoxy-Based Resin Filled with Mwcnts

Conventional, commonly produced epoxy-based resins are combined with a tiny dispersion of multi-wall carbon nanotubes to create multi-wall carbon nanotubes-filled epoxy-based resin. The stereolithography epoxy resin type used in the study was Watershed-TM-11120, which is commonly accessible commercially. Multi-wall carbon nanotubes (MWCNTs) were gradually incorporated into the stereolithography resins using mechanical blending and acoustic dispersing methods, preventing multi-wall carbon nanotubes from clumping. Multi-wall carbon nanotubes are prone to aggregation even without pre-treatment. This sample was stirred into packed glass jars before being placed in a conventional ultrasonic tank filled with water to reduce the formation of MULTI-WALL CARBON NANOTUBES agglomerates. The containers were subjected to three 60-second ultrasonication cycles. During this stage, the water temperature in the ultrasonic tank was regularly checked to ensure that the nanocomposite did not overheat. After reaching 303 K, the water in the tank was replaced with brand-new, fresh tap water with a temperature of 293 K. The multi-wall carbon nanotubes filler was created using ceramic oxide support, which was then refined using acid etching. After that, the final sample of multi-wall carbon nanotubes was created on a Cu mesh grid with a silicon monoxide/formvar coating. These polymer nanocomposites are still being improved in order to improve the nanocomposite SLA resins for their intended end-use application.

Nylon 6/Clay-Reinforced Nanocomposite

Clay nanoparticles make up five weight percent of the polyamide 6 that makes up the nylon 6/clay-reinforcement nanocomposite (The nanocomposites were created by milling polyamide 6 pellets into powders. The pellet was first put in a wire mesh basket with five to seven grams of charged pellets, and after five minutes in liquid nitrogen, the system was brought to equilibrium with liquid Nitrogen gas.

Secondly, the milling machine quickly emptied the cold pellets into it, where they were crushed for five minutes. An as pellet and crushed powders were pre-dried at 353 K for 24 hours under vacuum prior to the melt flow index and sintered investigations in order to eliminate any potential humidity from the atmosphere during the powdered grinding method (Kim et al., 2004). The powders were then sintered in a specially designed aluminium sintered cell. A modified aluminium cylindrical block with a conventional oven was used to monitor the sintering behaviour, while a differential scanning calorimeter was used to evaluate the temperature transition. The nylon 6-clay-reinforcement nanocomposites are still being developed in order to advance their end-use applications.

Commercially Available Additive Manufactured High-Performance Polymer Nanocomposites

Arevolabs is one of the companies that specializes in developing, manufacturing, and marketing high-performance polymers and polymer composites for additive manufacturing. The company manufactures and uses high-performance polymers such as poly-arylether ketone, poly-etherether ketone, and poly-acrylamide as the base high-performance polymeric materials to create innovative and improved high-performing polymer nanocomposites. Then, specialized filler particles such as carbon nanotubes are added. Poly-arylether ketone is a high-performance polymer substance of choice in aircraft structures due to its low weight, high specific strength, and non-corrosion properties. Furthermore, the automobile industry chose this material due to its durability. This substance is used in medicine because of its inertness and biocompatibility. On the other hand, poly-etherether ketone is a high-performance polymeric material with extraordinary properties that outperform those of standard engineering resin materials such as nylon and allow it to replace a wide range of materials. The petroleum industries prefer poly-etherether ketone for use in extreme environments where temperatures can exceed 523 K and a long dynamic fatigue life is required.

Similarly, the semiconductor industries use poly-etherether ketone to fabricate semiconductor wafers due to its exceptional resistance to abrasive produce chemicals. When compared to other engineering thermoplastics, poly-arylamide has superior strength and stiffness. Poly-arylamide has an exceptionally low coefficient of thermal expansion and creep due to its unique chemistry, while still retaining a flawlessly smooth finish. In situations where regular polyamides are too technical, inadequate poly-arylamide is a better choice for single-use medical products. Poly-arylether ketone, including some high-performance polymer materials, has been mixed with carbon nanotubes for the electrostatically dissipative substance featured in Quantevo TM-ESD products. These high-performance polymer nanocomposites are particularly well suited for the fabrication of jigs and fixtures in the aviation, pharmaceutical, and telecommunications industries. Arevolabs, in addition to carbon nanotubes, offers a variety of filler materials as reinforcing agents for its high-performance base polymers designed for use in specific applications such as carbon fibre and glass fibre (High-Performance Additive Manufacturing Materials, 2011).

In addition to the businesses that make high-performance polymer nanocomposite materials, it is important to note the important function that these materials currently play in the automobile industry. Given their higher mechanical qualities, improved stiffness and specific stiffness, enhanced scratching and abrasion resilience, outstanding thermal and processing characteristics, decreased components surface roughness, and excellent impact rigidity, high-performance polymer nanocomposite materials are regarded as a new class of substances. Due to these characteristics, the high-pressure polymer nanocomposites can be used as substitutes of metals in the automobile industry (Nanowerk.Com Reports on Polymer Nanocomposites Drive Opportunities in the Automotive Sector, 2011). The decrease in vehicle weight, enhanced engine efficiency, decrease in carbon dioxide emissions, and excellent quality (more security, greater convenience, and better driving dynamics) are the main factors driving the usage of high-performance polymer nanocomposite materials in the automobile industries. The automobile sector is becoming more motivated to use high-performance polymer nanocomposite materials for automobile components in conjunction with additive manufacturing. Despite these benefits, high-performance polymer nanocomposite materials fabrication still faces a number of obstacles and constraints, including making, suitability, distribution, and emulsification between nanomaterial and polymer matrix; expense, the manufacturing of nanocomposite materials on mass production at feasible rates (as polymer matrix price is dependent on oil price levels and carbon nanotubes price is also high); consistency and reliability in batch sizes, it is inconsistent and unreliable in large scale production. Nevertheless, stability and dependability in volume manufacturing are gradually declining due to the usage of tested high-performance polymer nanocomposite materials and additive manufacturing procedures that use these components. This has given fresh motivation for combining high-performance polymer nanocomposite materials with 3D printing once more.

MOTIVATION AND CONCLUDING REMARKS

It anticipates that additive manufacturing will develop and eventually replace our existing production system. The technology has witnessed numerous advancements ever since its inception in 1970, including an expansion of manufacturing techniques, post-processing, and the spectrum of substances that can be used. High-performance polymer nanocomposite, one of these novel materials, is anticipated to broaden the use of AM in applications that demand extremely resistant components for harsh environments. While there are many investigations on high-performing polymers, nanofillers, and their composite structures in the research journals, there are few publications on their use in additive manufacturing. This is anticipated to shift as additive manufacturing technological advancements and more organizations see the potential benefits for their operations.

FUTURE SCOPE

1. Utilizing both fiber and particle kind reinforcing agents is the current trend in high-performance polymer composite developments. Although it complicates the situation in terms of wearing mechanisms and wear-debris production, this technology provides the potential to boost the strength and durability of high-performance polymers. In that regard, a more quantitative approach to individual relationships and the overall effect on wear and friction behavior is required.
2. Other aspect of this topic that needs more study is the impact of vibrations, both naturally occurring and framework. Vibrations may alter the manner in which the mated components make contact, causing wear and friction to increase or decrease proportionally. It is important to use due caution in that area, which is frequently neglected.

3. Discrepancies exist in our comprehension of the elevated tribological behavior of high performance polymers and their composites since some researchers have claimed that adding particles often does not improve the wear rate of high-performance polymer composites, especially at high temperatures. Additionally, the price of nano-high-performance-filler and its comparatively weighed benefits would be taken into account.

REFERENCES

Abu Al-Rub, R. K. (2012). On the aspect ratio effect of multi-walled carbon nanotube reinforcements on the mechanical properties of cementitious nanocomposites. *Construction & Building Materials*, *35*, 647–655.

Additive Manufacturing Materials. (2017). Retrieved January 20, 2023, from https://www.additivemanufacturing.media/kc/what-is-additive-manufacturing/am-materials.

Additive Manufacturing State of the Industry. (2015). Retrieved January 20, 2023, from https://www.sme.org/technologies/latest-news/

Advanced Additive Solutions. (2012). Retrieved January 20, 2023, from https://oldworldlabs.com/

Athreya, S., Kalaitzidou, K., & Das, S. (2010). Processing and characterization of a carbon black filled electrically conductive Nylon-12 nanocomposite produced by selective laser sintering. *Materials Science and Engineering A*, *527*, 2637–2642.

Aviation, G. E. (2015, April 6). *Additive Manufacturing Machines & Materials | GE Additive*. Retrieved January 20, 2023, from https://www.ge.com/additive/

Beaman, J. J., Barlow, J. W., Bourell, D. L., Crawford, R. H., Marcus, H. L., & McAlea, K. P. (1996). *Solid Freeform Fabrication: A New Direction in Manufacturing*. Springer.

Callister, W. D. J. (2007). *Materials Science and Engineering: An Introduction* (7th ed.). John Wiley & Sons, Inc.

Camargo, P. H. C., Satyanarayana, K. G., & Wypych, F. (2009). Nanocomposites: Synthesis, structure, properties and new application opportunities. *Materials Research*, *12*, 1–39.

Columbus, L. (2015, April 28). *Roundup of 3D printing market forecasts and estimates*. Retrieved January 20, 2023, from https://www.forbes.com/sites/louiscolumbus/2015/03/31/2015-roundup-of-3d-printing-market-forecasts-and-estimates/

Crump, S. S. (1991). Fast, precise, safe prototype with FDM. *ASME, PED.*, *50*, 53–60.

Deckard, C., & Beaman, J. J. (1988). Process and control issues in selective laser sintering. *ASME, PED*, *33*, 91–197.

Di Ventra, M. (2004). *Introduction to Nanoscale Science and Technology*. Kluwer Academic Publisher.

Duoss, E. B., Weisgraber, T. H., Hearon, K., Zhu, C., Small, W., Metz, T. R., & Spadaccini, C. M. (n.d.). Three-dimensional printing of elastomeric, cellular architectures with negative stiffness. *Advanced Functional Materials, 24*, 4905–4913.

Elkins, K., Nordby, H., Janak, C., Gray, R. W. IV, Bohn, H. H., & Baird, D. G. (1997). Soft elastomers for fused deposition modeling. *Proc. 8th Solid Freeform Fabrication Symposium.*

Evert, R. F. (2006). Eas Esau's Plant Anatomy — Meristems, Cells, and Tissues of the Plant Body: Their Structure, Function, and Development (3rd ed.). Wiley & Son.

Feygin, M., & Hsieh, B. (1991). Laminated object manufacturing (LOM): A simpler process. *Proc. 2nd Solid Freeform Fabrication Symposium*, 123–130.

Frank, K. (2014). *Introduction to Nanofiber Materials*. Cambridge University Press.

Gibson, I., Rosen, D., & Stucker, B. (2015). *Additive Manufacturing Technologies: 3D Printing, Rapid Prototyping, and Direct Digital Manufacturing* (2nd ed.). Springer. doi:10.1007/978-1-4939-2113-3

Grunewald, S. (2021, October 18). *Worldwide spending on 3D printing to grow from nearly $11 billion to $26.7 billion by 2019 - 3DPrint.com: The Voice of 3D printing / additive manufacturing*. Retrieved January 20, 2023, from https://3dprint.com/116562/3d-printing-26-billion-by-2019/

Halloran, Tomeckova, & Gentry. (2011). Photopolymerization of powder suspensions for shaping ceramics. *J. Eur. Ceram. Soc., 31*, 2613–2619.

Hergenrother, P. (2003). The use, design, synthesis, and properties of high performance/ high temperature polymers: An overview. *High Performance Polymers, 15*, 3–45.

Hernández, R., Calister, I., Corsen, R., Rodríguez, A., Plaate, R., Imperial, C., ... Gutierrez, P. E. (2007). *Podio Europeo del jazz episodio 10-2007*. Academic Press.

High-performance additive manufacturing materials. (2016). Retrieved January 20, 2023, from https://www.arevo.com/

Huang, S. H., Liu, P., Mokasdar, A., & Hou, L. (2013). Additive manufacturing and its societal impact: A literature review. *International Journal of Advanced Manufacturing Technology, 67*, 1191–1203.

Hwa-Hsing, T., Ming-Lu, C., & Hsiao-Chuan, Y. (2011). Slurrybased selective laser sintering of polymer-coated ceramic powders to fabricate high strength alumina parts. *Journal of the European Ceramic Society, 31*, 1383–1388.

Iancu, C., Iancu, D., & Stancioiu, A. (2013). From CAD model to 3D print via "STL" file format. *Fiabilitate Durab, 1*, 73.

Kalin, M., Zalaznik, M., & Novak, S. (2014). Wear and friction behaviour of poly-ether-etherketone (PEEK) filled with graphene, WS2 and CNT nanoparticles. *Wear*, 855–862.

Kamrani, A. K., & Nasr, E. A. (2010). *Engineering Design and Rapid Prototyping*. Springer.

Kim, G. D., & Oh, Y. T. (2008). A benchmark study on rapid prototyping processes and machines: Quantitative comparisons of mechanical properties, accuracy, roughness, speed, and material cost. *Proceedings - Institution of Mechanical Engineers*, *22*, 201–215.

Kim, J., & Creasy, T. (2004). Selective laser sintering characteristics of nylon 6/clay-reinforced nanocomposite. *Polymer Testing*, *23*, 629–636.

Laske, S. (2015). *Polymer Clay Nanocomosite* (1st ed.). Elsevier.

Le, H. P. (1998). Progress and trends in ink-jet print technology. *The Journal of Imaging Science and Technology*, *42*, 49–62.

Lee, K. S., Kim, R. H., Yang, D. Y., & Park, S. H. (2008). Advances in 3D nano/microfabrication using two-photon initiated polymerization. *Progress in Polymer Science*, *33*, 631–681.

Lu, X., & Chen, Z. (2005). Curved pi-conjugation, curved pi-conjugation, aromaticity, and the related chemistry of small fullerenes (bC60) and single-walled carbon nanotubes. *Chemical Reviews*, *105*, 3643–3696.

Materials and filaments for industrial 3D printing. (2015). Retrieved January 20, 2023, from https://www.stratasys.com/en/materials/materials-catalog/

Maxey, K. (2013, September 3). *Infographic: The history of 3D printing*. Retrieved January 20, 2023, from https://www.engineering.com/story/infographic-the-history-of-3d-printing

Melchels, F. P. W., Feijen, J., & Grijpma, D. W. (2010). A review on stereolithography and its applications in biomedical engineering. *Biomaterials*, *31*, 6121–6130.

Mohanty, A. K. (2005). *Natural Fibers, Biopolymers and Biocomposites*. Taylor & Francis Group.

Mota, C., Puppi, D., Dinucci, D., Gazzarri, M., & Chiellini, F. (2013). Additive manufacturing of star poly(ε-caprolactone) wet-spun scaffolds for bone tissue engineering applications. *Journal of Bioactive and Compatible Polymers*, *28*, 320–340.

Mudge, R. P., & Wald, N. R. (2007). Laser engineered net shaping advances additive manufacturing and repair. *Weld. J. N. Y.*, *86*, 44–48.

Murr, L. E., Gaytan, S. M., Ramirez, D. A., Martinez, E., Hernandez, J., Amato, K. N., Shindo, P. W., Medina, F. R., & Wicker, R. B. (2012). Metal fabrication by additive manufacturing using laser and electron beam melting technologies. *Journal of Materials Science and Technology*, *28*, 1–14.

Muth, J. T., Vogt, D. M., Truby, R. L., Mengüç, Y., Kolesky, D. B., Wood, R. J., & Lewis, J. A. (2014). Embedded 3D printing of strain sensors within highly stretchable elastomers. *Advanced Materials*, *26*, 6307–6312.

Neto, A. H. (2009). The electronic properties of graphene. *Reviews of Modern Physics*, *81*, 109–162.

Noorani, R. (2006). Rapid Prototyping—Principles and Applications. John Wiley & Sons, Inc.

Paul, D., & Robenson, L. (2008). Polymer nanotechnology: Nanocomposites. *Polymer*, *49*, 3187–3204.

Pham, D. T., & Ji, C. (2000). Design for stereolithography. *Proceedings - Institution of Mechanical Engineers*, *214*, 635–640.

Report, P. T. (2016, February 26). Retrieved January 20, 2023, from https://www.hubs.com/get/trends/

Reports on polymer nanocomposites drive opportunities in the automotive sector. (2016, February 16). Retrieved January 20, 2023, from https://www.nanowerk.com/category-spotlight.php

Sachs, E., Cima, M., & Cornie, J. (1990). Three-dimensional printing: Rapid tooling and prototypes directly from a CAD model. *CIRP Ann. Manuf. Technol.*, *39*, 201–204.

Sherman, L. (2015, February 25). *Additive manufacturing: Materials for 'real-world' parts*. Retrieved January 20, 2023, from https://www.ptonline.com/articles/additive-manufacturing-materials-for-real-world-parts

Singh, R. (2011). Process capability study of poly jet printing for plastic components. *Journal of Mechanical Science and Technology*, *25*, 1011–1015.

Stansbury, J., & Idacavage, M. (2015). 3D printing with polymers: Challenges among expanding options and opportunities. *Dental Materials*, *32*(1), 54–64. doi:10.1016/j.dental.2015.09.018 PMID:26494268

The strongest choice for 3D printing. (2014). Retrieved January 20, 2023, from https://www.stratasys.com/en/resources

The Third Industrial Revolution. (2012, April 21). Retrieved January 20, 2023, from https://www.economist.com/leaders/2012/04/21/the-third-industrial-revolution

Therriault, D., Shepherd, R. F., White, S. R., & Lewis, J. A. (2005). Fugitive inks for direct-write assembly of three-dimensional microvascular networks. *Advanced Materials*, *17*, 395–399.

Therriault, D., White, S. R., & Lewis, J. A. (2003). Chaotic mixing in three-dimensional microvascular networks fabricated by direct-write assembly. *Nature Materials*, *2*, 347.

Tranchard, S., & Rojas, V. (2015, May 5). *Manufacturing our 3D future*. Retrieved January 20, 2023, from https://www.iso.org/news/2015/05/Ref1956.html

Yin, J., Zhu, G. Z., & Deng, B. (2013). Multi-walled carbon nanotubes (MWNTs)/polysulfone (PSU) mixed matrix hollow fiber membranes for enhanced water treatment. *Journal of Membrane Science*, *437*, 237–248.

Chapter 14
Machine Learning Applications for 3D-Printed Polymers and Their Composites

Mamta B. Savadatti
New Horizon College of Engineering, India

Mukesh Thakur
NMDC DAV Polytechnic, India

Kiran Kumar N.
Vemana Institute of Technology, India

R. Suresh Kumar
BMS College of Engineering, India

Jaya Christiyan K. G.
M.S. Ramaiah Institute of Technology, India

Richard Lincoln Paulraj
New Horizon College of Engineering, India

Amithkumar Gajakosh
Dayananda Sagar College of Engineering, India

Madhusudhana H. K.
KLE Technological University, India

ABSTRACT

Although the number of applications for 3D printing has substantially risen over the past several years, it is required to calibrate the AM processing settings. Various methods of AL are being applied in today's world in order to improve the parameters of 3D printing and to forecast the quality of components that have been 3D printed. An application of ML in the prediction of the properties and performance of 3D-printed components has been demonstrated in the current work. This research begins with an introduction to machine learning and continues with a summary of its uses in the 3D printing process. The majority of this chapter is dedicated to discussing the applications of ML in the forecasting of essential properties of 3D-printed components. In order to accomplish this objective, prior research studies that studied the application of ML in the characterisation of polymeric and polymer composites have been reviewed and addressed.

DOI: 10.4018/978-1-6684-6009-2.ch014

INTRODUCTION

Products are being replaced and upgraded more frequently now than in preceding decades, and manufacturer competitiveness encourages the creation of new technology. Rapid prototyping techniques like additive manufacturing (AM) and three-dimensional (3D) printing use a consecutive layer of materials to build three-dimensional things underneath computer control. The benefits of additive manufacturing over traditional procedures have been demonstrated to include the opportunity to function with a large variety of materials and few limits in the manufacture of complicated shapes. Numerous industries, including aviation, automobile, communications, construction, and health monitoring (Kong L et al., 2016, Leal R et al., 2017, Khosravani M R et al., 2020, Murr L E et al., 2019, Machment T et al., 2019, Nasiri et al., 2020), have employed this production technique. The American Society for Testing Materials (ASTM) has categorized additive manufacturing into 7 procedures: vat photopolymerisation, binders spraying, sheet laminations, direct energy deposition, materials' extrusion, and powders' bed fusion (ASTM, 2012). Various engineering elements were researched in this area basis on applications of additive manufacturing. For example, current research studies looked into durability, tensile strength, ecological consequences, and various welding purposes (Colorado H.A et al., 2020, Palanikumar et al., 2020, Khosravani MR and Reinicke T, 2020, Karavel E and Bozkurt Y, 2020). Various options were used to optimize these variables and forecast the mechanical characteristics of printed parts (Ren D et al., 2019, Munprom R, Limtasiri S, 2019, Goh GD et al., 2020, Khosravani MR and Reinicke T, 2020, Saeed K et al., 2021, Pei S et al., 2021) because additive manufacturing operating variables (such as particle size of powders, deposition rate, film thickness, laser power, and raster angel) get a significant influence on the network authenticity and mechanical characteristics of three dimensional-printed components. For instance, the rigidity and strengths of three-dimensional-printed polymeric composite parts were recently assessed in (Saeed K et al., 2021) using a sequence of tensile tests. Additionally, it was shown that fiber orientations have an impact on the studied components' mechanical behavior. The article referenced (Pei S et al., 2021) looked studied the micro and macro influences of process variables on three-dimensional-printed composites manufacturing. In this instance, short materials extrusion-based carbon-fiber reinforcing polymeric composite materials were manufactured. Morphology characterization was done using an image-based statistical approach (e.g., fiber volume fractions). A Monte Carlo sampling technique was also employed to enhance the datasets. The observed results demonstrated that processing factors are important for void production and void volume concentration dispersion.

According to an analysis of relevant literature, mathematical modeling and other Artificial Intelligence (AI) methods have been utilized to explore the performance requirements of three-dimensional-printed parts (Yang J et al., 2017, Li Y et al., 2021, Trivedi A and Gurrala PK, 2021, Pazhamannil RV et al., 2020) in addition to experimentations. For example, three-dimensional FEM modeling has been used in (Chen Q et al., 2017) to assess the impact of process variables on the melt pool profile and beads forms in ceramic materials three-dimensional printing. A fundamental quantitative concept was established at the very same time to examine residual stresses in metal components made using additive manufacturing (Fergani O et al., 2017). To do this, the evaluation of the operation's thermo signature was done using the predicted temperature distributions. According to research, thermal stresses were employed as an input when calculating residual stresses. Those earlier investigations showed that models were exclusively focused on one or two facets of the additive manufacturing procedure. The use of data-driven algorithms, also known as Machine Learning (ML) concepts, has become commonplace because it is hard to anticipate all mechanical qualities and the entire thing of various production processes rapidly and effectively (Ulas M et al., 2020, Khanzadeh M et al., 2018, Xu D et al., 2019, Awoyers PO et al., 2020, Sizemore NE et al., 2015). A branch of artificial intelligence, machine learning (ML) is an interdisciplinary field that supports reduced-cost computing utilizing algorithm learning (Jordan MI and Mitchell TM, 2015). A long list of physical-based equations is not necessary when using machine learning approaches; instead, historical facts are utilized. The adoption of machine learning approaches in the areas of additive manufacturing has varied depending on their benefits (Yao X et al., 2017, Tootooni M et al., 2017, Zohdi TI, 2018, Delli U and Chang S, 2018, Zhu Z et al., 2018, Amini M et al., 2019, Qi X et al., 2019, Wang C et al., 2020, Hertlein N et al., 2020, Goh GD et al., 2021). A hybrid machine learning system, for instance, was published in (Yao X et al., 2017) to suggest unique features for three-dimensional-printed components. Through the creation of three-dimensional-printed automotive parts, the suggested approach was evaluated. Untrained artists might apply the provided strategy throughout the design stage. The abilities of the proposed technique were shown using the specified three-dimensional design features. Research (Delli U and Chang S, 2018) used machine learning to supervise the three-dimensional printing methodology. The machine learning, image analysis, and webcam integrations in the suggested methodology allowed it to judge the effectiveness of the three-dimensional-printed polymeric components. The conceptual methodology was shown using two polymer composites. The advantages of automatic production tracking in performance analysis were proven by the outcomes that were documented. Uses of neural-network-based machine learning in additive manufacturing were developed and presented in (Qi X et al., 2019). Simulation models, in-situ surveillance, and quality assurance are among these uses. Various forms of the original dataset could be used for in-situ surveillance (e.g., spectra, images, and tomography). Given the reported uses of neural-network-based machine learning in additive manufacturing, it was determined that this area has tremendous potential again from designing through post-processing stages. A later study (Scime L and Beuth J, 2019) employed machine to describe how powder spreads during the laser powder bed fusion procedure. Scientists investigated the shape of melt pools in detail using a high-speed camera, and they used machine learning to distinguish between the many melt pools they saw. The suggested method lays the way for the identification of multiple problems throughout the printing process because improper powder dispersal might result in a variety of defects.

The benefits of machine learning techniques in the mechanical characterization of three-dimensional-printed objects are addressed in the current paper. To achieve this, 3 distinct are taken into account: (a) process variables; (b) porosity; and (d) printed component flaws, all of which have an immediate impact on the mechanical behavior of three dimensional-printed components. The study attempts to compile the established machine learning-based technologies that are used in various additive manufacturing techniques. The present study illustrates the development, difficulties, and potential for expanding investigation in this exciting area. We anticipate that researchers who are considering using machine learning to improve their simulations will find this assessment to be useful. We outlined, clarified, and contrasted machine learning implementations in the aforementioned sub-domains in this study. The rest of this essay is structured as follows: A basic introduction to machine learning is given in the second domain. Uses of machine learning for forecasting the mechanical behavior of three-dimensional-printed objects are discussed in the third domain. In the fourth domain, benefits, difficulties, and opportunities are presented. Furthermore, the fifth domain has provided a conclusion.

MACHINE LEARNING

Since the real-world challenges nowadays are difficult, various artificial intelligence techniques demonstrated their effectiveness in various fields. There have been continuous efforts throughout the years to automate the process of learning using various connection-based methodologies (Messaoud S et al., 2020). Later, other techniques dependent on statistical learning theory were developed, including Decision Trees (DTs) and Support Vector Machines (SVMs) (Vapnik VN, 2013). A device or computer can train autonomously to make predictions without even being supervised learning using the artificial intelligence approach known as machine learning (ML) (Liu Y et al., 2017). In fact, machine learning seeks to accomplish a goal by discovering things from a specific data set. Three kinds of machine learning exist because distinct procedures depend on the information: (a) supervised learning, (b) unsupervised learning, and (c) reinforcement learning. Figure 1 displays a few typical machine learning methodologies. In supervised learning, the algorithm operates from labeled training data to aid in outcome prediction, whereas in unsupervised learning, the program uses unlabelled data to find correlations between desired properties. In reinforcement learning, the system can engage with its surroundings to learn from them and choose the best course of action that will provide the highest rewards.

Figure 1. Classification of Machine leaning Algorithms

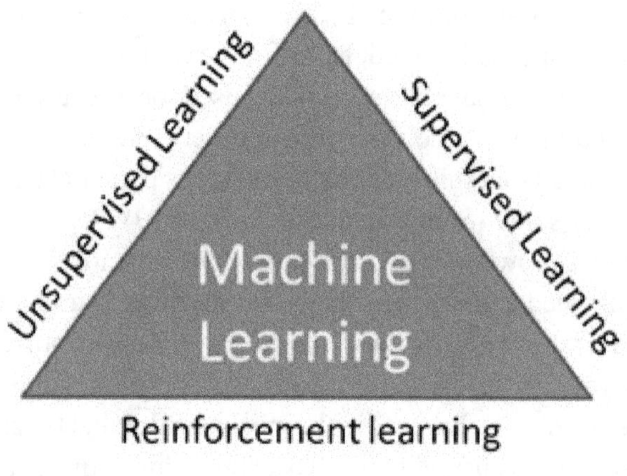

Regression, categorization, as well as other high-dimensional data-related demands, are suitable candidates for machine learning applications. The information might be in numerous types, including acoustic input, writing, or pictures, and the retraining is dependent on understanding from prior calculations (Li Z et al., 2019, Cheng MY et al., 2020, Shi B et al., 2016). This should be emphasized that the neural networks-based machine learning algorithms include Convolution Neural Network (CNN), Adaptives' Network-Fuzzy Interference Systems (ANFIS), Recurrent Neural Network (RNN), Self-Organising Maps (SOM), Deep Belief Networks (DBN), and Multi-Layer Perceptrons (MLP). Some other algorithms are part of the conventional machine learning methodology. Deep learning is the term for machine learning methods that use neural networks with much more than two hidden layers.

Choosing the appropriate ML algorithm is crucial when developing an ML system, since each algorithm has a significant influence on the correctness of the results. There is no algorithm that is suitable for all problems since each algorithm offers distinct advantages for a specific application. Regression, estimation, classification, and clustering are some of the most frequently used machine learning (ML) techniques in mechanical engineering. Primarily, regression, clustering, and classification techniques are used to predict material properties. Figure 2 depicts ML methods that are often utilized to handle diverse mechanical engineering difficulties.

Figure 2. frequently used ML algorithms

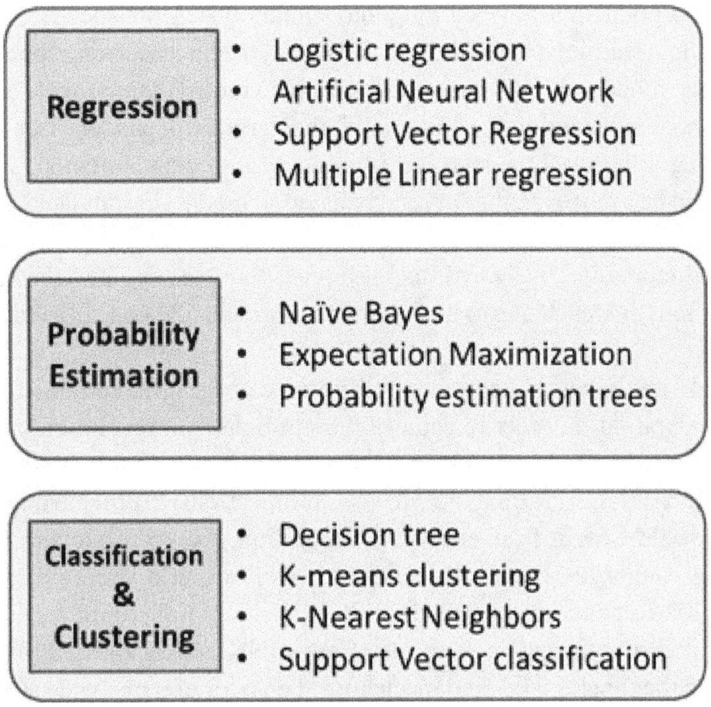

Data preprocessing, description choice, algorithm choice, tend to issue, and deployment is the 5 major machine learning procedures (Liu Y et al., 2020). Empirical data gathering, acceptable property predictions, and experimental verification make up the whole cycle of predicting the mechanical properties and characteristics of a structural element. The fundamental concept is based on the application of a probability model to the forecasting of material behavior and mechanical characteristics.

Because both numeric simulations and machine learning deal with representations, they are connected. Even though the inputs for the stochastic process in simulations are not fully recognized, the framework is. Prior to training in machine learning, the network is undefined, but the input is understood. Prior research has indicated an interaction between machine learning and simulations that should be taken into account in applications. Here is an overview of these interactions:

The methodologies of machine learning can replace conventional numerical modeling techniques. Machine learning (ML) seeks to predict system behavior by data collection and mathematical modelling (Rueden L et al., 2020), similar to simulations. As machine learning-based models can replicate high-dimensional phenomena far more quickly than the vast majority of physical simulations, they are becoming viewed as a feasible choice.

Application in industry and environmental sciences can be used to implement simulations into machine learning. Simulated and the results can be precisely implemented in numerous machine learning parts (e.g., training data, algorithm, and final hypothesis). According to the research, enhancing the training examples is how simulation findings are most frequently integrated into machine learning systems. As an illustration, numerical simulations were employed in (Rahman A et al., 2021, Bayliss C, 2021) to train data for the forecasting of structural rigidity and the urban route problems, correspondingly. The machine learning model can learn from prior numerical findings. The simulation studies can be employed as a data provider for the machine learning algorithms, much like the actual data.

It is possible to incorporate machine learning into simulations. More specifically, several simulation elements can use machine learning approaches. In particular, machine learning approaches can be applied to a model to minimize modeling order and create surrogate models that provide approximations of the real problem but with easier solutions. Additionally, the variable dependence of simulation outcomes can be studied by integrating machine learning approaches into the simulation. The machine learning model, for instance, can be used to identify several behavioral models in simulated data, which saves time and effort throughout the design phase. This is demonstrated in (Bohn B et al., 2013). The numerical solutions are obtained more effectively by integrating machine learning into the numerical simulation. In this situation, learning models that can be computed more quickly can solve model components that use a lot of resources.

Although ML and simulation integration is still in its early stages, certain considerations must be considered into consideration. In order to achieve the 4th industrial revolution's technological goal of combining the project planning phases, a current breed of computer-aided design software systems is needed to give extremely quick cycle times for process optimizations. In order to implement such systems, ML and simulation would benefit from enhanced integration. According to a review of the literature, machine learning has caught the interest of academics working in a variety of disciplines, including medicine (Smiti A, 2020), material science (Juan Y et al., 2021), the building industry (Bilal M and Ovedele LO, 2020), and mechanical engineering (Geetha NK and Bridjes P, 2020). Uses of machine learning for predicting the functionality and mechanical behavior of parts made additively are discussed in the section that follows (Decost BL et al., 2017, Jiang J et al., 2019, Shen X et al., 2004, Ye D et al., 2018, Gu GX et al., 2018, Lu ZL et al., 2010, Xiong J et al., 2014, Aoyagi K et al., 2019).

APPLICATION OF MACHINE LEARNING IN ADDITIVE MANUFACTURING

An Outline of Using Machine Learning in Additive Manufacturing

Designing, printing, post-processing, and component assessment are the 4 primary stages of additive manufacturing fabrication. The main components of such stages are depicted in Figure 3, wherein various machine learning approaches may be applied. The very first stage in the design phase for three-dimensional printing is to create a computer-aided design model of the component. The examination of the three-dimensional-printed component completes the manufacturing chain. Adhesive spraying, materials' extrusion, and powders' bed fusion are 3 extensively used machine learning processes among 7 different types. A popular printed method depending on the material extrusion is fused deposition modeling (FDM). Selective Laser Melting (SLM), Electron Beam Melting (EBM), Selective Laser Sintering (SLS), and other processes fall within the category of powder bed fusion.

Figure 3. Applications of Machine leaning methods in Additive Manufacturing

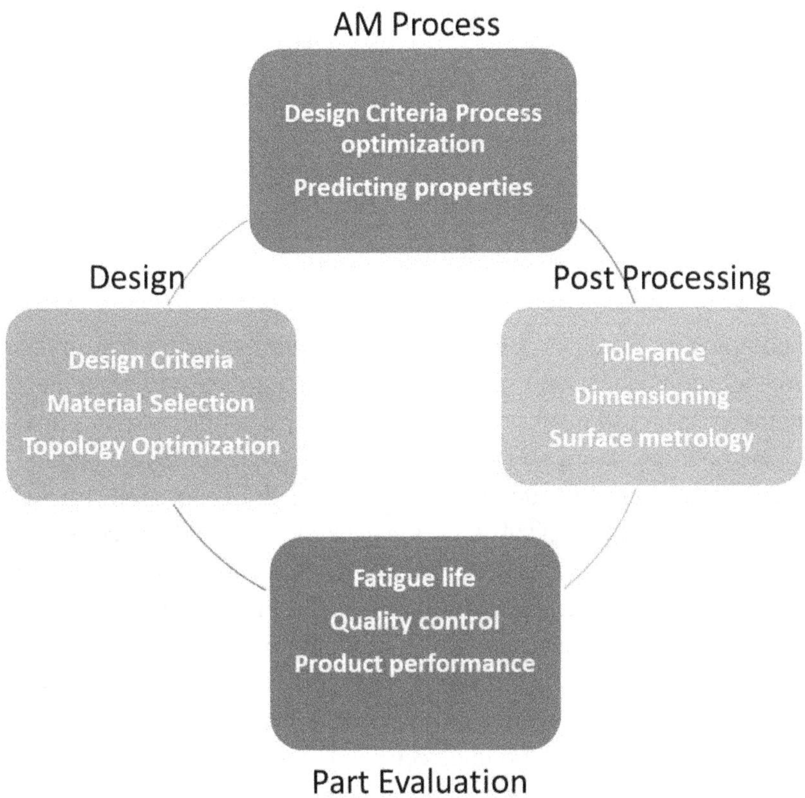

Machine Learning for Forecasting Mechanical Characteristics

Even though there are techniques to anticipate the mechanical behavior of components, such as FEM modeling, these might vary with experimental observations due to simple assumptions. As a result, past work has investigated the viability of machine learning techniques in forecasting the mechanical behavior of engineering structures. The review primarily concentrates on 3 categories where used machine learning approaches have an impact on the functionality and mechanical behavior of three-dimensional-printed parts. The following are these categories: a) Processing parameter optimization, b) porosity forecasting, and c) flaw identification in three-dimensional-printed objects. Because printing variables, porosity, and flaws can dramatically alter how well components made using additive manufacturing perform mechanically, optimizing and predicting these characteristics can help the items perform more mechanically. Three categories of methodologies are used to categorize machine learning strategies that use machine learning methodologies in the present investigation. These methodologies and the pertinent methods are shown in Figure 4.

Figure 4. Additive Manufacturing process involving machine learning

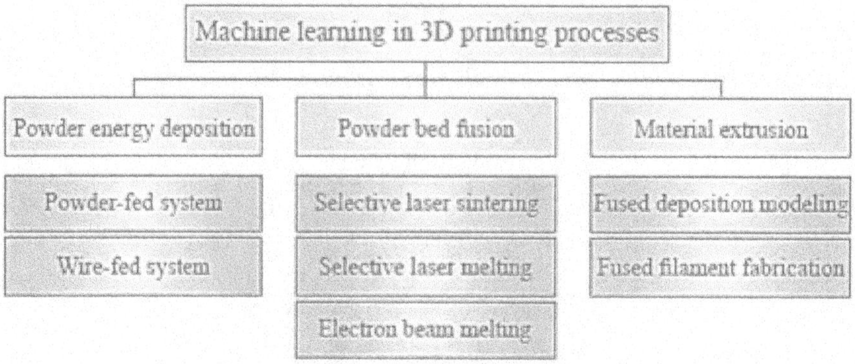

Various solid materials, including metallic materials, are classified as structure or function components, and the qualities of these components rely on the sort of binding that exists among their component atoms and how those atoms are arranged geometrically. While many ceramic materials display a combination of covalent and ionic bonding, metal and its alloys typically have metallic bonding. Through various techniques, atomic organization and structure can be controlled. Because the structure does have a significant impact on the characteristics of substances, a general framework is needed to assess a particular microstructural characteristic. A thorough analysis of the structural modeling of three-dimensional-printed metal items is provided in (J. H. K. T. and Sing SL et al., 2020). Knowing how structure and mechanical characteristics relate to one another is important for optimizing the mechanical behavior of metal components. Varieties of different stages including, grain size, details of their construction, form, and sizes are used to define the morphology of engineered materials. One-dimensional flaws like dislocations, zero-dimensional defects like point defects, and two-dimensional faults like grain boundaries are significant surface morphology characteristics that frequently regulate the ensuing characteristics.

Mechanical stresses affect the structure and durability of metals in metal engineering structures. The behavior of the components during aggregation depends on microstructural evolution. Various metals materials' micrographs, such as pores at grain boundaries, can cause morphological and structural damage and subsequently shorten lifespan based on the loading circumstances. Consequently, research on crystal structures is crucial. Component micrographs were studied using a variety of characterization approaches over the years (Zaldivar RJ et al., 2018, Sing SL et al., 2020, Li H et al., 2020, Sing SL et al., 2021). The impact of post-processing on the structure and thermal characteristics of three-dimensional printed objects is currently being researched in (Pascual-Gonz alez C et al., 2021).

Numerous morphology characteristics are connected to mechanical qualities. For instance, particle volume percent, preferential grain orientations (textured), phase transformations, and particle density, form, and volumes. Though machine learning approaches were used to forecast specific behaviors, characteristics, and fatigue resistance of three-dimensional-printed components (Khadilkar A et al., 2019, Zhang M et al., 2019, Herriott C and Spear AD, 2020, Zolfagharian A et al., 2021), the subject of the ongoing investigation is not directly related to previous researchers' efforts.

Processing Parameter Optimization

Various processing variables are essential to the effectiveness and functionality of parts produced when parts are made employing three-dimensional printing methods. As a result, the impacts of such variables have really been studied in numerous study projects (Khosravani MR et al., 2020, Chikkangoudar RN et al., 2020, Ramalingam PS et al., 2020, Pihlehvar S et al., 2021), however, experiments are expensive and time-consuming. Machine learning approaches outperformed a variety of choices in the optimization of three-dimensional printing processing variables. In (Tapia G et al., 2017), a system based on the Gaussian distribution has been used to forecast the depth of the melt pool during laser powder-bed fusion. Scanning rate and laser power were taken into account as variables in this situation, and an experiment regarding manufacturing 316L stainless steel was used. When the system was tested, it performed well and had a small mean absolute prediction error.

Melt pool structure has a substantial effect on the performance of Direct Energy Deposition (DED) three-dimensionally printed objects (e.g., shape, continuation, and homogeneity). In this context, machine learning was used to predict the width, depth, and height of the melt pool in different direct energy deposition processes (Li Y et al., 2018, Mozaffar M et al., 2018, Khanzadeh M et al., 2018, Johnson NS et al., 2020, Sampson R et al., 2020). Using an ultrafast camera, (Zhang Y et al., 2018) constructed a field of view for identifying the properties of the melt pool, plumes, and spatter in a Powder Bed Fusion (PBF) technique. On the basis of system information, the extracted attributes were input into a typical machine learning model. A computational neural network model was employed to explicitly detect quality abnormalities, and the method achieved a 92.70 percent success rate in predicting quality.

A relationship between both the inputs laser metal depositing processing variables and the final design parameters of the printed item was discovered using machine learning in the direct energy deposition method in (Caiazzo F and Caggiano A, 2018) at the very same period. The results of the research were utilized to educate the artificial neural network in two stages. The analysis indicates that neural network-based machine learning can predict process variables required to build a metal object with a particular shape precisely. Machine learning is helpful to optimize processing variables in the fused deposition modeling technology, according to a prior study (Mohamed OA et al., 2017). In this case, visco-elastic reactions are improved by optimizing film thickness, free space, building orientations, raster angle, and quantity of contour. In other studies (Sood AK et al., 2012, Sood AK et al., 2012), machine learning was employed to increase the three-dimensional-printed polymeric parts' compression strength and durability by optimizing the process parameters.

The significance of this process optimization has grown as three-dimensional printing's uses in the construction sector have greatly risen significantly. In this connection, parameters optimization for three-dimensional printing of cement-based materials using machine learning recently in (Liu Z et al., 2020). In order to explore the materials' flow strategy throughout extruded, scientists have created a mathematical method. To ascertain the effects of various variables on the flow mechanisms, they used an SVM-based machine learning technique. The network was evaluated using scientific results, and the SVM modeling findings showed that printing rate and materials yielding stress have a significant impact on the deformation of printed filaments, while plastic viscosity would not be a factor.

Most recently, in (Zhang Z et al., 2021), machine learning was used to create a data-driven model for predicting the temperatures of the melt pool in the direct energy deposition process. Extreme Gradient Boosting (XGBoost) and Long Short-Term Memory (LSTM), 2 machine learning techniques that are easily customizable and excellent at interpreting time-series data, were specifically utilized by the investigators. A nickel-based superalloy was also utilized to fabricate a thin-walled testing coupon and gauge the temperatures of the melt pool inside every layer. The acquired findings demonstrated that both machine learning systems successfully forecasted the melt pool temperatures. The machine learning also optimized three-dimensional extrusion bio-printing at the very same time (Ruberu K et al., 2021). In order to build a replicable three-dimensional scaffold, the Bayesian optimization method was used.

Various bioink proportions and print variables, such as extruded pressure, reservoir temperatures, platform temperatures, and scan speed, are fed into the solution space of the optimization. To suggest the printing configuration for the laboratory step, a probability modeling of the systems was used as output. To assess print quality, the authors created a classification scale for intercalation and filaments structure. A quantitative research method for extruder three-dimensional printing, the Bayesian optimization strategy was shown to be effective by the recorded outcomes. The suggested Bayesian technique may be utilized to enhance the print quality of various bioink technologies based on extrusion.

According to study in the academic literature, artificial neural networks are the most often used machine learning technique for improving the performance of diverse three-dimensional printing technologies. According to a comparison of supervised and unsupervised learning algorithms, unsupervised learning approaches showed the lowest accuracy when applied to AM process improvements. For instance, K means clustering was used to categorise pores in (Snell R et al., 2019), and the accuracy varied between 40 and 45 percent.

According to a review of the extant literature, the degrees of used machine learning techniques in diverse three-dimensional printing procedures are macro scale levels (such as mechanical qualities) and mesoscale levels (e.g., melt pool geometries). Employing machine learning is advantageous because the abovementioned levels' many variables have a major impact on the final product performance. More uses of this methodology in optimizing the process parameters of three-dimensional printing technologies are anticipated given the benefits of machine learning techniques (Mohamed OA et al., 2016, Vahabi E and Rahmati S, 2016, Asadi-Evdivand et al., 2016, Bayraktar O et al., 2017, Zhang W et al., 2017, Li Z et al., 2019, Singh A et al., 2012).

Porosity Forecasting

While one of the main goals of metal three-dimensional printing is to achieve complete solidity, the permeability of the pieces is very important. To forecast permeability, which has a direct impact on the mechanical behavior of the materials, machine learning approaches have been employed in this context. Back-Propagation Neural Network (BPNN) technology was used in (99) to three-dimensional printing bronze and estimate porosity by determining volume fraction. The classifier was tested using a variety of training algorithms, and predictive ability was improved using an optimization focused on the Genetic Algorithm (GA).

The part location and orientations were connected to the part porosity using random forest network machine learning (ML) in a similar manner (Kappes B et al., 2018). Each design is made up of parts that are oriented in relation to the build direction (polar angle). The study's findings showed that the orientations of the component had the biggest impact on the pore population. describes the development of a Gaussian process-based prediction model to forecast porosity in metal-based three-dimensional printing. The parameters of the model were precisely estimated using the Bayesian framework, and forecasts of the permeability at any specified site were made using these estimations. The relationship between scanning speed and laser power and the part's permeability was established. These variables have the most impact on the component's porosity. The researcher makes use of actual data from stainless steel selective laser melting (Tapia G et al., 2016).

In order to predict porosity and achieve optimal porosity by altering processing parameters in selective laser sintering printing technology, film thickness and scan speed were used as inputs for machine learning-based systems (Garg A et al., 2015, Garg A and Lam JSL, 2015). Specifically, a multi-gene dynamic programming technique was employed to generate the model in for porosity analysis in Additive manufacturing (Garg A et al., 2015). Since the issue of generalization hinders the implementation of multigene genetic codes, machine learning methods have been used. In the first step, the initialization was constructed by combining the genes using the least-squares method. Using the structure reduction method, the initial population was then approximated. The subsequent phase includes general processes for the establishment of a new generation. The studies revealed the system's benefits for boosting the efficiency of selective laser sintering.

Previous research indicated that supervised machine learning generates a black-box model for the probability distribution of porosity. The approach benefits from this model's independence from sample design and material properties. Because permeability has a substantial influence on the mechanical behaviour of three-dimensionally printed components, there is a great need to develop a system that can predict their permeability. This is due to the increasing utilisation of 3D-printed components as functioning end products (Montazeri M and Rao P, 2018, Khanzadeh M et al., 2018, Khanzadeh Met al., 2019, Jafari-Marandi et al., 2019, Zhang B et al., 2019, Gobert C et al., 2020).

Flaw Detection

There may still be processing-related faults (such as debonding, cracks, and rough surfaces) in these techniques, despite the significant advancements achieved in three-dimensional printing technology. Some faults have the capacity to propagate from one layer to another. From this viewpoint, machine learning (ML) might be seen as a solution-oriented technology. K-mean clustering was utilized to differentiate between normal and abnormal printing processes in (Wu H et al., 2016). The alteration in the audio signal pattern would indicate that the printing technology had changed from regular to atypical settings, according to investigators who employed acoustic waves as a sensing tool. A variety of empirical practices on the fused deposition modeling technique were carried out, and the reported outcomes showed that this approach can lower manufacturing waste. In a similar vein, flaws were found and located using K-means clustering (Grasso M et al., 2017). K-means clustering was employed to identify the flaw caused by the overheating since the brightness patterns of melt pool image pixels in overheated areas are vastly different from normal melting conditions. Additionally, in order to locate trouble spots of the print layers, the research employed a statistic classifier to image information. In experimental experiments, hollowed cylinders and structures depending on the selective laser melting technique were printed in stainless steel. The test results demonstrated that the proposed method operated effectively in both simple and complicated geometries. In parallel, machine learning was used in (Wu M et al., 2017) to develop a vision-based system for the detection of intentional assaults in three-dimensional printing.

In fact, unique infills were found when real-world gray-scale photographs and simulations of the additive manufacturing process were examined. 3 machine learning algorithms—KNN, random forest, and anomaly detection—were specifically applied, and the suggested strategy was used to assess honeycomb pieces manufactured using the fused deposition modeling process. The study's findings suggested that outlier detection performed better than KNN or random forest. Additionally, the results showed that mounting a webcam on a 3D printer frame causes unnecessary movements, which decreased the effectiveness of detection.

The next year, investigators used a multi-scale CNN to train the network and find flaws in the PBF procedure through the use of modern computer vision techniques (Scime L and Beuth J,2018). To do this, square picture patches from images taken during several builds were chosen by a human expert.

To allow the algorithms to learn important context data and the appearances of the powder bed abnormalities, the CNN's input layer was updated. The anomaly classification can be seen in several formats, as researchers stated in their prior paper (Scime L and Beuth J, 2018). The experimental test revealed that the system performed better than expected. ML and acoustic signals were simultaneously employed to identify flaws in the SLM process (Ye D et al., 2018). To accomplish this, a Deep Belief Network (DBN) was used, and the recommended technique employed a simplified classification structure without signal processing and feature extraction. The DBN is a probabilistic visual model with many layers of latent variables (hidden units). On a 3D-printed steel component, researchers conducted many tests and collected acoustic data from five distinct patterns. The presented findings demonstrated that DBN was effective at recognising various difficulties. Similarly, supervised machine learning was used to develop a metallic 3D printing technique for in situ fault identification (Gobert C et al., 2018). Using the SVM approach and input characteristics including layer-wise images and ex-situ CT scan data, certain defects were identified. Manual inspection or automated analysis identified incomplete fusion and fractures, and the researcher's utilized 3D CT scan data to train the binary classifiers. After training, validation tests demonstrated the effectiveness of the proposed technique.

The next year, Deep Convolutional Neural Network (DCNN) was used for online defect identification in selective laser melting (Caggiano A et al., 2019). The experimental assessment demonstrated that the proposed method is successful in identifying defects resulting from process nonconformities. Using the selective laser melting method described in (Okaro IA et al., 2019), tensile test bars were made, and a semi-supervised ML algorithm was utilized to identify faults. Specifically, a nickel-based alloy was used to print the samples. Based on ultimate tensile strength testing and the Gaussian Mixture Model, samples were separated into excellent and poor 3D-printed components (GMM). In semi-supervised learning, labeled and unlabeled data from whole data sets are used. Each extracted bar feature was subjected to semi-supervised learning. If the chance of a problem was more than 0.50, the samples were judged flawed. The semi-supervised GMM was trained using unlabeled points chosen at random. According to the collected data, defective samples were recognised with a high rate of success. Similarly, deep learning was used to predict distortion in 3D printing using lasers (Francis J and Bian L, 2019). Traditional neural networks and artificial neural networks were used to evaluate thermal pictures and pertinent process parameters, respectively. The output of conventional neural networks and artificial neural networks was combined and trained further to get the final prediction of distortion. Conventional neural networks consist of two pooling layers to lower the spatial dimension and, thus, the computational complexity. As a case study, a disc was manufactured, and experimental results confirmed the suggested method's accuracy in predicting distortions. Using machine learning (ML) (Wu M et al., 2019), harmful infill structures generated by the fused deposition modelling technique were concurrently identified. Cyber security and physical data machine learning have been merged in order to improve the accuracy of detection. In this case, KNN and random forest were used, and 3D printing of polymeric components was taken into account while assessing the suggested method. During the printing process, researchers used both actual camera pictures and 3D printing software simulations. Using a camera-based vision detection system, the picture classification strategy was also evaluated. A real-time detection prototype was developed to alert administrators of dangerous situations. The collected data demonstrated the system's greater accuracy in spotting printing process flaws. Later, in (Khan MF et al., 2020), ML was used to identify defects in the fused deposition modelling process and reduce the requirement for human quality assurance. The proposed technique used the extraction of characteristics of manufacturing-discovered geometrical anomalies and was based on standard neural networks. Weak infills, uneven extrusion, and a lack of reinforcements may all lead to the occurrence of irregularities. Experiment results indicated that the suggested algorithm was capable of detecting faults and interrupting the printing process in the midst of them.

In a performed investigation, machine learning techniques were employed to determine the error in 3D printing through extrusion (Jin Z et al., 2021). In this instance, two customized machine learning algorithms, semantic segmentation model and localization model, were used to capture real-time pictures utilizing an extrusion-mounted camera. Using the semantic segmentation approach, the criteria (such as high quality or anomalies), the location, and the amount of faults were found. In addition, an algorithm was used to identify defects in both the current and previous layers. By defining a threshold, highly sensitive areas would be seen on the input photographs. Various hues are used to indicate model layers with distinct functionalities. Using the specified evaluation methods for 3D printing machines revealed the intervention's success.

BENEFITS, CHALLENGES, AND MOTIVATION

Even while experiments produce trustworthy results, the necessary equipment can make them occasionally uneconomical. Mathematical computations are less expensive than experimentation but could not be as accurate. In this situation, machine learning surpasses the restrictions on a study in the field of additive manufacturing. Earlier studies looked at the effectiveness of the machine learning-assisted experimental technique. The experiments proved that applying machine learning to the field of three-dimensional printing can result in simulations that are completed more quickly than those produced by purely physical simulations. A requirement for enhancing a material's characteristics is a comprehension of the impacts of morphology factors. It is interesting that ML algorithms can be applied to the systems to produce the necessary microstructure that complies with the necessary requirements. It might be viewed as one of the key benefits of the machine learning approach in comparison to other widely used methods. This opens the door for creating microstructures with unique features.

Significant advancements in machine learning applications in this area are anticipated since additive manufacturing, can be used to fabricate Functionally Graded Materials. Even though Functionally Graded Materials have been extensively used in contemporary industrial applications, there are still some problems that call for more advancements. For instance, the inclusion of uncertainties during the manufacturing process causes the intended and actual Functionally Graded Materials to diverge. In this regard, machine learning can be utilized to create a reliable analytic system that is suitable. Additionally, machine learning can be quite critical in analyzing how Functionally Graded Materials structures behave mechanically under various circumstances.

Given the different difficulties, further study is needed to concentrate on machine learning applications in three-dimensional printing technology. A review of earlier research projects revealed that many databases had been created using instruments, methods, and concepts from various disciplines. In reality, mechanical engineering, computer science, manufacturing, and data science have collaborated to accelerate production breakthroughs and enhance product quality. Despite the excellent achievements, the materials community lacks a strong data-sharing culture. In order to exchange knowledge and improve manufacturing procedures, it is necessary to create a standard database environment for storing data generated by diverse academic and industrial research institutions.

CONCLUSION

Though three-dimensional printing technologies were used extensively across a variety of industries across the globe, but are still in the early stages of development and encounter a number of production-related issues. Various experiments have been conducted in this regard to ascertain the influence of printing processing parameters on the mechanical behavior of finished goods. Other procedures that are precise and efficient in terms of cost have been used in this discipline because experimentation practices are time-consuming and expensive ways. Owing to its high qualities, machine learning has recently generated a lot of study interest in three-dimensional printing. Machine learning algorithms are helpful in a variety of applications for three-dimensional printing. Despite the fact that machine learning can be applied to a variety of tasks in three-dimensional printing, including the design process, optimization, morphology description, and quality analysis, the present program consists of three machine learning applications in three aspects that have an impact on the mechanical behavior of finished 3D-printed components. The following are some domains: flaw identification, porosity forecasting, and processing parameters optimization. Uses of machine learning-based systems can boost productivity and speed up industrial breakthroughs because the abovementioned categories have a major impact on the mechanical characteristics of the finished goods. In the current paper, a general review of machine learning (ML) has been provided, and prior reports using machine learning to anticipate the mechanical behavior of three-dimensional-printed components have been reviewed. Furthermore, we discussed our predictions for machine learning applications in three-dimensional printing going forward. The stated difficulties examined in research articles, and taken into account will all be then used to create new machine learning (ML)-based systems in the future. In order to meet the needs, there are a number of topics that need to be studied and methods that need to be taken to enhance the use of machine learning three-dimensional printing technology.

REFERENCES

Amini, M., Chang, S.I., & Rao, P. (2019). A cybermanufacturing and AI framework for laser powder bed fusion (LPBF) additive manufacturing process. *Manuf. Lett., 21.*

Aoyagi, K., Wang, H., Sudo, H., & Chiba, A. (2019). Simple method to construct process maps for additive manufacturing using asupport vector machine. *Addit. Manuf., 27.*

Asadi-Eydivand, M., Solati-Hashjin, M., Fathi, M., Padashi, M., & Abu Osman, N.A. (2016). Optimal design of a 3D-printed scaffold using intelligent evolutionary algorithms. *Appl Soft Comput, 39.*

ASTM. (2012). *F2792 -12 standard terminology for additive manufacturing technologies, standard.* American Society for Testing Materials.

Awoyera, P.O., Kirgiz, M.S., Viloria, A., & Ovallos Gazabon, D. (2020). Estimating strength properties of geopolymer selfcompacting concrete using machine learning techniques. *J. Mater. Res. Technol., 9.*

Bayliss, C. (2021). Machine learning based simulation optimization for urban routing problems. *Applied Soft Computing, 105*, 107269.

Bayraktar, O., Uzun, G., Cakiroglu, R., & Guldas, A. (2017). Experimental study on the 3D-printed plastic parts and predicting the mechanical properties using artificial neural networks. *Polym Adv Technol, 28.*

Bilal, M., & Oyedele, L. O. (2020). Guidelines for applied machine learning in construction industry - a case of profit margins estimation. *Advanced Engineering Informatics, 43*, 101013.

Bohn, B., Garcke, J., Iza Teran, R., Paprotny, A., Peherstorfer, B., & Schepsmeier, U. (2013). Analysis of car crash simulation data with nonlinear machine learning methods. *Proc. Comput. Sci., 18.*

Caggiano, A., Zhang, J., Alfieri, V., Caiazzo, F., Gao, R., & Teti, R. (2019). Machine learning-based image processing for on-line defect recognition in additive manufacturing. *CIRP J. Manuf. Sci. Technol., 68.*

Caiazzo, F., & Caggiano, A. (2018). Laser direct metal deposition of 2024 Al alloy: trace geometry prediction via machine learning. *Materials, 11.*

Chen, Q., Guillemot, G., Gandin, C.A., & Bellet, M. (2017). Three dimensional finite element thermo-mechanical modeling of additive manufacturing by selective laser melting for ceramic materials. *Addit. Manuf., 16.*

Cheng, M. Y., Kusoemo, D., & Gosno, R. A. (2020). Text mining-based construction site accident classification using hybridsupervised machine learning. *Automation in Construction, 118*, 103265.

Chikkangoudar, R.N., Sachidananda, T.G., & Pattar, N. (2020). Influence of 3D printing parameters on the dimensional stability of polypropylene/clay printed parts using laser scanning technique. *Mater. Today: Proc.*

Colorado, H.A., Velasquez, E.I.G., & Monteiro, S. (2020). Sustainability of additive manufacturing: the circular economy of materials and environmental perspectives. *J. Mater. Res. Technol., 9.*

Decost, B.L., Jain, H.A., Rollett, A., & Holm, E.A. (2017). Computer vision and machine learning for autonomous characterization of AM powder feedstocks. *J Occup Med, 69.*

Delli, U., & Chang, S. (2018). Automated process monitoring in 3D printing using supervised machine learning. *Procedia Manuf, 26.*

Fergani, O., Berto, F., Welo, T., & Liang, S.Y. (2017). Analytical modelling of residual stress in additive manufacturing. *Fatig Fract Eng Mater Struct, 40.*

Francis, J., & Bian, L. (2019). Deep learning for distortion prediction in laser-based additive manufacturing using big data. *Manuf. Lett., 20.*

Garg, A., & Lam, J.S.L. (2015). Measurement of environmental aspect of 3-D printing process using soft computing methods. *Measure, 75.*

Garg, A., Lam, J.S.L., & Savalani, M.M. (2015). A new computational intelligence approach in formulation of functional relationship of open porosity of the additive manufacturing process. *Int J Adv Manuf Technol, 80.*

Geetha, N. K., & Bridjes, P. (2020). Overview of machine learning and its adaptability in mechanical engineering. *Materials Today: Proceedings.*

Gobert, C., Kudzal, A., Sietnis, J., Mock, C., Sun, J., & McWilliams, B. (2020). Porosity segmentation in X-ray computed tomography scans of metal additively manufactured specimens with machine learning. *Addit. Manuf.*, *36*, 101460.

Gobert, C., Reutzel, E.W., Petrich, J., Nassar, A.R., & Phoha, S. (2018). Application of supervised machine learning for defect detection during metallic powder bed fusion additive manufacturing using high resolution imaging. *Addit. Manuf., 21.*

Goh, G.D., Dikshit, V., An, J., & Yeong, W.Y. (2020). Process-structureproperty of additively manufactured continuous carbon fiber reinforced thermoplastic: An investigation of mode I interlaminar fracture toughness. *Mech Adv Mater Struct, 27.*

Goh, G.D., Sing, S.L., & Yeong, W.Y. (2021). A review on machine learning in 3D printing: applications, potential, and challenges. *Artif Intell Rev, 54.*

Grasso, M., Laguzza, V., Semeraro, Q., & Colosimo, B. M. (2017). In-process monitoring of selective laser melting: Spatial detection of defects viaimagedataanalysis. *Journal of Manufacturing Science and Engineering*, *139*, 051001.

Gu, G.X., Chen, C.T., Richmond, D.J., & Buehler, M.J. (2018). Bioinspired hierarchical composite design using machine learning: simulation, additive manufacturing, and experiment. *Mater. Horiz., 5.*

Herriott, C., & Spear, A. D. (2020). Predicting microstructure-dependent mechanical properties in additively manufactured metals with machine- and deep-learning methods. *Computational Materials Science*, *175*, 109599.

Hertlein, N., & Deshpande, S. (2020). Prediction of selective laser melting part quality using hybrid Bayesian network. *Addit. Manuf.*, *32*, 101089.

J. H. K. T. & Sing. (2020). Microstructure modelling for metallic additive manufacturing: a review. *Virtual Phys Prototyp, 15.*

Jafari-Marandi, R., Khanzadeh, M., Tian, W., Smith, B., & Bian, L. (2019). From in-situmonitoring toward high-throughput process control: cost-driven decision-making framework for laser-based additive manufacturing. *J Manuf Syst, 51.*

Jiang, J., Hu, G., Li, X., Xu, X., Zhang, P., & Stringer, J. (2019). Analysis and prediction of printable bridge length in fused deposition modelling based on back propagation neural network. *Virtual Phys Prototyp, 14.*

Jin, Z., Zhang, Z., Ott, J., & Gu, G. X. (2021). Precise localization and semantic segmentation detection of printing conditions in fused filament fabrication technologies using machine learning. *Addit. Manuf.*, *37*, 101696.

Johnson, N. S., Vulimiri, P. S., To, A. C., Zhang, X., Brice, C. A., & Kappers, B. B. (2020). Invited review: Machine learning for materials developments in metals additive manufacturing. *Addit. Manuf.*, *36*, 101641.

Jordan, M.I., & Mitchell, T.M. (2015). Machine learning: trends, perspectives, and prospects. *Science, 349.*

Juan, Y., Dai, Y., Yang, Y., & Zhang, J. (2021). Accelerating materials discovery using machine learning. *J Mater Sci Technol, 79.*

Kappes, B., Moorthy, S., Drake, D., Geerlings, H., & Stebner, A. (2018). Machine learning to optimize additive manufacturing parameters for laser powder bed fusion of Inconel 718. *Proceedings of the 9th international symposium on superalloy 718 & derivatives: energy, aerospace, and industrial applications.*

Karayel, E., & Bozkurt, Y. (2020). Additive manufacturing method and different welding applications. *J. Mater. Res. Technol., 9.*

Khadilkar, A., Wang, J., & Rai, R. (2019). Deep learning-based stress prediction for bottom-up SLA 3D printing process. *Int J Adv Manuf Technol, 102.*

Khan, M.F., Alam, A., Siddiqui, M.A., Alam, M.S., Rafat, Y., & Salik, N. (2020). Real-time defect detection in 3D printing using machine learning. *Mater. Today: Proc.*

Khanzadeh, M., Chowdhury, S., Marufuzzaman, M., Tschop, M.A., & Bian, L. (2018). Porosity prediction: supervised-learning of thermal history for direct laser deposition. *J Manuf Syst, 47.*

Khanzadeh, M., Chwodhury, S., Tschopp, M.A., Doude, H.R., Marufuzzaman, M., & Bian, L. (2019). In-situ monitoring of melt pool images for porosity prediction in directed energy deposition processes. *IISE Trans, 51.*

Khanzadeh, M., Rao, P., Jafari Mardani, R., Smith, B.K., Tschopp, M.A., & Bian, L. (2018). Quantifying geometric accuracy with unsupervised machine learning: using self-organizing map on fused filament fabrication additive manufacturing parts. *J Manuf Sci Eng, 140.*

Khanzadeh, M., Tian, W., Yadollahi, A., Doude, H.R., Tschop, M.A., & Bian, L. (2018). Dual process monitoring of metal-based additive manufacturing using tensorde composition of thermal image streams. *Addit. Manuf., 23.*

Khosravani, M. R., Berto, F., Ayatollahi, M. R., & Reinicke, T. (2020). Fracture behavior of additively manufactured components: A review. *Theoretical and Applied Fracture Mechanics*, *109*, 102763.

Khosravani, M. R., & Reinicke, T. (2020). 3D-printed sensors: Current progress and future challenges. *Sensors and Actuators. A, Physical*, *305*, 111916. doi:10.1016/j.sna.2020.111916

Khosravani, M. R., & Reinicke, T. (2020). On the environmental impacts of 3D printing technology. *Appl. Mater. Today*, *20*, 100689.

Khosravani, M.R., & Reinicke, T. (2020). Effects of raster layup and printing speed on strength of 3D-printed structural components. *Procedia Struct. Integrity, 28.*

Kong, L., Ambrosi, A., Nasir, M.Z.M., Guan, J., & Pumera, M. (2019). Selfpropelled 3D-printed "Aircraft Carrier" of light-powered smart micromachines for large-volume nitroaromatic explosives removal. *Adv Funct Mater, 19*, 1-9.

Leal, R., Barreiros, F.M., Alves, L., Romeiro, F., Vasco, J.C., & Santos, M. (2017). Additive manufacturing tooling for the automotive industry. *Int J Adv Manuf Technol, 92*, 1671-6.

Li, H., Liu, Y., Liu, Y., Zeng, Q., Hu, K., & Lu, Z. (2020). Effect of debinding temperature under an argon atmosphere on the microstructure and properties of 3D-printed alumina ceramics. *Materials Characterization, 168*, 110548.

Li, Y., Li, X., Zhang, G., Horvath, I., & Han, Q. (2021). Interlayer closed-loop control of forming geometries for wire and arc additive manufacturing based on fuzzy-logic inference. *J Manuf Process, 63.*

Li, Y., Sun, Y., Han, Q., Zhang, G., & Horvath, I. (2018). Enhanced beads overlapping model for wire and arc additive manufacturing of multi-layer multi-bead metallic parts. *J Mater Process Technol, 252.*

Li, Z., Liu, R., & Wu, D. (2019). Data-driven smart manufacturing: tool wear monitoring with audio signals and machine learning. *J Manuf Process, 48.*

Li, Z., Zhang, Z., Shi, J., & Wu, D. (2019). Prediction of surface roughness in extrusion-based additive manufacturing with machine learning. *Robot Comput Integrated Manuf, 57.*

Liu, Y., Niu, C., Wang, Z., Gan, Y., Zhu, Y., & Sun, S. (2020). Machine learning in materials genome initiative: a review. *J Mater Sci Technol, 57.*

Liu, Y., Zhao, T., Ju, W., & Shi, S. (2017). Materials discovery and design using machine learning. *J Mater, 3.*

Liu, Z., Li, M., Weng, Y., Qian, Y., Wong, T. N., & Tan, M. J. (2020). Modelling and parameter optimization for filament deformation in 3D cementitious material printing using support vector machine. *Composites. Part B, Engineering, 193*, 108018.

Lu, Z.L., Li, D.C., Lu, B.H., Zhang, A.F., Zhu, G.X., & Pi, G. (2010). The prediction of the building precision in the laser engineered net shaping process using advanced networks. *Opt Laser Eng, 48.*

Machment, T., Sanjayan, J., & Xia, M. (2019). Method of enhancing interlayer bond strength in construction scale 3D printing with mortar by effective bond area amplification. *Materials & Design, 169*, 107684. doi:10.1016/j.matdes.2019.107684

Messaoud, S., Bradai, A., Bukhari, S. H. R., Quang, P. T. A., Ahmed, O. B., & Atri, M. (2020). A survey on machine learning in Internet of Things: Algorithms, strategies, and applications. *Internet Things, 12*, 100314.

Mohamed, O.A., Masood, S.H., & Bhowmik, J.L. (2017). Influence of processing parameters on creep and recovery behavior of FDM manufactured part using definitive screening design and ANN. *Rapid Prototyp J, 23.*

Mohamed, O.A., Masood, S.H., & Bhowmik, J.L. (2016). Investigation of dynamic elastic deformation of parts processed by fused deposition modeling additive manufacturing. *Adv. Prod. Eng. Manage., 13.*

Montazeri, M., & Rao, P. (2018). Sensor-based build condition monitoring in laser powder bed fusion additive manufacturing process using a spectral graph theoretic approach. *Journal of Manufacturing Science and Engineering*, *140*, 091002.

Mozaffar, M., Paul, A., Al Bahrani, R., Wolff, S., Choudhary, A., & Agrawal, A. (2018). = Data-driven prediction of the highdimensional thermal history indirected energy deposition processes via recurrent neural networks. *Manuf. Lett., 18.*

Munprom, R., & Limtasiri, S. (2019). Optimization of stereolithographic 3D printing parameters using Taguchi method for improvement in mechanical properties. *Mater. Today: Proc., 17.*

Murr, L.E. (2019). Metallurgy principles applied to powder bed fusion 3D printing/additive manufacturing of personalized and optimized metal and alloy biomedical implants: An overview. *J. Mater. Res. Technol., 9.*

Nasiri, S., & Khosravani, M. R. (2020). Progress and challenges in fabrication of wearable sensors for health monitoring. *Sensors and Actuators. A, Physical*, *312*, 112105. doi:10.1016/j.sna.2020.112105

Okaro, I.A., Jayasinghe, S., Stucliffe, C., Black, K., & Paoletti, P. (2019). Automatic fault detection for laser powder-bed fusion using semi-supervised machine learning. *Addit. Manuf., 27.*

Palanikumar, K., Mudhukrishnan, M., & Prahba, P.S. (2020). Technologies in additive manufacturing for fiber reinforced composite materials: a review. *Curr. Opin. Chem. Eng., 28*, 51-9.

Pascual-Gonzalez, C., Martı´n, P.S., Lizarralde, I., Fernandez, A., Leon, A., & Lopes, C.S. (2021). Post-processing effects on microstructure, interlaminar and thermal properties of 3D printed continuous carbon fibre composites. *Composites, Part B, 210.*

Pazhamannil, R.V., Govindan, P., & Sooraj, P. (2020). Prediction of the tensile strength of polylactic acid fused deposition models using artificial neural network technique. *Mater. Today: Proc.*

Pei, S., Wang, K., Chen, C.B., Li, J., Li, Y., & Zeng, D. (2021). Process structure- property analysis of short carbon fiber reinforced polymer composite via fused filament fabrication. *J Manuf Process, 64.*

Pihlehvar, S., Arnhof, M., Erichsen, A., Valentini, L., & Kjoniksen, A. (2021). Investigation of severe lunar environmental conditions on the physical and mechanical properties of lunar regolith geopolymers. *J. Mater. Res. Technol., 12.*

Qi, X., Chen, G., Li, Y., Cheng, X., & Li, C. (2019). Applying neural-networkbased machine learning to additive manufacturing: current applications, challenges, and future perspectives. *Engineering, 5.*

Rahman, A., Deshpande, P., Raude, M., Odegard, G. M., Gowtham, S., & Ghosj, S. (2021). A machine learning framework for predicting the shear strength of carbonnanotube polymerr interfaces based on molecular dynamics simulation data. *Composites Science and Technology*, *207*, 108627.

Ramalingam, P.S., Mayandi, K., Balasubramanian, V., Chandrasekar, K., Stalany, V.M., & Munaf, A. (2020). Effect of 3D printing process parameters on the impact strength of onyx-Glass fiber reinforced composites. *Mater. Today: Proc.*

Ren, D., Li, S., Wang, H., Hou, W., Hao, Y., & Jin, W. (2019). Fatigue behavior of Ti-6Al-4V cellular structures fabricated by additive manufacturing technique. *J Mater Sci Technol, 35*, 285-94.

Ruberu, K., Senadeera, M., Rana, S., Gupta, S., Chung, J., & Yue, Z. (2021). Coupling machine learning with 3D bioprinting to fast track optimisation of extrusion printing. *Appl. Mater. Today, 22*, 100914.

Rueden, L., Mayer, S., Sifa, R., Bauckhage, C., & Garcke, J. (2020). Combining machine learning and simulation to a hybrid modelling approach: current and future directions. *Proceedings of the 18th international symposium on intelligent data analysis.*

Saeed, K., McIlhagger, A., Harkin Jones, E., Kelly, J., & Archer, E. (2021). Predication of the in-plane mechanical properties of continuous carbon fibre reinforced 3D printed polymer composites using classical laminated-plate theory. *Composite Structures, 259*, 113226.

Sampson, R., Lancaster, R., Stucliie, M., Carswell, D., Hauser, C., & Barras, J. (2020). An improved methodology of melt pool monitoring of direct energy deposition processes. *Optics & Laser Technology, 127*, 106194.

Scime, L., & Beuth, J. (2019). Using machine learning to identify in-situ melt pool signatures indicative of flaw formation in a laser powder bed fusion additive manufacturing process. *Addit. Manuf., 25.*

Scime, L., & Beuth, J. (2018a). A multi-scale convolutional neural network for autonomous anomaly detection and classification in a laser powder bed fusion additive manufacturing process. *Addit. Manuf., 24.*

Scime, L., & Beuth, J. (2018b). Anomaly detection and classification in a laser powder bed additive manufacturing process using a trained computer vision algorithm. *Addit. Manuf., 19.*

Shen, X., Yao, J., Wang, Y., & Yang, J. (2004). Density prediction of selective laser sintering parts based on artificial neural network. *Proceedings of the international symposium on neural networks.*

Shi, B., Bai, X., & Yao, C. (2016). An end-to-end trainable neural network for image-based sequence recognition and its application to scene text recognition. *IEEE Trans Pattern Anal Mach Intell, 39.*

Sing, S. L., Huang, S., Goh, G. D., Goh, G. L., Tey, C. F., & Tan, J. H. K. (2021). Emerging metallic systems for additive manufacturing: Insitu alloying and multi-metal processing in laser powder bed fusion. *Progress in Materials Science, 119*, 100795.

Sing, S. L., Huang, S., & Yeong, W. Y. (2020). Effect of solution heat treatment on microstructure and mechanical properties of laser powder bed fusion produced cobalt-28cromium-6 molybdenum. *Materials Science and Engineering A, 769*, 138511.

Singh, A., Cooper, D., Blundell, N., Gibbons, G., & Pratihar, D. (2012). Modelling of direct metal laser sintering of EOS DM20 bronze using neural networks and genetic algorithms. *Proceedings of the 37th international MATADOR conference.*

Sizemore, N.E., Nogueira, M.L., Greis, N.P., & Davies, M.A. (2020). Application of machine learning to the prediction of surface roughness in diamond machining. *Procedia Manuf, 48.*

Smiti, A. (2020). When machine learning meets medical world: Current status and future challenges. *Computer Science Review, 37*, 100280.

Snell, R., Tammas-Williams, S., Chechik, L., Lyle, A., Hernandez-Nava, E., & Boig, C. (2019). Methods for rapid pore classification in metal additive manufacturing. *J Occup Med, 72.*

Sood, A.K., Equbal, A., Toppo, V., Ohdar, R.K., & Mahapatra, S.S. (2012). An investigation on sliding wear of FDM built parts. *CIRP J. Manuf. Sci. Technol., 5.*

Sood, A.K., Ohdar, R.K., & Mahapatra, S.S. (2012). Experimental investigation and empirical modelling of FDM process for compressive strength improvement. *J Adv Res, 3.*

Tapia, G., Elwany, A.H., & Sang, H. (2016). Prediction of porosity in metalbased additive manufacturing using spatial Gaussian process models. *Addit. Manuf., 12.*

Tapia, G., Khairallah, S., Matthews, M., King, W.E., & Elwany, A. (2017). Gaussian process-based surrogate modeling frame work for process planning in laser powder-bed fusion additive manufacturing of 316L stainless steel. *Int J Adv Manuf Technol, 94.*

Tootooni, M., Dsouza, A., Donovan, R., Rao, P. K., Kong, Z., & Borgesen, P. (2017). Classifying the dimensional variation in additive manufactured parts from laser-scanned three journal dimensional point cloud data using machine learning approaches. *Journal of Manufacturing Science and Engineering, 139*, 091005.

Trivedi, A., & Gurrala, P.K. (2021). Fuzzy logic based expert system for prediction of tensile strength in fused filament fabrication (FFF) process. *Mater. Today: Proc., 44.*

Ulas, M., Aydur, O., Gurgnec, T., & Ozel, C. (2020). Surface roughness prediction of machined aluminum alloy with wire electrical discharge machining by different machine learning algorithms. *J. Mater. Res. Technol., 9.*

Vahabi, E., & Rahmati, S. (2016). Application of an RBF neural network for FDM parts' surface roughness prediction for enhancing surface quality. *Int J Precis Eng Manuf, 17.*

Vapnik, V. N. (2013). *The nature of statistical learning theory*. Springer.

Wang, C., Tan, X. P., Tor, S. B., & Lim, C. S. (2020). Machine learning in additive manufacturing: State-of-the-art and perspectives. *Addit. Manuf., 36*, 101538.

Wu, H., Yu, Z., & Wang, Y. (2016). A new approach for online monitoring of additive manufacturing based on acoustic emission. *Proceedings of the ASME 2016 international manufacturing science and engineering conference.*

Wu, M., Song, Z., & Moon, Y.B. (2019). Detecting cyber-physical attacks in cyber manufacturing systems with machine learning methods. *J Intell Manuf, 30.*

Wu, M., Zhou, H., Lin, L. L., Silva, B., Song, Z., & Cheung, J. (2017). Detecting attacks in cyber manufacturing systems: Additive manufacturing example. *MATEC, 108*, 06005.

Xiong, J., Zhang, G., Hu, J., & Wu, L. (2014). Bead geometry prediction for robotic GMAW-based rapid manufacturing through a neural network and a second-orderregression analysis. *J Intell Manuf, 25.*

Xu, D., Liu, P.F., Li, J.G., & Chen, Z.P. (2019). Damage mode identification of adhesive composite joints under hygrothermal environment using acoustic emission and machine learning. *Compos Struct, 211.*

Yang, J., Chen, Y., Huang, W., & Li, Y. (2017). Survey on artificial intelligence for additive manufacturing. *Proceedings of the 23rd international conference on automation and computing.*

Yao, X., Moon, S.K., & Bi, G. (2017). A hybrid machine learning approach for additive manufacturing design feature recommendation. *Rapid Prototyp J, 23.*

Ye, D., Fuh, J.Y.H., Zhang, Y., Hong, G.S., & Zhu, K. (2018). In situ monitoring of selective laser melting using plume and spatter signatures by deep belief networks. *ISA Trans, 81.*

Ye, D., Hong, G.S., Zhang, Y., Zhu, K., & Fuh, J.Y.H. (2018). Defect detection in selective lasermelting technology by acoustic signalswith deep belief networks. *Int J AdvManuf Technol, 96.*

Zaldivar, R.J., Mclouth, T.D., Ferrelli, G.L., Patel, D.N., Hopkins, A.R., & Witkin, D. (2018). Effect of initial filament moisture content on the microstructure and mechanical performance of ULTEM 9085 3D printed parts. *Addit. Manuf., 24.*

Zhang, B., Liu, S., & Shin, Y.C. (2019). In-process monitoring of porosity during laser additive manufacturing process. *Addit. Manuf., 28.*

Zhang,, M., Sun,, C.N., Zhang,, X., Goh,, P.C., & Wei,, J. (2019). High cycle fatigue life prediction of laser additive manufactured stainless steel: A machine learning approach. *International Journal of Fatigue, 128*, 105194.

Zhang, W., Mehta, A., Deasi, P. S., & Higgs, C. (2017). Machine learning enabled powder spreading process map for metal additive manufacturing (AM). *Proceedings of the 28th annual international solid freeform fabrication symposium.*

Zhang, Y., Hong, G.S., Ye, D., Zhu, K., & Fuh, J.Y.H. (2018). Extraction and evaluation of melt pool, plume and spatter information for powder-bed fusion AM process monitoring. *Mater Des, 156.*

Zhang, Z., Liu, Z., & Wu, D. (2021). Prediction of melt pool temperature in directed energy deposition using machine learning. *Addit. Manuf., 37*, 101692.

Zhu, Z., Answer, N., Huang, Q., & Mathieu, L. (2018). Machine learning in tolerancing for additive manufacturing. *CIRP Ann Manuf Technol, 67.*

Zohdi, T.I. (2018). Dynamic thermomechanical modeling and simulation of the design of rapid freeform 3D printing processes with evolutionary machine learning. *Comput Methods Appl Mech Eng, 331.*

Zolfagharian, A., Durran, L., Gharaie, S., Rolfe, B., Kaynak, A., & Bodaghi, M. (2021). 4D printing soft robots guided by machine learning and finite element models. *Sensors and Actuators. A, Physical, 328*, 112774.

Compilation of References

A., J., & G., M. (2019). Applications of 3D printing in dentistry - A review. Journal of Pharmaceutical Sciences *and Research, 11(5), 1670–1675.*

Ab*do*, H. S., Elzatahry, A. A., Alharbi, H. F., & Khalil, K. A. (2017). Electrical conductivity behavior of biopolymer composites. In Biopolymer Composites in Electronics (pp. 13–2*5). Elsevier. doi:10.1016/B978-0-12-809*261-3.00002-4

Abu Al-Rub, R. K. (2012). On the aspect ratio effect of multi-walled carbon nanotube reinforcements on the mechanical properties of cementitiou*s nanocomposites. Constructio*n *& B*uilding Materials, 35, 647–655.

Additive Manufacturing Materials. (2017). Retrieved January 20, 2023, from ht*tps://www.additivemanuf*actu*r*ing.media/kc/what-is-additive-manufacturing/am-materials.

Additive Manufacturing State of the Industry. (2015). Retrieved January 20, *2023, from https://www.*sm*e*.org/technologies/latest-news/

Advanced Additive Solutions. (2012). Retrieved January 20, 2023, from https://oldworldlabs.com/

Ahmad, F., Choi, H. S., & Park, M. *K. (2015). A r*eview: Natural fiber composites selection in view of mechanical, light weight, and economic proper*ties. Macromolecular Materials* a*n*d Engineering, 300(1), 10–24. doi:10.1002/mame.201400089

Ahmed, S., & Ikram, S. (2016). Chitosan based scaffolds and their applications in wound healing. Achievements in the Lif*e Sciences, 10(1), 27–37. doi:10.1016/j.als.20*1*6.04.*001

Al-alkawi, H. J., Al-fattal, D. S., & Ali, A. H. (2012). Fatigue behavior of woven glass fiber reinforced polyester under variable temperature. Exilir Mech. Engg., 53, 12045–12050.

Al-Maadeed, M. *A., & Labidi, S. (2014). Recycled polym*ers *i*n natural fibre-reinforced polymer composites. In A. Hodzic & R. Shanks (Eds.), Natural fibre composites Materials, Processes and properties (pp. 103–114). Woodhead Publishing Limited.

Allcock, H. R. (1990). Polyphosphazenes as new biomedical and bioactive materials. Drugs and the Pharmaceutical Sciences, 45, 163-193.

Allen, R. J. A., & Trask, R. S. (2015). An experimental demonstration of effective Curved Layer Fused Filament Fabrication utilising a parallel deposition robot. Additive Manufacturing, 8, 78–87. doi:10.1016/j*.addma.2015.09.001*

Ambrose, C. G., *H*artline, B. E., Clanton, T. O., Lowe, W. R., & McGarvey, W. C. (2015). Polymers in orthopaedic surgery. In Advanced Polymers in Medicine (pp. 129–145). Springer.

Ambrosi, A., & Pumera, M. (2016). 3D-printing technologies for electrochemical applicatio*ns. Chemical Society Reviews, 45(10), 2740–2755. doi*:1*0.*1039/C5CS00714C PMID:27048921

Ameta, Solanki, Haque, Singh, Devi, & Chundawat. (2022). Critical appraisal and systematic review of 3D & 4D printing in sustainable and environment-friendly smart manufacturing technologies. Sustainable *Materials and* Technologies.

Amini, M., Chang, S.I., & Rao, P. (2019). A cybermanufacturing and AI framework for laser powder bed fusion (LPBF) additive manufacturing process. Manuf. Lett., 21.

An, Y., Myung, J. H., Yoon, J., & Yu, W. *(2022). Three-dimensional printing of* continuous carbon fiber-reinforced polymer composites via in-situ pin-assisted melt impregnation. Additive Manufacturing, 55, 102860. doi:10.1016/j.addma.2022.102860

Anandjiwala, R. D., & Blouw, S. (2007). Composites from bast fibres-prosp*ects and potential in* the changing market environment. Journal of Natural Fibers, 4(2), 91–109. doi:10.1300/J395v04n02_07

Ang, H. Y., Bulluck, H., Wong, P., Venkatram*an, S. S., Huang, Y., &* Fo*in*, N. (2017). Bioresorbable stents: Current and upcoming bioresorbable technologies. International Journal of Cardiology, 228, 931–939. doi:10.1016/j.*ijcard.2016.11.258 PMID:27912202*

*Ao*yagi, K., Wang, H., Sudo, H., & Chiba, A. (2019). Simple method to construct process maps for additive manufacturing using asupp*ort vecto*r *m*achine. Addit. Manuf., 27.

Arakawa, C. K., & DeForest, C. A. (2017). Polymer Design and Development. In Biology and Engineering of Stem Cell Niches (pp. 295-314). Academic Press. doi:10.1016/B978-0-12-802734-9.00019-6

Armillotta, A. (2006). Assessment of surface quality on textu*red FDM protot*ypes. Rapid Prototyping Journal, 12(1), 35–41. doi:10.1108/13552540610637255

Asadi-Eydivand, M., Solati-Hashjin, M., Fathi, M., Padashi, M., & Abu Osman, N.A. (2016). Optimal design of a 3D-printed scaffold using intelligent ev*olutionary algorithms. App*l Soft Comput, 39.

ASTM. (2012). F2792 -12 standard terminology for additive manufacturing technologies, standard. American Society for Testing Materials.

Athreya, S., Kalaitzidou, K., & Das, S. (2010). Processing and characterization of *a carbon black filled electrically conductive Nylo*n*-12* nanocomposite produced by selective laser sintering. Materials Science and Engineering A, 527, 2637–2642.

Aviation, G. E. (2015, April 6). Additive Manufacturing Machines & Materials | GE Additive. Retrieved January 20, 2023, from https://www.ge.com/additive/

Awan, G.H., Ali, L., & Ghauri, K.M. (2009). *Effect of various forms of glass fi*ber reinforcements on tensile properties of polyester matrix composite. J Faculty Eng Techno, 16, 33–39.

Awoyera, P.O., Kirgiz, M.S., Viloria, A., & Ovallos Gazabon, D. (2020). Estima*ting strength properti*es *o*f geopolymer selfcompacting concrete using machine learning techniques. J. Mater. Res. Technol., 9.

Axpe, E., & Oyen, M. L. (2016). Applications of alginate-based bioinks in 3D bioprinting. Internat*ional Journal of Molecular* Sciences, 17(12), 1976. doi:10.3390/ijms17121976 PMID:27898010

Babaie, E., & Bhaduri, S. B. (2018). Fabrication aspects of porous biomaterials in orthopedic applications: A review. ACS Biomaterials Science & Engineering, *4(1), 1–39. doi:10.1021/acsb*io*mate*rials.7b00615 PMID:33418675

Backes, E. H., de Nóbile Pires, L., Selistre-de-Araujo, H. S., Costa, L. C., Passador, F. R., & Pessan, L. A. (202*1). Development and charact*eri*z*ation of printable PLA/β-TCP bioactive composites for bone tissue applications. Journal of Applied Polymer Science, 138(5), 49759. *doi:10.1002/*app*.*49759

Badekila, A. K., Kini, S., & Jaiswal, A. K. (2021). Fabrication techniques of biomimetic scaffolds in three-dimensional cell culture: A review. Journal of Cellular Physiology, 236(2), 741–762. doi:10.1002/jcp.29935 PMID:32657458

Badouard, C., Traon, F., Denoual, C., Mayer-L*aigle, C., Paës, G., & Bourmaud, A. (2019). Explor*ing *me*chanical properties of fully compostable flax reinforced composite filaments for 3D printing applications. Industrial Crops and Products, 135, 246–250.

Baer, G. M. (2007). Fabrication and in vitro deployment of a laser-activated shape memory polymer vascular stent. Biomedical Engin*eering Online, 6(1), 43. https:// doi.*org/10.1186/1475-925X-6-43

Baer, G. M. (2009). Thermomechanical properties, collapse pressure, and expansion of shape memory polymer neurovascular stent prototypes. Journal of Biomedical Materials Research. Part B, Applied Biomaterials, 90(1), 421–429.

Balla, V. K., Kate, K. H., Satyavolu, J., Singh, P., & Tadimet*i, J. G. D. (2019). Additive manufacturing of na*tural fiber reinforced polymer composites: Processing and prospects, Compos. B Eng., 174, 106956.

Bandyopadhyay*, A., & Heer, B. (2018). Additive manufac*turing of multi-material structures. Materials Science and Engineering R Reports, 129, 1–16. doi:10.1016/j.mser.2018.04.001

Bartholomew, R. S. (1981). PDS (polydioxanone suture): A new synthetic absorbable suture in cataract surg*ery. Ophthalm*ologica, 183(2), 81–85. doi:10.1159/000309144 PMID:6795557

Bartkowiak, T., & Walkowiak-Śliziuk, A. (2018). 3D printing technology in orthodontics – Review of current applications. Journal of Stomatology, 71(4), 356–364. doi:10.5114/jos.2018.83410

Batyrbekov, E. O., Iskakov, R., & Zhubanov, B. A. (1998, February). Synthetic and natural polymers as drug car*riers for tuberculosis tr*eatment. Macromolecular Symposia, 127(1), 251–256. doi:10.1002/masy.19981270132

Bayliss, C. (2021). Machine learning based simulation opti*mization for urban routing problems.* Applied Soft Computing, 105, 107269.

Bayraktar, O., Uzun, G., Cakiroglu, R., & Guldas, A. (2017). Experimental study on the 3D-printed plastic parts and predicting the mecha*nical properties using artificial* neur*al* networks. Polym Adv Technol, 28.

Beaman, J. J., Barlow, J. W., Bourell, D. L., Crawford, R. H., Marcus, H. L., & McAlea, K. P. (1996). Solid Freeform Fabrication: A New Direction in Manufacturing. Springer.

Behl, M., Razzaq, M. Y., & Lendlein, A. (2010). Multifunctional shape-memory polymers. Advanced Materials, 22(31), 3388–3410.

Bellaire, G. & Talmi, K., Oezgu*er, E., & Kos*ch*an*, A. (1998). Object recognition: Obtaining 2-D reconstructions from color edges. . doi:10.1109/IAI.1998.666884

Berman, B. (2012). 3-D printing: The new industrial revolution. Business Horizons, 55(2), 155–162.

Berthomieu, C., & Hienerwadel, R. (2009). Fourier transf*orm infrared (FTI*R) *s*pectroscopy. Photosynthesis Research, 101(2), 157–170. doi:10.100711120-009-9439-x PMID:19513810

Bhatt, P. M., Malhan, R. K., Rajendran, P., & Gupta, S. K. (2020). Building free-form thi*n shell parts using suppor*tl*ess* extrusion-based additive manufacturing. Additive Manufacturing, 32, 101003. doi:10.1016/j.addma.2019.101003

Bilal, M., & Oyedele, L. O. (2020). Guidelines for applied machine learning in construction industry - a case of profit *margins estimation. Advanced Eng*ineering Informatics, 43, 101013.

Bledzki, A. K., & Gassan, J. (1999). Composites reinforced with cellulose based fibres. Progress in Polymer Science, 24(2), 221–274.

Blok, L. *G., Longana, M. L., Yu, H.*, & Woods, B. K. S. (2018). An investigation into 3D printing of fibre reinforced thermoplastic composites. Additive Manufacturing, 22, 176–186. doi:10.1016/j.addma.2018.04.039

Bodkhe, S., Turcot, G., Gosselin, F. P., & Therriault, D. (2017). One-step solvent evaporation-assisted 3D pr*inting of piezoel*ectric PVDF nanocomposite structures. ACS Applied Materials & Interfaces, 9(24), 20833–20842.

Boerio, F. J., & Koenig, J. L. (1971). Raman spectroscopy of polymers. In Polymer Characterization Interdisciplinary Approaches (pp. 1–13). Springer. doi:10.1007/978-1-4684-1929-0_1

Bogoeva-Gaceva, G., Avella, M., Malinconico, M., Buzarovska, A., Grozdanov, *A., Gentile, G.*, & Errico, M. E. (2007). Natural fiber eco-composites. Polymer Composites, 28(1), 98–107. doi:10.1002/pc.20270

Bogue, R. (2014). Smart materials: A review of capabilities and applications. Assembly Automation.

Bohn, B., Garcke, *J., Iza Teran, R., Paprotn*y, *A.*, Peherstorfer, B., & Schepsmeier, U. (2013). Analysis of car crash simulation data with nonlinear machine learning methods. Proc. Comput. Sci., 18.

Bongiovanni Abel, S., Montini Ballarin, F., & Abraham, G. A. (20*20). Combination* of *e*lectrospinning with other techniques for the fabrication of 3D polymeric and composite nanofibrous scaffolds with improved cellular interactions. Nanotechnology, 31(17), 172002. doi:10.1088/1361-6528/ab6ab4 PMID:31931493

Boopathi, S. (2019). Experimental investigation and parameter analysis of LPG refrig*eration system using Taguchi method.* In SN Applied Sciences (Vol. 1, Issue 8). doi:10.100742452-019-0925-2

Boopathi, S. (2021). An investigation on gas emission concentration and relative emission rate of the near-dry wire-cut electrical discharge machining process. Environmental Science and Pollution Research Internatio*nal, 1–10. doi:10.100711356-021-176*58*-1* PMID:34837614

Boopathi, S. (2021). Improving of Green Sand-Mould Quality using Taguchi Technique. Journal of Engineering Research. doi:10.36909/jer.14079

Boopathi, S. (2022a). An experimental investigation of Quench Polish Quench (QPQ) coating on AISI 4150 stee*l. Engineering Resear*ch Express, 4(4), 045009. doi:10.1088/2631-8695/ac9ddd

Boopathi, S. (2022a). Cryogenically treated and untreated stainless steel grade 317 in sustainable wire electrical discharge machining process: *A comparativ*e *s*tudy. Environmental Science and Pollution Research International, 1–10. doi:10.100711356-022-22843-x PMID:36057706

Boopathi, S. *(2022a). Experimental investigation and multi*-objective optimization of cryogenic Friction-stir-welding of AA2014 and AZ31B alloys using MOORA technique. Materials Today. Communications, 33, 104937. doi:10.1016/j. mt*comm.2022.104937*

*Boopathi, S. (2022b). An Extensive Re*view on Sustainable Developments of Dry and Near-Dry Electrical Discharge Machining Processes. Journal of Manufacturing Science and Engineering, 144(5), 1–37. doi:10.1115*/1.4052527*

*Boo*pa*thi*, S. (2022b). Performance Improvement of Eco-Friendly Near-Dry Wire-Cut Electrical Discharge Machining Process Using Coconut Oil-Mist Dielectric Fluid. Journal of Advanced Manufacturing Systems, 1–20. Advance online publication. do*i:10.1142/S02196867235001*78

Boopathi, S., & Myilsamy, S. (2021). Material removal rate and surface roughness study on Near-dry wire electrical discharge Machining process. Materials Today: Proceedings, 45, 8149–8156. doi:10.1016/j.matpr.2021.02.267

Boopathi, S., Balasubramani, V., Kumar, R. S., & Singh, G. R. (2021). The influence of human hair on kenaf and Grewia fiber-based hybrid natural composite material: An experimental study. Functional Composites and Structures, 3(4), 045011. doi:10.1088/2631-6331/ac3afc

Boopathi, S., Haribalaji, V., Mageswari, M., & Asif, M. M. (2022). Influences of Boron Carbide Particles on the Wear Rate and Tensile Strength of Aa2014 Surface Composite Fabricated By Friction-Stir Processing. Materiali in Tehnologije, 56(3), 263–270. doi:10.17222/mit.2022.409

Boopathi, S., Jeyakumar, M., Singh, G. R., King, F. L., Pandian, M., Subbiah, R., & Haribalaji, V. (2022). An experimental study on friction stir processing of aluminium alloy (AA-2024) and boron nitride (BNp) surface composite. Materials Today: Proceedings, 59, 1094–1099. doi:10.1016/j.matpr.2022.02.435

Boopathi, S., Lewise, K. A. S., Subbiah, R., & Sivaraman, G. (2022). Near-dry wire-cut electrical discharge machining process using water—air-mist dielectric fluid: An experimental study. Elsevier: Materials Today: *Proceedings, 50(5),* 1885–1890. doi:10.1016/j.matpr.2021.08.077

Boopathi, S., Sureshkumar, M., Jeyakumar, M., Kumar, R. S., & Subbiah, R. (2022). Influences of Fabrication Parameters on Natural Fiber Reinforced Polymer Composite (NFRPC) Material: A Review. Materials Science *Forum, 1075, 115–124.* doi:10.4028/p-095f0t

Boopathi, S., Thillaivanan, A., Pandian, M., Subbiah, R., & Shanmugam, P. (2021). Friction stir processing of boron carbide reinforced aluminium surface (Al-B4C) composite: Mechanical characteristics analysis. Materials Today: Proceedings, 50(5), 2430–2435. doi:10.1016/j.matpr.2021.10.261

Boopathi, S., Venkatesan, G., & Anton Savio Lewise, K. (2023). *Mechanical Pro*perties Analysis of Kenaf–Grewia–Hair Fiber-Reinforced Composite. In Lecture Notes in Mechanical Engineering (pp. 101–110). Springer. doi:10.1007/978-981-16-9057-0_11

Bose, S., Vahabzadeh, S., & Bandyopadhyay, A. (2013). Bone tissue engineering using 3D printing. *Materials Today*, *16*(12), 496–504. doi:10.1016/j.mattod.2013.11.017

Bourmaud, A., Beaugrand, J., Shah, D. U., Placet, V., & Baley, C. (2018). Towards the design of high-performance plant fibre composites. Progress in Materials Science, 97, 347–408. doi:10.1016/j.pmatsci.2018.05.005

Boyer, C. J., Boktor, M., Samant, H., White, L. A., Wang, Y., Ballard, D. H., & Alexander, J. S. (2019). 3D printing for bio-synthetic biliary stents. Bioengineering (Basel, Switzerland), 6(1), 16. doi:10.3390/bioengineering6010016 PMID:30744131

Bracaglia, L. G., Yu, L., Hibino, N., & Fisher, J. P. (2014). Reinforced pericardium as a hybrid material for cardiovascular applications. Tissue *Engineering. Part A*, *20*(21-22), 2807–2816. doi:10.1089/ten.tea.2014.0516 PMID:25236439

Brown, G. B., Currier, G. F., Kadioglu, O., & Kierl, J. P. (2018). Accuracy *of 3-dimensional pri*nted dental models reconstructed from digital intraoral impressions. American Journal of Orthodontics and Dentofacial Orthopedics, 154(5), 733–739. doi:10.1016/j.ajodo.2018.06.009 PMID:30384944

Bugay, D. E., & Martogolio Smith, P. A. (2008). In S. Jickells & A. Negrusz (Eds.), Clarke's Analytical Forensic Toxicology (3rd ed., pp. 455–468). Pharmaceutical Press.

Burgert, I. (2006). *Exploring the micro*mechanical design of plant cell walls. American Journal of Botany, 93(10), 1391–1401. doi:10.3732/ajb.93.10.1391 PMID:21642086

Burkoth, A. K., & Anseth, K. S. (2000). A review of photocrosslinked polyanhydrides: In situ forming degradable networks. Biomaterials, 21(23), 2395–*2404. doi:10.1016/S0142-9612(00)00107-1* PMID:11055287

Burnham-Fay, E. D., Le, T., Tarbutton, J. A., & Ellis, J. D. (2017). Strain characteristics of additive manufactured polyvinylidene fluoride (PVDF) actuators. Sensors and Actuators. A, Physical, 266, 85–92.

Caggiano, A., Zhang, J., Alfieri, V., Caiazzo, F., Gao, R., & Teti, R. *(2019). Machine learn*ing-based image processing for on-line defect recognition in additive manufacturing. CIRP J. Manuf. Sci. Technol., 68.

Caiazzo, F., & Caggiano, A. (2018). Laser direct metal deposition of 2024 Al alloy: trace geometry prediction via machin*e learning. Materials, 11.*

Calles, J. A., Bermúdez, J., Vallés, E., Allemandi, D., & Palma, S. (2015). Polymers in ophthalmology. In Advanced Polymers in Medicine (pp. 147–176). Springer.

Callister, W. D. J. (2007). Materials Science and Engineering: An Introduction *(7th ed.). John Wiley & So*ns, Inc.

Camargo, P. H. C., Satyanarayana, K. G., & Wypych, F. (2009). Nanocomposites: Synthesis, structure, properties and new application opportunities. Materials Research, 12, 1–39.

Campbell, I., Bourell, D., & Gibson, I. (20*12). Additive manufacturing: Rapid prototyping* comes of age. Rapid Prototyping Journal, 18(4), 255–258. doi:10.1108/13552541211231563

Camposeco-Negrete, C. (2020). Optimization of printing parameters in fused deposition modeling for improving part quality and process sustainability. Internatio*nal Journal of Advanced Manufacturing Techno*logy, 108(7–8), 2131–2147. doi:10.100700170-020-05555-9

Canal, C., Aparicio, R. M., Vilchez, A., Esquena, J., & García-Celma, M. J. (2012). D*rug delivery properties of macropo*rous polystyrene solid foams. Journal of Pharmacy & Pharmaceutical Sciences, 15(1), 197–207. doi:10.18433/J3X884 PMID:22365097

Cantrell, J. T., Rohde, S., Damiani, D., Gurnani, R., DiSandro, L., Anton, J., You*ng, A., Jerez, A., Stei*nbach, D., Kroese, C., & Ifju, P. G. (2017). Experimental characterization of the mechanical properties of 3D-printed ABS and polycarbonate parts. Rapid Prototyping Journal, 23(4), 811–824. doi:10.1108/RPJ-03-2016-0042

Cao, H., & Kuboyama, N. (2010*). A biodegradable por*ous composite scaffold of PGA/β-TCP for bone tissue engineering. Bone, 46(2), 386–395. doi:10.1016/j.bone.2009.09.031 PMID:19800045

Cao, X. V., Ismail, H., Rashid, A. A., Takeichi, T., & Vo-Huu, T. (2012). *Maleated natural rubber as a coupling agent for re*cycled high density polyethylene/natural rubber/kenaf powder biocomposites. Polymer-Plastics Technology and Engineering, 51(9), 904–910. doi:10.1080/03602559.2012.671425

Carmali, S., & Br*occhini, S. (2014). Polyacetals. In Na*tural and Synthetic Biomedical Polymers (pp. 219-233). Elsevier.

Carneiro, O. S., Silva, A. F., & Gomes, R. (2015). Fused deposition modeling with polypropylene. Materials & Design, 83, 768–776. doi:10.1016/j.matdes.2015.06.053

Carrell, J., Gruss, *G., & Gomez, E. (2020)*. Four-dimensional printing using fused-deposition modeling: A review. Rapid Prototyping Journal.

Car*riedo, G. A., García Alonso, F. J., González, P. A., & Menéndez, J. R.* (1998). Infrared and Raman spectra of the phosphazene high polymer [NP(O2C12H8)]n. Journal of Raman Spectroscopy: JRS, 29(4), 327–330. doi:10.1002/(SICI)1097-4555(199804)29:4<327::AID-*JRS245*>3.*0.C*O;2-5

Caulfield, B., McHugh, P., & Lohfeld, S. (2007). analyzing the effect of build parameters on polyamide components' mechanical properties. Journal of Materials Processing Technology, 182(1-3), 477–488.

Caulfield, B., McHugh, P. E., & Lohfeld, S. *(2007). Dependen*ce of mechanical properties of polyamide components on build parameters in the SLS process. Journal of Materials Processing Technology, 182(1-3), 477–488. do*i:10.1016/j. jmatprotec.2006.09.007*

*Cestari, F., Pet*re*tt*a, M., Yang, Y., Motta, A., Grigolo, B., & Sglavo, V. M. (2021). 3D printing of PCL/nano-hydroxy-apatite scaffolds derived from biogenic sources for bone tissue engineering. Sustainable Materials and Technologies, 29, e00318. doi:10.1016/j.susmat.2021.e00318

Cha*cón, Caminero, Núñez, García-Plaza, García-Moreno, & Reverte. (*2019). Additive manufacturing of continuous fibre reinforced thermoplastic composites using fused deposition modelling: Effect of process parameters on mechanical properties. Composites Science and Technology, 181(8).

Chamerski, K.*, Lesniak, M., Sitarz, M.,* St*op*a, M., & Filipecki, J. (2016). An investigation of the effect of silicone oil on polymer intraocular lenses by means of PALS, FT-IR and Raman spectroscopies. Spectrochimica Acta. Part A: Molecular and B*iomolecular Spec*tr*os*copy, 167, 96–100. doi:10.1016/j.saa.2016.05.036 PMID:27261889

Chand, N., Naik, A., & Neogi, S. (2000). Three-body abrasive wear of short glass fibre polyester composite. Wear, 242(1–2), 38–46. doi:10.1016/S0043-164*8(00)00398-7*

*Ch*ang, C., Tran, V. H., Wang, J., Fuh, Y.-K., & Lin, L. (2010). Direct-write piezoelectric polymeric nanogenerator wit*h high energy conversion efficiency.* Na*n*o Letters, 10(2), 726–731.

Charalampous & Kostavelis. (2020). In Situ Visual Quality Control in 3D Printing. IVAPP 2020 - 11th International Conference on Information Visualization Theory and Application, 3, 317-324.

*Charalampous, P., Kostavelis, I., Kontod*ina*,* T., & Tzovaras, D. (2021). Learning-based error modeling in FDM 3D printing process. Rapid Prototyping Journal. . doi:10.1108/RPJ-03-2020-0046

Chavan, S. A., & Anwar, P. M. A., & R, *P. S. B. (2017). Review of Effect* o*f Pro*cess Parameters on Mechanical Properties of 3D Printed Components. Internation Journal of Advance Research in Science and Engineerng*, 6(5), 461–467.*

Chen, G,. & Luo, H. (2020). Effects of node with discontinuous hierarchical fibers on the tensile fracture behaviors of natural bamboo. Sustainable Materials and Technologies, 26.

Chen, H., Yang, X., Chen, L., Wang, Y., & Sun, Y. (2016). Application of FDM three-dimensional *printing technology in the digital m*anu*f*acture of custom edentulous mandible trays. Scientific Reports, 6(January), 1–6. doi:10.1038rep19207 PMID:26763620

Chen, Liu, Tian, Zhu, Yan, Shi, Kong, Qi, & Zhou. (2022). 3D-Printed Anisotropic Polymer Materials fo*r Functional Applications. Advanced* M*a*terials, 34(5).

Chen, Q., Guillemot, G., Gandin, C.A., & Bellet, M. (2017). Three dimensional finite element thermomechanical modeling of additive manufac*turing by selec*t*ive* laser melting for ceramic materials. Addit. Manuf., 16.

Chen, Q., Maalihan, R. D., Ren, J., da Silva, Í. G., Dugos, N. *P., Caldona, E.* B., & Advincula, R. C. (2021). 3D printing of biomedically relevant polymer materials and biocompatibility. MRS Communications, 11(2), 197–212. doi:10.155743579-021-000*38-8 PMID:33936866*

Chen, R. S., Ab Ghani, M. H., Salleh, M. N., Ahmad, S., & Tarawneh, M. A. (2015). Mechanical, water absorption, and morphology of recycled poly*mer ble*nd *ric*e husk flour biocomposites. Journal of Applied Polymer Science, 132(8), 41494.

Chen, R. S., Ahmad, S., & Gan, S. (2016). Characterization *of Rice Husk-Incorporated Recycled Thermoplastic Ble*nd Composites. BioResources, 11(4), 8470–8482.

Chen, R. S., Ahmad, S., & Gan, S. (2016). Rice husk bio-filler reinforced polymer blends of recycled HDPE/PET: Three-dimensional st*ability under water immersion and* mechanical performance. Polymer Composites, 39, 2695–2704.

Chen, R. S., Salleh, M. N., Ab Ghani, M. H., Ahmad, S., & Gan, S. (2015). Biocomposites Based on Rice Husk Flour and Recycled Polymer Blend: Effects of Interfacial Modification and High Fibre Loading. BioResources, 10(4), 6872–6885.

Chen, Z., Liu, X., Lu, R., & Li, T. (2006). Mechanical and tribological properties of PA66/PPS blend. III. Reinforced with GF. Journal of Applied Polymer Science, 102(1), 523–529. doi:10.1002/app.24253

Cheng, M. Y., Kusoemo, D., & Gosno, R. A. (2020). Text mining-base*d construction site accident classificati*on *u*sing hybridsupervised machine learning. Automation in Construction, 118, 103265.

Cherng, J. Y., Hou, T. Y., Shih, M. F., Talsma, H., & Hennink, W. E. (2013). Polyurethane-based drug d*elivery systems. Inter*national Journal of Pharmaceutics, 450(1-2), 145–162. doi:10.1016/j.ijpharm.2013.04.063 PMID:236322*62*

*Chia, H. N., & Wu, B. M. (2015). Rec*ent advances in 3D printing of biomaterials. Journal of Biological Engineering, 9(1), 1–14. doi:10.118613036-015-0001-4 PMID:25866560

Chikkangoudar, R.N., Sachidananda, T.G., *& Pattar, N. (2020).* In*fl*uence of 3D printing parameters on the dimensional stability of polypropylene/clay printed parts using laser scanning technique. Mater. Today: Proc.

Childers, E. P., Wang, M. O., *Becker, M. L., Fisher, J. P., & Dean, D. (2015). 3D printing* of resorbable poly (propylene fumarate) tissue engineering scaffolds. MRS Bulletin, 40(2), 119–126. doi:10.1557/mrs.2015.2

Choi, J., Kim, K., Kim, T., Liu, G., Bar-Shir, *A., Hyeon, T., & Gilad, A*. A. (2011). Multimodal imaging of sustained drug release from 3-D poly (propylene fumarate) (PPF) scaffolds. Journal of Controlled Release, 156(2), 239–245. doi:10.1016/j.jconrel.2011.06.035 PMID:21763735

Choi, S. H., & *Samavedam, S. (2002). Modelling and o*ptimisation of rapid prototyping. Computers in Industry, 47(1), 39–53. doi:10.1016/S0166-3615(01)00140-3

Chung, H., & Das, S. (2006). Processing and p*roperties of gla*ss bead particulate-filled functionally graded Nylon-11 composites produced by selective laser sintering. Materials Science and Eng*ineering A, 437(2), 226–23*4.

Cipurkovic, A., Horozic, E., Donlagic, N., Maric, S., Saletovic, M., & Ademovic, Z. (2018). Biodegradable Polymers: Production, properties and application in medicine. Academic Press.

Colorado, H.A., Velasquez, E.I.*G., & Monteiro, S. (2020).* Sustainability of additive manufacturing: the circular economy of materials and environmental perspectives. J. Mater. Res. Technol., 9.

Colthup, N. B., Daly, L. H., & Wiberly, S. E. (1975). *Introduction to i*nfrared and Raman spectroscopy. Academic Press.

Columbus, L. (2015, April 28). Roundup of 3D printing market forecasts and estimates. Retrieved January 20, 2023, from https://www.forbes.com/sites/louiscolumbus/2015/03/31/2015-roundup-of-3d-printing-market-forecasts-and-esti*mates/*

*Comb, J. W., Prie*dema*n*, W. R., & Turley, P. W. (1994). FDM Technology Process Improvements. Proceedings of Solid Freeform Fabrication Symposium, 42–49.

Cousley, R. R. (2020). Introducing 3D printing in your orthodontic practice. Journal *of Orthodontics, 47*(3), 265–272. doi*:10.1177/1465312520936704 PMID:32627658*

*Croisier, F., & Jérôme, C. (2013). Chit*osan-based biomaterials for tissue engineering. European Polymer Journal, 49(4), 780–792. doi:10.1016/j.eurpolymj.2012.12.009

Crump, S. S. (1991). Fast, precise, safe prototype with FDM. ASME, PED., 50, 53–60.

Cui, J., Kratz, K., & *Lendlein, A. (2010). Adjusting shape-memory* properties of amorphous polyether urethanes *and radio-opaque composites thereof by variation of physical* parameters during programming. Smart Materials and Structures, 19(6), 065019. https://doi.org/10.1088/0964-1726/19/6/065019

Daimler Chrysler. (2019). Awarded for banana fibre use in Mercedes a class - Net-Compos*ites. https://netcompos*ites.com/news/2005/august/19/daimlerchrysler-awarded-for-banana-fibre-use-in-mercedes-a-class/

Daly, A. C., Cunniffe, G. M., Sathy, B. N., Jeon, O., Alsberg, E., & Kelly, D. J. (2016). *3D Bioprinting: 3D Bioprin*ting of Developmentally Inspired Templates for Whole Bone Organ Engineering. Advanced Healthc*are Materials, 5(18), 2352. doi:10.1002/adh*m.201670100 PMID:27281607

Daly, A. C., Riley, L., Segura, T., & Burdick, J. A. (2020). Hydrogel microparticles for biomedical applications. Nature Reviews. Materials, 5(1), 20–43. doi:10.103841578-01*9-0148-6 PMID:34123409*

*Damodar*an, V. B., Bhatnagar, D., & Murthy, N. S. (2016). Biomedical polymers: Synthesis and processing. Springer.

Damodaran, V. B., Joslin, J., & Reynolds, M. M. (2012). Preparing biocompatible materials for non-permanent medical devices. European Pharmaceutical Review, *17(5), 71–77.*

*Das, R. S., & Agr*awal, Y. K. (2011). Raman spectroscopy: Recent advancements, techniques and applications. Vibrational Spectroscopy, 57(2), 163*–176. doi:10.1016/j.vibspec.*201*1*.08.003

Datta, P., Dey, M., Ataie, Z., Unutmaz, D., & Ozbolat, I. T. (2020). 3D bioprinting for reconstituting the cancer microenvironment. NPJ Precision Oncology, 4(1), 1–13. doi:10.1038*41698-020-0121-2 PMID:32793806*

*Davachi, S. M., & Kaffashi, B. (2015). Po*lylactic acid in medicine. Polymer-Plastics Technology and Engineering, 54(9), 944–967. doi:10.1080/03602559.2014.979507

Davies, M. C., Binns, J. S., Melia, C. D., Hendra, P. J., Bourgeois, D., Church, *S. P.,* & *St*ephenson, P. J. (1990). FT Raman spectroscopy of drugs in polymers. International Journal of Pharmaceu*tics, 66(1-3), 223–232. doi:10.1016/0378-5173(90)*90403-Q

Dawood, A., Marti, B. M., Sauret-Jackson, V., & Darwood, A. (2015). 3D printing in dentistry. British Dental Journal, 219(11), 521–529. doi:10.1038j.bd*j.2015.914 PMID:26657*435

Dean, D., Wolfe, M. S., Ahmad, Y., Totonchi, A., Chen, J. E. K., Fisher, J. P., Cooke, M. N., Rimnac, C. M., Lennon, D. P., Caplan, A. I., Topham, N. S., & Mikos, A. G. (2005). Effect of *transforming growth fa*ctor β2 on marrow-infused foam poly (propylene fumarate) tissue-engineered constructs for the repair of critical-size cranial defects in rabbits. *Tissue Engineering,* 11(*5*-6), 923–939. doi:10.1089/ten.2005.11.923 PMID:15998232

Deckard, C., & Beaman, J. J. (1988). Process and control issues in selective laser sintering. ASME, PED, 33, 91–197.

Decost, B.L., Jain, H.A., *Rollett, A., & Holm*, E.A. (2017). Computer vision and machine learning for autonomous characterization of AM powder feedstoc*ks. J Occup Med, 69.*

*Delli, U., & Chang, S. (2018). Automat*ed process monitoring in 3D printing using supervised machine learning. Procedia Manuf, 26.

De Oliv*eira, A. S., de Macedo*, N. J. R., & Rosa, D. S. (2019). Eco-efficiency of Poly (lactic acid)-Starch-C*otton composite with high natural cotton fiber content: Environmen*tal and functional value. Journal of Cleaner Production, (217), 32–41.

Depuydt, D., Balthazar, M., *Hendrickx, K., Six*, W., Ferraris, E., Desplentere, F., & Ivens, J., Van Vuure, & A.W. (2019). Production and characterizati*on of bamboo and flax fiber* reinforced polylactic acid filaments for fused deposition modeling (FDM). Polymer Composites, 40, 1951–1963.

Derby, B. (2012). Printing and prototyping of tissues and scaffolds. Science, 338(6109), 921–926. doi:10.112*6cience.1226340 PMID:23161*993

Derr, K., Zou, J., Luo, K., Song, M. J., Sittampalam, G. S., Zhou, C., Michael, S., Ferrer, M., & Derr, P. (2019). Fully Three-Dimensional Bioprinted Skin Equivalent Constructs w*ith Validated Morphology and Ba*rr*ier* Function. Tissue Engineering. Part C, Methods, 25(6), 334–343. doi:10.1089/ten.tec.2018.0318 PMID:*31007132*

*Devesse, W., De Ba*ere, D., Hinderdael, M., & Guillaume, P. (2017). Modelbased temperature feedback control of laser cladding using highresolution hyperspectral imaging. IEEE/ASME *Transactions on Mechat*ronics, 22(6), 2714–2722. doi:10.1109/TMECH.2017.2754550

Dey, A., & Yodo, N. (2019). A systematic survey of FDM process parameter optimization and their influence on part characteristics. Journal of Manufacturin*g and Materials Processing, 3(3),* 64. Advance online publication. doi:10.3390/jmmp3030064

Dhandapani, R., Krishnan, P. D., Zennife*r, A., Kannan, V., Manigandan, A., Arul, M. R., & Set*huraman, S. (2020). Additive manufacturing of biodegradable porous orthopaedic screw. Bioactive Materials, 5(3), 458–467. doi:10.1016/j.bioactmat.2020.03.009 PMID:32280835

Dicker, M. P. M., Duckworth, P. *F., Baker, A. B., F*rancois, G., Hazzard, M. K., & Weaver, P. M. (2014). Green composites: A review of material attributes and complem*entary applications*. Composites. Part A, Applied Science and Manufacturing, 56, 280–289. doi:10.1016/j.compositesa.2013.10.014

Dilberoglu, U. M., Gharehpapagh, B., Yaman, U., & Dolen, M. (2017). *The Role of Additive M*anufacturing in the Era of Industry 4.0. Procedia Manufacturing, 11, 545–554. doi:10.1016/j.promfg.2017.07.148

Ding, D., Pan, Z., Cuiuri, D., & Li, H. (2014). A tool-path generation strategy for wire and arc additive manufacturing. I*nternational J*our*n*al of Advanced Manufacturing Technology, 73(1–4), 173–183. doi:10.100700170-014-5808-5

Diouf-Lewis, A., Farahani, R. D., Iervolino, F., Pierre, J., Abderrafai, Y., Lévesque, M., Piccirelli, N., & Therriault, D. (2022). Design and characterization of carbon fiber-reinforced PEEK/PEI blends for Fused Filament Fabrication additive man*ufacturing. Materials Today. C*ommunications, 31, 103445. doi:10.1016/j.mtcomm.2022.103445

Di Ventra, M. (2004). Introduction to Nanoscale Science and Technology. Kluwer Academic Publisher.

Domingos, *M., Intranuovo, F., Russo, T., D*e Santis, R., Gloria, A., Ambrosio, L., & Bartolo, P. (2013). The first systematic analysis of 3D rapid prototyped poly (ε-caprolactone) scaffolds manufactured through BioCell printing: The effect of pore size and geometry on compressive mechanical behaviour and in vitro hMS*C viability. Biofabrication, 5(4*), 045004. doi:10.1088/1758-5082/5/4/045004 PMID:24192056

Donnaloja, F., Jacchetti, E., Soncini, M., & Raimondi, M. T. (2020). Natural and Synthetic Polymers for Bone Scaffolds Optimization. Polymers, 12(4*), 905. doi:10.3390/polym12040905 PMID:32295115*

Dou, J., Zhang, Q., Ma, M., & Gu, J. (2012). Fast fabrication of epoxy-functionalized magnetic polymer core-shell microspheres using glycidyl methacrylate as monomer via p*hoto-initiated miniemulsion* polymerization. Journal of Magnetism and Magnetic Materials, 324(19), 3078–3082. doi:10.1016/j.jmmm.2012.05.005

Duigou, A. L., Correa, D., Ueda, M., Matsuzaki, R., & Castro, M. (2020). A review of 3D and 4D pri*nting of natural fibre biocomposites. Materials & Design,* 194(108911), 1–26.

Duoss, E. B., Weisgraber, T. H., Hearon, K., Zhu, C., Small, W., Metz, T. R., & Spadaccini, C. M. (n.d.). Three-dimensional printing of elastomeric, cellular architectures with negative stiffness. *Advanced Functional Materials, 24*, 4905–4913.

Durfee, W. K., & Iaizzo, P. A. (2019). Medical applications of 3D printing. In Engineering in medicine (pp. 527–543). Academic Press. doi:10.1016/B978-0-12-813068-1.00021-X

Duy, T. T., Dang, N. M., Thuc, C., Thuc, H., & Dang, T. T. (2013). Study of Mechanical Properties of Composite Material Based on Polypropylene and Vietnamese Rice Husk Filler. Journal of Chemistry, 2013, 1–6.

Eder, M., & Burgert, I. (2010). *Natural fibres - function in nature.* In J. Mussig (Ed.), Industrial Applications of Natural Fibres: Structure, Properties and Technical Applications (pp. 23–40). John Wiley & Sons, Ltd.

Eichhorn, S. J., & Young, R. J. (2004). Composite micromechanics of hemp fibres and epoxy resin microdroplets. *Composites Science and Technology, 64*(5), 767–772. doi:10.1016/j.compscitech.2003.08.002

Eichhorn, S. J., Baillie, C. A., Zafeiropoulos, N., Maikambo, L. Y., Ansell, M. P., Dufresne, A., & … . (2001). Current international research into cellulosic fibres and composites. Journal of *Materials Science, 36*(9), 2107–2131.

Einmahl, S., Behar–Cohen, F., D'Hermies, F., Rudaz, S., Tabatabay, C., Renard, G., & Gurny, R. (2001). A new poly (ortho ester)-based drug delivery system as an adjunct treatment in filtering surgery. Investigative *Ophthalmology & Visual Science, 42*(3), 695–700. PMID:11222529

Ekaputra, A. K., Zhou, Y., Cool, S. M., & Hutmacher, D. W. (2009). Composite electrospun scaffolds for engineering tubular bone grafts. Tissue Engineering. Part A, 15(12), 3779–3788. doi:10.1089/ten.tea.2009.0186 PMID:19527183

El-Tayeb, N. S. M., & Yousif, B. F. (2007). Evaluation of glass fiber reinforced polyester composite for multi-pass abrasive wear applications. Wear, 262(9–10), 1140–1151. doi:10.1016/j.wear.2006.11.015

El-Tayeb, N. S. M., Yousif, B. F., & Yap, T. C. (2008). An investigation on worn surfaces of chopped glass fibre reinforced polyester through SEM observations. Tribology International, 41(5), 331–340. doi:10.1016/j.triboint.2007.07.007

Elkins, K., Nordby, H., Janak, C., Gray, R. W. IV, Bohn, H. H., & Baird, D. G. (1997). *Soft elastomers for fused deposition modeling.* Proc. 8th Solid Freeform Fabrication Symposium.

Elsawy, M. A., Kim, K. H., Park, J. W., & Deep, A. (2017). Hydrolytic degradation of polylactic acid (PLA) and its composites. *Renewable & Sustainable Energy Reviews*, 79, 1346–1352.

Equbal, A., Sood, A. K., Ohdar, R. K., & Mahapatra, S. S. (2011). Prediction of dimensional accuracy in fused deposition modelling: A fuzzy logic approach. International *Journal of Productivity and Quality Management, 7*(1), 22–43. doi:10.1504/IJPQM.2011.037730

Erden, S., Sever, K., Seki, Y., & Sarikanat, M. (2010). Enhancement of the mechanical properties of glass/polyester composites via matrix modification glass/polyester composite siloxane matrix modification. Fibers and Polymers, 11(5), 732–737. doi:10.100712221-010-0732-2

Ergul, N. M., Unal, S., Kartal, I., Kalkandelen, C., Ekren, N., Kilic, O., & Gunduz, O. (2019). 3D printing of chitosan/poly (vinyl alcohol) hydrogel containing synthesized hydroxyapatite scaffolds for hard-tissue engineering. *Polymer Testing*, 79, 106006. doi:10.1016/j.polymertesting.2019.106006

Ertay, D. S., Yuen, A., & Altintas, Y. (2018). Synchronized material deposition rate control with path velocity on fused filament fabrication machines. Additive *Manufacturing, 19*, 205–213. doi:10.1016/j.addma.2017.05.011

Espalin, D. (2014). A review of melt extrusion additive manufacturing processes: I. Process design and modeling. Rapid Prototyping Journal, 20(3), 192–204. doi:10.1108/RPJ-01-2013-0012

*Espalin, D., Mus*e, D. W., MacDonald, E., & Wicker, R. B. (2014). 3D Printing multifunctionality: Structures with electronics. International Journal of *Advanced Manu*facturing Technology, 72(5-8), 963–978.

Etemad-Shahidi, Y., Qallandar, O. B., Evenden, J., Alifui-Segbaya, F.*, & Ahmed, K. E. (2020). Acc*uracy of 3-dimensionally printed full-arch dental model*s: A systematic review. Journal of Clinical Medici*ne, 9(10), 1–18. doi:10.3390/jcm9103357 PMID:33092047

Evert, R. F. (2006). Eas Esau's Plant Anatomy — Meristems, Cells, and Tissues of the Plant Body: Their Structure, Function, and *Development (3rd e*d.*).* Wiley & Son.

F. O., A., P. O., A., & O. O., O. (2013). Mechanical Properties of a Polyester Fibre Glass Composi*te. International Journal of* Composite Materials, 2(6), 147–151. doi:10.5923/j.cmaterials.20120206.06

Facca, A. G., Kortschot, M. T., & Yan, N. (2006). Predicting the tensile strength of natural fibre reinforced thermoplast*ics. Composites. Part A, Applied Science and Manufacturing, 37(*10), 1660–1671. doi:10.1016/j.compositesa.2005.10.006

Fan, J., Gonzalez, D., Garcia, J., Newell, B., & Nawrocki, R. A. (2020). The Effects of Additive Manufacturing and Electric Poling Techniques *on PVdF Thin Films: Towards 3D Printed Functi*onal Materials, in: Smart Materials. Adaptive Structures and Intelligent Systems, 84027.

Farah, S., Anderson, D. G., & Langer, R. (2016). Physical and mechanical properties of PLA, and their functions in widespread applications—A comprehensive review. Advanced Drug Delivery Reviews, 107, 367–392. doi:10*.1016/j.addr.2016.06.012* PM*ID*:27356150

Faruk, O., Bledzki, A. K., Fink, H. P., & Sain, M. (2012). Biocomposites reinforced with natural fibers: 2000-2010. Progress in Polymer Science, 37(11), 1552–1596. doi:10.1016/j.progpolymsci.2012.04.003

Faulstich de Paiva, J. M., & *Frollini, E. (2006). Unmodified and modifi*ed *su*rface sisal fibers as reinforcement of phenolic and lignophenolic matrices composites: Thermal analyses of fibers and composites. Macromolecular Materials and Engineering, 291(4), 405–417.

Felton, S. M. (2013). Self-folding with shape memory composites. Soft Matter*, 9(32), 7688–7694.*

*F*eng, Y., Zhu, Sh., Mei, D., Li, J., Zhang, J., Yang, Sh., & Guan, Sh. (2020). Application of 3D Printing Technology in Bone Tissue Engineering: A Review. Curr*ent Drug Delivery, 18(7),* 847–861. doi:10.2174/18755704MTExsNDcy2 PMID:33191886

Fergani, O., Berto, F., Welo, T., & Liang, S.Y. (2017). Analytical modelling of residual stre$_s$s$_i$n ad$_d$it*ive manufacturing. Fatig Fract Eng Ma*ter Struct, 40.

Ferreira, R. T. L., Amatte, I. C., Dutra, T. A., & Bürger, D. (2017). Experimental characterization and micrography of 3D printed PLA and PLA reinforced with short carbon fibers. Composites. Part B, Engineering, 124, 8*8–100. doi:10.1016/j.compositesb.2017.05.0*13

*F*eygin, M., & Hsieh, B. (1991). Laminated object manufacturing (LOM): A simpler process. Proc. 2nd Solid Freeform Fabrication Symposium, 123–130.

Fidan, I., & … . (2019). The trends and chal*lenges of fiber reinforced additive manuf*acturing. International Journal of Advanced Manufacturing Technology, 102(5–8), 1801–1818.

Filgueira, D., Holmen, S., Melbø, J. K., Moldes, D., Echtermeyer, A. T., & Chinga-Carrasco, G. (2017). Enzymatic-assisted modification of thermomechanical pulp fibers to im*prove the interfacial adhesion with po*ly (lactic acid) for 3D printing. ACS Sustainable Chemistry & Engineering, 5(10), 9338–9346.

Fisher, J. P., Dean, D., & Mikos, A. G. (2002). Photocrosslinking characteristics and mechanical properties of die*thyl fumarate/poly(propylene fumarate) biomaterials. Biomaterials, 23(22),* 4333–4343. doi:10.1016/S0142-9612(02)00178-3 PMID:12219823

FluorX. (n.d.). https://go.arkema.com/fluorx-filament

Francis, J., & Bian, L. (2019). Deep learning *for* d*ist*ortion prediction in laser-based additive manufacturing using big data. Manuf. Lett., 20.

Frank, K. (2014). Introduction to Nanofiber Materials. Cambridge University Press.

Fu, C.-C. (2009). Tunable Nan*owrinkles on Sha*pe Memory Polymer Sheets. Advanced Materials, 21(44), 4472–4476. https://doi.org/10.1002/adma.200*902294*

*Fu, C.-C. (2010). Bimetallic nanopetals for thousand-fold fluorescence enhancements. Appl*ied Physics Letters, 97(20), 203101. https://doi.org/ 10.1063/1.3495773

Fu, Y., Kan, Y., Fan, X., Xuan, Sh., & Yao, X. (2022). Novel designab*le strategy and multi-sca*le analysis of 3D printed thermoplastic fabric composites. Composites Science and Technology, 222, 109388. doi:10.1016/j.compscitech.2022.109388

Fuller, S. M., Butz, D. R., Vevang, *C. B., & Makhlouf, M. V. (2014). Application of 3-dimensional pri*nting in hand surgery for production of a novel bone reduction clamp. The Journal of Hand Surgery, 39(9), 1840–1845. doi:10.1016/j.jhsa.2014.06.009 PMID*:25042538*

Fuqua, M. A., Huo, S., & Ulven, C. A. (2012). Natural fiber reinforced composites. Polymer Reviews (Philadelphia, Pa.), 52, 259–320.

Gackowski, B. M., Phua, H., *Sharma, M., & Idapalap*ati, S. (2022). Hybrid additive manufacturing of polymer composites reinforced with buckypapers and short carbon fibres. Composites. Part A, Applied Science and Manufacturing, 154, 106794. doi:10.1016*/j.compositesa.202*1.106794

Gackowski, B. M., Sharma, M., & Idapalapati, S. (2022). Processing and characterization of tailorable 3D-printed acrylonitrile butadiene styrene *composites with hybrid bu*ck*ypa*pers. Composites Communications, 32, 101162. doi:10.1016/j.coco.2022.101162

Gao, W., Zhang, Y., Nazzetta, D. C., Ramani, K., & Cipra, R. J. (2015). RevoMaker: Enabling multi-directional and *functionally-embe*dded 3D printing using a rotational cuboidal platform. UIST 2015 - Proceedings of the 28th Annual ACM Symposium on User Interface Software and Technology, 437–446. 10.1145/2807442.2807476

Gao, W., Zhang, Y., Ramanujan, D., Ramani, K., Ch*en, Y., Williams, C. B., Wang, C. C. L.*, Shin, Y. C., Zhang, S., & Zavattieri, P. D. (2015). The status, challenges, and future of additive manufacturing in engineering. CAD Comp*uter Aided D*esi*g*n, 69, 65–89. doi:10.1016/j.cad.2015.04.001

Garcia, C. R., Correa, J., Espalin, D., Barton, J. H., Rumpf, R. C., Wicker, R., & Gonzalez, V. (2012). 3D printing of anisotropic metamaterials. Progress in Ele*ctromagnetics Rese*arch Letters, 34, 75–82. doi:10.2528/PIERL12070311

Garcia, C. R., Correa, J., Espalin, D., Barton, J. H., Rumpf, R. C., Wicker, R., & Gonzalez, V. (2012). Anisotropic metamaterials' 3D printing. Prog Electromagn Res, 3*4, 75–82.*

*Gard*an, J. (2019). Smart materials in additive manufacturing: State of the art and trends. Virtual and Physical Prototyping, 14(1), 1–18.

Garg, A., *& Lam, J.S.L. (2015). Measurement* of *en*vironmental aspect of 3-D printing process using soft computing methods. Measure, 75.

Garg, A., Lam, J.S.L., & Savalani, M.M. (2015). A new computational *intelligence approach in formulation of* functional relationship of open porosity of the additive manufacturing process. Int J Adv Manuf Technol, 80.

Gebelein, C. G. (1985). Medical applications of polymers. In Applied polymer science (vol. 2, pp. 535–556). American Chemical Society. doi:10.1021/bk-1985-*0285.ch023*

Geetha, N. K., & Bridjes, P. (2020). Overview of machine learning and its adaptability in mech*anical engineering. Mate*rials Today: Proceedings.

Geng, P., Zhao, J., Wu, W., Ye, W., Wang, Y., Wang, S., & Zhang, S. (2019). Effects of extrusion speed and printing speed on the 3D printing stability of extruded PEEK filament. Journal of *Manufacturing Processes, 37(December),* 266–273. doi:10.1016/j.jmapro.2018.11.023

George, J., Sreekala, M. S., & Thomas, S. (2001). A review on in*terface modification and characterization of natural fiber reinforced plas*tic composites. Polymer Engineering and Science, 41(9), 1471–1485.

Gershenfeld, N. A. (2005). Fab: The coming revolution on your desktop–from personal computers to personal fabrication. *Basic Books.*

Ghosh, S. K. (1991). Manufacturing engineering and technology. Jour*nal of Materials Processing Technology, 25(1),* 112–113. doi:10.1016/0924-0136(91)90107-P

Giannit*elli, S. M., Accoto, D., Trombetta, M., & Rainer, A.* (2014). Current trends in the design of scaffolds for computer-aided tissue engineering. Acta Biomaterialia, 10(2), 580–594. doi:10.1016/j.actbio.2013.10.024 PMID:24184176

Gibson, I., Rosen, D., & Stucker, B. (2015). Additive Manufacturing Te*chnologies: 3D Printing, Rapid Prototyping, and Dir*ect Digital Manufacturing (2nd ed.). Springer. doi:10.1007/978-1-4939-2113-3

Gibson, I.*, Rosen, D. W., & Stuc*ker, B. (2014). Additive manufacturing technologies. Springer.

Gibson, Rosen, & Stucker. (2014). Additive Manufacturing Te*chnologies*: *3D* Printing, Rapid Prototyping, and Direct Digital Manufacturing. Springer.

Gierek, M., Kuśnierz, K., Lampe, P., Ochała, G., Kurek, J., Hekner, B., & Majewski, J. (2018). Absorbable sutures in general surgery–revie*w, available materials, and* o*ptim*um choices. Polish Journal of Surgery, 90(2), 34–37. doi:10.5604/01.3001.0010.5632 PMI*D:29773758*

Giles, H. F. Jr, Mount, E. M. III, & Wagner, J. R. Jr. (2004). Extrusion: The Definitive Processing Guide and Handbook. William Andrew Publishing.

Gobert, C., Kudzal, A., Sietnis, J., Mock, C., Sun, J., & McWilliams, B. (2020). Porosity segmentation in X-ray computed tomography scans of metal additively manufactured specimens with machine learning. Addit. Ma*nuf., 36, 101460.*

*Gobert, C., R*eutzel, E.W., Petrich, J., Nassar, A.R., & Phoha, S. (2018). Application of supervised machine learning for defect detection during metallic powder bed fusion additive manufacturi*ng using high resolution imagi*ng. Addit. Manuf., 21.

Godoi, B. Prakash, & Zhang. (2018). Fundamentals of 3D food printing and applications. Academic Pres*s.*

*Godoy & Carmo. (2020). S*tudy of Defects and Geometric Anomalies on Monolayer Parts Obtained by Fused Deposition Modeling Process. Proceedings.

Goh, G.D., Dikshit, V., An, J., & Yeong, W.Y. (2020). Process-*structureproperty of* add*itively* manufactured continuous carbon fiber reinforced thermoplastic: An investigation of mode I interlaminar fracture to*ughness. Mech Adv Mater Struct, 27.*

*Goh, G.*D., Sing, S.L., & Yeong, W.Y. (2021). A review on machine learning in 3D printing: applications, potential, and challenges. Artif Intell Rev, 54.

Goh, G. D., Yap, Y. L., Agarwala, S., & Yeong, W. Y. (2019). Rec*ent progress in additive manufacturing of* fiber reinforced polymer composite. Adv. Mater. Technol., 4(1), 1800271.

Gonzalez, M., Galego, F., Ortiz, D. T., & Rapado, M. (2007). Raman spe*ctroscopy of poly (3-h*ydr*ox*ybutyrate) modified with poly (vinyl acetate) by radiation-induced copolymerization. Nucleus.

Gonzalez, R. C., & Woods, R. E. (2002). Digital image processing. Prentice-Hall.

Goonoo, N., Jeetah, R., Bhaw-Luximon, A., & Jhurry, D. (2015). Polydioxanone-ba*sed bio-materials for tissue* engineering and drug/gene delivery applications. European Journal of Pharmaceutics and Biopharmaceutics, 97, 37*1–391. do*i:*10.*1016/j.ejpb.2015.05.024 PMID:26614558

Goswami, P., & O'Haire, T. (2016). Developments in the use of green (biodegradable), recycled and biopolymer materials in tec*hnical nonwoven*s. In Advances in Technical Nonwovens (pp. 97–114). Woodhead Publishing. doi:10.1016/B978-0-08-100575-0.00003-6

*Grainger, D. W. (*2020). 1.3.2C - Fluorinated Biomaterials. In W. R. Wagner, S. E. Sakiyama-Elbert, G. Zhang, & M. J. Yaszemski (Eds.), Biomaterials Science (4th ed., pp. 125–138). Academic Press.

Grand View Research. (n.d.). Natural fiber composites (NFC) market size, shar*e & trends analysi*s *re*port by raw material, by matrix, by technology (injection molding, compression mold*ing, pu*ltr*us*ion), by application, and segment forecasts. Academic Press.

Grasso, M., Laguzza, V., Semeraro, Q., & Colosimo, B. M. (2017). In-process monitoring of selective laser melting: Spatial detection of defects viaimagedataanalysis. Journal of Manufacturing Science and Engineering, 139, *051001.*

Graupner, N., Herrmann, A. S., & Müssig, J. (2009). Natural and man-made cellulose fibre-reinforced poly(lactic acid) (PLA) composites: An overview about mechanical characteristics and application areas. Composites. Part A, Applied Science and Manufacturing, *40(6), 810–821.*

Griffith, P. R., & de *Ha*seth, J. A. (1986). Fourier transform infrared spectroscopy. Wiley.

Gross, R. A., & Kalra, B. (2002). Biodegradable polymers for the environment. Science, 297(5582), 803–807. doi*:10.1126cience.297.5582.803 PMID:12161646*

*Grunew*al*d*, S. (2021, October 18). Worldwide spending on 3D printing to grow from nearly $11 b*illion to $26.7 billion by 2019 - 3DPrint.com: T*he Voice of 3D printing / additive manufacturing. Retrieved January 20, 2023, from https://3dprint.com/116562/3d-printing-26-billion-by-2019/

Gu, G.X., Chen, C.T., Richmond, D.J., & Buehler, M.J. (2018). Bioinspired *hierarchical composite design using machine learning*: *si*mulation, additive manufacturing, and experiment. Mater. Horiz., 5.

Gu, H., Ma, C., Gu, J., Guo, J., Yan, X., Huang, J., Zhang, Q., & Guo, Z. (2016). An overview of multifunctional *epoxy nanocomposites.* Jo*urnal of Materials Chemistry. C, Materials for Optical and Electronic Devices, 4(25), 5890–5906. doi:10.1039/C6TC01210H

Gu, H., Ma, C., Gu, J., Guo, J., Yan, X., Huan*g, J., Zhang, Q., & Guo, Z. (2016). Multifunctional epoxy* nan*o*composites: A brief overview. Journal of Materials Chemistry. C, Materials for Optical and Electronic Devices, 4(25), 5890–5906.

Gu, J., Yang, X., Lv, Z., Li, N., Liang, C., & Zhang, Q. (2016). Functionalized graphite nanoplatelets/epoxy resin nanocomposites with high thermal conductivity. Internation*al Journal of Heat and Mass Transfe*r, 92, 15–22. doi:10.1016/j.ijheatmasstransfer.2015.08.081

Gu, S.-Y., Jin, S.-P., Gao, X.-F., & Mu, J. (2016). Polylactide-based polyurethane shape memory nanocomposites (Fe3O4/*PLAUs) wi*th fast magnetic responsiveness. Smart Materials and Structures, 25(5), 055036.

Guen, L. M., Hill, S., Smith, D., Theobald, B., Gaugler, E., Barakat, A., & Mayer-Laigle, C. (2019). Influence of Rice Husk and Wood Biomass Properties on the Manufacture of Filaments for *Fused Deposition Modeling. Frontiers in Ch*emi*st*ry, 7(735), 1–10.

Gugliandolo, A., Diomede, F., Cardelli, P., Bramanti, A., Scionti, D., Bramanti, P., & Mazzon, E. (2018). Transcriptomic analysis of gingival mesenchymal stem cells *cultured on 3 d bio*print*ed* scaffold: A promising strategy for neuroregeneration. Journal of Biomedical Materials Research. Part A, 106(1), 126–137. doi:10.1002/jbm.a.36213 PMID:28879677

Gunasekaran, K., Boopathi, S., & Sureshkumar, M. *(2022). Analysis of a Cryogenic*ally Cooled Near-Dry Wedm Process Using Different Dielectrics. Materiali in Tehnologije, 56(2), 179–186. doi:10.17222/mit.2022.397

Guo, R., Ren, Z., Bi, H., Song, Y., & Xu, M. (2018). Effect of *toughening agents o*n *the* properties of poplar wood flour/poly (lactic acid) composites fabricated with Fused Deposition Mo*deling. European Polymer Journal, 107, 34–45.*

Guo, Z., Poot, A. A., & Grijpma, D. W. (2021). Advanced polymer-based composites and structures for biomedical applications. European Polymer Journal, 149, 110388. doi:10.1016/j.eurpolymj.2021.110388

Gupta, B., Revagade, N., & Hilborn, J. (2007). Poly (lactic acid) *fiber: An overview. Progress i*n Polymer Science, 32(4), 455–482. doi:10.1016/j.progpolymsci.2007.01.005

Gupta, M. K., & Bansil, R. (1981). Laser Raman spectros*copy of polyacrylamide. Journal o*f *Po*lymer Science. Polymer Physics Edition, 19(2), 353–360. doi:10.1002/pol.1981.180190214

Gupta, P. K., Raghunath, S. S., Prasanna, D. V., Venkat, P., Shree, V., Chithananthan, C., ... Geetha, K. (2019). An update on overview of cellulose, its structure an*d applications. Cellulose (London, England), 8*46–1297.

Hakobyan, D., Kerouredan, O., Remy, M., Dusserre, N., Medina, C., Devillard, R., Fricain, J. C., & Oliveira, H. (2020). Laser-Assisted Bioprinting for Bone *Repair. In Methods in Molecular Biology (Vol.* 2140, pp. 135–144). Springer., doi:10.1007/978-1-0716-0520-2_8

Halloran, Tomeckova, & Gentry. (2011). Photopolymerization of pow*der suspensions for shaping ceramics.* J. *E*ur. Ceram. Soc., 31, 2613–2619.

Han, J., Lazarovici, P., Pomerantz, C., Chen, X., Wei, Y., & Lelkes, P. I. (2011). Co-electrospun blends of PLGA, gelatin, an*d ela*s*tin* as potential nonthrombogenic scaffolds for vascular tissue engineering. Biomacromolecules, 12(2), 399–408. doi:10.1021/bm101149r PMID:21182235

Han, X., Sharma, N., Xu, Z., Scheideler, L., Geis-Gerstorfer, J., *Rupp, F., Thieringer, F. M.*, & Spintzyk, S. (2019). An in vitro study of osteoblast response on fused-filament fabrication 3D printed PEEK for dental and Cranio-Maxillofacial implants. Journal of Clinical Medicine, 8(6), 771. Advance *online publication. doi:10.3390/jcm8060771 PMID:31159171*

*Hari*balaji, V., Boopathi, S., Asif, M. M., Jeyakumar, M., Subbiah, R., & Lewise, K. A. S. (2021). Influences of Friction stir tool parameters for joining two similar AZ61A alloy plates. Materials Today: Proceedings, 50(5), 2547–2553. doi:10.1016/j.matpr.*2021.12.074*

Haribalaji, V., Venkatesan, G., Asif, M. M., Pandian, M., Subbiah, R., & Boopathi, S. (2022). Investigation on corrosion and tensile Characteristics: Friction stir welding of AA7075 and AA2014. Materials Today: Proceedings, 66, 743–748. doi:10.1016/j.matpr.2022.04.037

Harrison & Ounaies. (2002). Piezoelectric polymers. Encyclopedia Polym. Sci. Technol., 3.

He, Zhang, & Hong. (2018). Profile monitoring based quality control method for fused deposition modeling process. Journal of Intelligent Manufacturing, (May). Advance online publication. doi:10.100710845-018-1424-9 21

Hector Sandoval, J., & Wicker, R. B. (2006). Functionalizing stereolithography resins: Effects of dispersed multi-walled carbon nanotubes on physical properties. Rapid Prototyping Journal, 12(5), 292–303.

Heidari-Rarani, Rafiee-Afarani, & Zahedi. (2019). Mechanical characterization of FDM 3D printing of continuous carbon fiber reinforced PLA composites. Composites Part B: Engineering, 175(15).

Heljak, M. K., Kurzydlowski, K. J., & Swieszkowski, W. (2017). Computer aided design of architecture of degradable tissue engineering scaffolds. Computer Methods in Biomechanics and Biomedical Engineering, 20(15), 1623–1632. doi:10.1080/10255842.2017.1399263 PMID:29106807

Heljak, M. K., Moczulska-Heljak, M., Choińska, E., Chlanda, A., Kosik-Kozioł, A., Jaroszewicz, T., Jaroszewicz, J., & Swieszkowski, W. (2018). Micro and nanoscale characterization of poly(DL-lactic-co-glycolic acid) films subjected to the L929 cells and the cyclic mechanical load. Micron (Oxford, England), 115, 64–72. doi:10.1016/j.micron.2018.09.004 PMID:30253318

Hench, L., & Jones, J. (Eds.). (2005). Biomaterials, artificial organs and tissue engineering. Elsevier.

Herbert, N., Simpson, D., Spence, W. D., & Ion, W. (2005). A preliminary investigation into the development of 3-D printing of prosthetic sockets. Journal of Rehabilitation Research and Development, 42(2), 141. doi:10.1682/JRRD.2004.08.0134 PMID:15944878

Hergenrother, P. (2003). The use, design, synthesis, and properties of high performance/ high temperature polymers: An overview. High Performance Polymers, 15, 3–45.

Hernández, R., Calister, I., Corsen, R., Rodríguez, A., Plaate, R., Imperial, C., ... Gutierrez, P. E. (2007). Podio Europeo del jazz episodio 10-2007. Academic Press.

Herrera-Franco, P. J., & Drzal, L. T. (1992). Comparison of methods for the measurement of fibre/matrix adhesion in composites. Composites, 23(1), 2–27.

Herriott, C., & Spear, A. D. (2020). Predicting microstructure-dependent mechanical properties in additively manufactured metals with machine- and deep-learning methods. Computational Materials Science, 175, 109599.

Hertlein, N., & Deshpande, S. (2020). Prediction of selective laser melting part quality using hybrid Bayesian network. Addit. Manuf., 32, 101089.

Hifumi, H., Ewing, A. V., & Kazarian, S. G. (2016). ATR-FTIR spectroscopic imaging to study the drying and dissolution of pharmaceutical polymer-based films. International Journal of Pharmaceutics, 515(1-2), 57–68. doi:10.1016/j.ijpharm.2016.09.085 PMID:27720876

High-performance additive manufacturing materials. (2016). Retrieved January 20, 2023, from https://www.arevo.com/

Hollister, S. J. (2005). Porous scaffold design for tissue engineering. Nature Materials, 4(7), 518–524. doi:10.1038/nmat1421 PMID:16003400

Holmes-Siedle, A., Wilson, P., & Verrall, A. (1983). PVdF: An electronically-active polymer for industry. Materials & Design, 4(6), 910–918.

Hong, B. T., Shin, K. S., & Kim, D. S. (2005). Ultraviolet-curing behavior of an epoxy acrylate resin *system. Journal of Applied Polymer Scie*nce, 98(3), 1180–1185. doi:10.1002/app.22130

Hong, K. Z. (1996). Poly (vinyl chloride) in medical device and packaging applications. Journal of Vinyl and Additive Technology, 2(3), 193–197. doi:10.1002/*vnl.10123*

*Honi*gmann, P., Sharma, N.*, Okolo, B., Popp, U., Msallem, B.,* & Thieringer, F. M. (2018). Patient-specific surgical implants made of 3D printed PEEK: Material, tec*hnology, and scope of* surgical application. BioMed Research International, 2018, 1–8. Advance online publication. doi:10.1155/2018/4520636 PMID:29713642

Howard, D.*, Partridge, K., Yang,* X., Clarke, N. M., Okubo, Y., Bessho, K., & Oreffo, R. O. (2002). Immunoselection and adenoviral genetic modulation of human osteoprogeni*tors: In vivo bone formation on PLA sca*ffold. Biochemical and Biophysical Research Communications, 299(2), 208–215. doi:10.1016/S0006-291X(02)02561-5 PMID:12437971

Hu, L., Sun, Y., & Wu, Y. (2013). Advan*ces in chitosan-based drug delive*ry *ve*hicles. Nanoscale, 5(8), 3103–3111. doi:10.1039/c3nr00338h PMID:23515527

Hua, L., Lei, T., Qian, H., Zhang, Y., Hu, Y., & Lei, P. (2021). 3D-printed porous tantalum: Recent application in various drug delivery systems t*o repair hard tissue defects. Expert Opinion on Drug* Deli*v*ery, 18(5), 625–634. doi:10.1080/17425247.2021.1860015 PMID:33270470

Huang, B., & Singamneni, S. B. (2015). Curved layer adaptive slicing (CLAS) for fused deposition modelling. Rapid Prototyping Journal, 21(4), 354–367. doi:10.11*08/RPJ-06-2013-0059*

*Huan*g, *S*. H., Liu, P., Mokasdar, A., & Hou, L. (2013). Additive manufacturing and its societal impact: A literature review. International Journal of Advanced Manufacturing Technology, 67(5-8), 1191–1203.

Huang, Y., Zhang, J., Hu, X., S*ong, G., Liu, Z., Yu, L., & Liu, L. (2016). FrameFab: Robotic Fabrication of Frame Shapes. ACM Tra*ns. Graph. Article, 3516(1112), 978–1. doi:10.1145/2980179.2982401

Hutmacher, D. W. (2000). Scaffolds in tissue engineering bone and cartilage. Biomaterials, 21(24), 2529–2543. doi:10.1016/S0142-9612(00)00121-6 PMID:11071603

Hwa-Hsing, T., Ming-Lu, C.*, & Hsiao-Chuan, Y. (2011).* Sl*u*rrybased selective laser sintering of polymer-coated ceramic powders to fabricate high strength alumina parts. Journal of the European Ceramic Society, 31, 1383–1388.

Hölzl, K., Lin, S.*, Tytgat, L., Van Vlierberghe, S., Gu, L., &* Ov*s*ianikov, A. (2016). Bioink properties before, during and after 3D bioprinting. Biofabrication, 8(3), 32002. doi:10.1088/1758-5090/8/3/032002 PMID:27658612

Iancu, C., Iancu, D., & S*tancioiu, A. (2013).* Fr*o*m CAD model to 3D print via "STL" file format. Fiabilitate Durab, 1, 73.

Ibrahim, H. M., & El-Zairy, *E. M. R. (2015). Chitosan as a* bio*m*aterial—structure, properties, and electrospun nanofibers. Concepts, compounds and the alternatives of antibacterials, 81-101.

IEEE R*obotics and* Automation Society, National Science Foundation (U.S.), & Institute of Electrical and Electronics Engineers. (2019). 2019 IEEE 15th International Conference on Automation Science and Enginee*ring (CASE). Author.*

*Ingbe*r, D. E., Mow, V. C., Butler, D., Niklason, L., Huard, J., Mao, *J., Yannas, I., Kaplan,* D., & Vunjak-Novakovic, G. (2006). Tissue engineering and developmental biology: Going biomimetic. Tissue Engineering, 12(12), 3265–3283. doi:10.1089/ten.2006.12.3265 PMID:17518669

Isaacson, A., *Swioklo, S., & Connon, C. J.* (2018). 3D bioprinting of a corneal stroma equivalent. Experimental Eye Research, 173, 188–193. doi:10.1016/j.exer.2018.05.010 PMID:29772228

Ishak, I., & Larochelle, P. (2019). Mo*toMaker: A robot FDM platform for multi*-plane and 3D lattice structure printing. Mechanics Based Design of Structures and Machines, 47(6), 703–720. doi:10.1080/15397734.2019.1615943

Isobe, T., Tanaka, T., Nomura, T., & Yuasa, R. (2018). Compariso*n of strength of 3D printing ob*jec*t*s using short fiber and continuous long fiber*. IOP Conference Series. Materials Science and Engineering, 406(1), 012042. Advance onlin*e publication. doi:10.1088/1757-899X/406/1/012042

Ivey, M., Melenka, G. W., Ca*rey, J. P., & Ayranci, C. (2017). Characte*ri*z*ing short-fiber-reinforced composites produced using additive manufacturing. Advanced Manufacturing: Polymer and Composites Science, 3(3), 81–91. doi:10.1080/20550340.2017.1341125

J. H. K. T. & Sin*g. (2020). Micros*truc*t*ure modelling for metallic additive manufacturing: a review. Virtual Phys Prototyp, 15.

Jaber, *S. T., Hajeer, M. Y., Khattab, T. Z., & Mahaini, L. (2021). Evaluation of the fused deposition model*ing and the digital light processing techniques in terms of di*mensional accuracy of printing dent*al models used for the fabrication of clear aligners. Clin*ical and Experimental Dental Research, 7(4), 591–600. doi:10.1002/cre2.366 PMID:33258297*

Jacob, M., Francis, B., Thomas, S., & Varughese, K. T. (2006). Dynamical mechanical analysis of sisal/oil palm hybrid fiber-rei*nforced natural rubber composites. Polymer Composites,* 27(6), 671–680. doi:10.1002/pc.20250

Jafari-Marandi, R., Khanzadeh, M., Tian, W., Smith, B., & Bian, L. (2019). From in-situ monitoring toward high-throughput process control: cost-driven decision-making framework for laser-based additive manufacturing. J *Manuf Syst, 51.*

Jain, A., Duvvuri, L. S., Farah, S., Beyth, N., Domb, A. J., & Khan, W. (2014). Antimicrobial polymers. Advanced Healthcare Materials, 3(12), 1969–1985. doi:10.1002/adhm.201400418 PMID:25408272

Jain, S., Mo*ndal, K., & Gupta*, M. N. (2006). Applications of alginate in bioseparation of proteins. Artificial Cells, Blood Substitutes, and Biotechnology, 34(2), 127-*144.*

*Jammalamadaka & Tappa. (2018). Recent Advances in Biomaterials for 3D Printing and Tissue Engineeri*ng. Journal of Functional Biomaterials, 9(1), 22.

Jamróz, W., Szafraniec, J., Kurek, M., & Jachowicz, R. (2018). 3D printing in pharmaceutical and medical applications–recent achievements and challenges. Pharmaceutical Research, 35(*9), 1–22. doi:10.100711109*5-018-2454-x PMID:29998405

Jawaid, M., Khalil, H. P. S. A., Hassan, A., Dungani, R., & Hadiyane, A. (2013). Effect of jute fibre loadi*ng on the mechanica*l and thermal properties of oil palm-epoxy composites. Composites. Part B, Engineering, 45(1), 619–624. doi:10.1016/j.compositesb.2012.04.068

Je*nsen, M. L., Mahshid, R.*, D'Angelo, G., Walther, J. U., Kiewning, M. K., Spangenberg, J., Hansen, H. N., & Pedersen, D. B. (2019). Toolpath strategies for 5DOF and 6DOF extrusion-ba*sed additive manufacturing. Applied Sciences (Switzerland), 9(19), 4168. Advance online publication. doi:10.3390/app919*4168

Jeyakymar, M., Munusami, V., Shanmugam, P., Boopalan, N., Boopathi, S., & Sureshkumar, M. (2022). An investigation on wear loss and hardness of nitro-ca*rburizing coated stainless-stee*l grade-316. Materials Today: Proceedings, 66, 1398–1404. doi:10.1016/j.matpr.2022.05.223

Jeznach, O., Kołbuk, D., & Sajkiewicz, P. (2018). Injectable hydrogels and nanocomposite hydrogels for cartilage regener*ation. Journal of Bi*omedical Materials Research. Part A, 106(10), 2762–2776. doi:10.1002/jb*m.a.36449 PMID:29726104*

Jiang, J., Hu, G., Li, X., Xu, X., Zhang, P., & Stringer, J. (2019). Analysis and prediction of printable bridge length in fused deposition modelling based on back propagation neural network. Virtual Phys Prototyp, 14.

Jiang, J., Xu, X., & Stringer, J. (2019). Optimisation of multi-part production in additive manufacturing for reducing support waste. Virtual and Physical Prototyping, 14(3), 219–228. doi:10.1080/17452759.2019.1585555

Jiang, J., Xu, X., & Stringer, J. (2019). Optimization of Process Planning for Reducing Material Waste in Extrusion Based Additive Manufacturing. Robotics and Computer-integrated Manufacturing, 59, 317–325. doi:10.1016/j.rcim.2019.05.007

Jiang, Y., Shiono, S., Hamada, H., Fujita, T., Higuchi, K., & Maenaka, K. (2010). Low-frequency energy harvesting using a laminated PVDF cantilever with a magnetic mass (Vol. 375378). Power MEMS.

Jiao, Z., Wang, C., Yang, Q., & Wang, X. (2018). Preparation and characterization of UV-curable urethane acrylate oligomers modified with cycloaliphatic epoxide resin. Journal of Coatings Technology and Research, 15(2), 251–258. doi:10.100711998-017-9994-6

Jim, W. (2015). The science and technology of composite materials - Curious. Australian Academy of Science. https://www.science.org.au/curious/technology-future/composite-materials

Jin, G. Q., Li, W. D., & Gao, L. (2013). An adaptive process planning approach of rapid prototyping and manufacturing. Robotics and Computer-integrated Manufacturing, 29(1), 23–38. doi:10.1016/j.rcim.2012.07.001

Jin, Y., an, He, Y., & Fu, J. (2014). Optimization of tool-path generation for material extrusion-based additive manufacturing technology. Additive Manufacturing, 1, 32–47. doi:10.1016/j.addma.2014.08.004

Jin, Y., He, Y., Fu, G., Zhang, A., & Du, J. (2017). A non-retraction path planning approach for extrusion-based additive manufacturing. Robotics and Computer-Integrated Manufacturing, 48(August), 132–144. doi:10.1016/j.rcim.2017.03.008

Jin, Y., Xia, N., & Gerhardt, R. A. (2016). Enhanced dielectric properties of polymer matrix composites with BaTiO3 and MWCNT hybrid fillers using simple phase separation. Nano Energy, 30, 407–416.

Jin, Z., Zhang, Z., Ott, J., & Gu, G. X. (2021). Precise localization and semantic segmentation detection of printing conditions in fused filament fabrication technologies using machine learning. Addit. Manuf., 37, 101696.

Johnson, N. S., Vulimiri, P. S., To, A. C., Zhang, X., Brice, C. A., & Kappers, B. B. (2020). Invited review: Machine learning for materials developments in metals additive manufacturing. Addit. Manuf., 36, 101641.

Johnson, T. (2018). History of Composites: The Evolution of Lightweight Composite Materials. ThoughtCo. https://www.thoughtco.com/history-of-composites-820404#:~:text=The

Jordan, M.I., & Mitchell, T.M. (2015). Machine learning: trends, perspectives, and prospects. Science, 349.

Joseph Francis, P. J., Arun, K. J., Navas, A. A., & Joseph, I. (2018). Biomedical Applications of Polymers-An Overview. Current Trends in Biomedical Engineering & Biosciences, 15(2), 555909.

Joshi, S. V., Drzal, L. T., Mohanty, A. K., & Arora, S. (2004). Are natural fiber composites environmentally superior to glass fiber reinforced composites? Composites. Part A, Applied Science and Manufacturing, 35(3), 371–376. doi:10.1016/j.compositesa.2003.09.016

Juan, Y., Dai, Y., Yang, Y., & Zhang, J. (2021). Accelerating materials discovery using machine learning. J Mater Sci Technol, 79.

Jung, Y. C., & Cho, J. W. (2010). Appli*cation of shape memory polyu*rethane in orthodontic. J. Mater. Sci. - Mater. Med., 21(10), 2881–2886. s10856-008-3538-7 doi:10.1007/

Kabir, S. F., Mathur, K., & Seyam, A.-F. M. (2020). A critical review *on 3D printed continuo*us fiber-reinforced composites: History, mechanism, materials and properties. Composite Structures, 232, 111476.

Kalin, M., Zalaznik, M., & Novak, S. (2014). Wear and friction behaviour of poly-ether-etherketone (PEEK) filled with graphene, WS2 and CNT nanoparticl*es. Wear, 855–862.*

*Kalishwa*ral*a*l, K., Jeyabharathi, S., Sundar, K., Selvamani, S., Prasanna, M., & Muthukumaran, A. (2018). A novel biocompatible chitosan–Selenium nanoparticles (SeNPs) film with electrical conductivity for cardiac tissue engineering application. Materials Science and Engine*ering C, 92, 151–160. doi:10.1016*/j.msec.2018.06.036 PMID:30184738

Kamei, M., Brian Saunders, W., Bayless, K. J., Dye, L., Davis, G. E., & Weinstein, B. M. (2006). Endothelial tubes assemble from intracellular vacuoles in vivo. Nature, 442(7101), 453–456. doi:10.1038/nature*04923 PMID:16799567*

Kamrani, A. K., & Nasr, E. A. (2010). Engineering Design and Rapid Prototyping. Springer.

Kang, H. W., Lee, *S. J., Ko, I. K., Kengla, C., Yoo,* J. J., & Atala, A. (2016). A 3D bioprinting system to produce human-scale tissue constructs with structural integrity. *Nature Biotechnology, 34(3), 312–3*19. doi:10.1038/nbt.3413 PMID:26878319

Kannan, E., Trabelsi, Y., Boopathi, S., & Alagesan, S. (2022). Influences of cryogenically treated work material on near-dry wire-cut electrical discharge machining process. Surface Topograph*y: Metrology and Properti*es, 10(1), 015027. Advance online publication. doi:10.1088/2051-672X/ac53e1

Kappes, B., Moorthy, S., Drake, D., Geerlings, H., & Stebner, A. (2018). Machine learning to *optimize additive manufacturing par*ameters for laser powder bed fusion of Inconel 718. Proceedings of the 9th international symposium on superalloy 718 & derivatives: energy, aerospace, a*nd industrial applications.*

Karayel, E., & Bozkurt, Y. (2020). Additive manufacturing method and different welding applications. J. Mater. Res. Technol., 9.

Kaur, H. S. (2006). Instrumental Methods of Chemical Analysis (3rd ed.). Pragati Prakashan.

Kavitha, C., Geetha Malini, P. S., Charan Kantumuchu, V., Manoj Kumar, N., Verma, A., & Boopathi, S. (2022). An experimental study on the hardness and wear rat*e of carbonitride coated stai*nless steel. Materials Today: Proceedings. Advance online publication. doi:10.1016/j.matpr.2022*.09.524*

Kawai, H. (1969). The piezoelectricity of poly (vinylidene fluoride). Japanese Journal of Applied Physics, 8(7), 975.

Kazmi, S. J. A., & Mitchell, A. G. (1973). Cellophane as a dialysis m*embrane for studying noni*o*nic* surfactants. Journal of Pharmaceutical Sciences, 62(8), 1299–1302. doi:10.1002/jps.2600620814 PMID:4725174

Keating, S.*, & Oxman, N. (2013). Compound fabricat*ion: A multi-functional robotic platform for digital design and fabrication. Robotics and Computer-integrated Manufacturing, 29(6), 439–448. doi:10*.1016/j.r*cim.2013.05.001

Keckes, J., Burgert, I., Fruhmann, K., Müller, M., Kollin, K., & Hamilton, M. (2003). Cell-wall recovery after irreversible deformation of wood. Nature Materials, 2(12), *810–814. doi:10.1038/nmat1019 P*MI*D:1*4625541

Kenett, R. S., Swarz, R. S., & Zonnenshain, A. (2019). Systems Engineering in the Fourth Industrial Revolution: Big Data. *Novel Technol*ogie*s*, and Modern Systems Engineering. doi:10.1002/9781119513957

Kennedy, Z. (2017). 3D-printed poly (vinylidene fluoride)/carbon nanotube composites as a tunable, low-cost *chemical vapour sensing platform. Nan*oscal*e*, 9(17), 5458–5466.

Kerimoglu, O., & Alarçin, E. (2012). Poly (lactic-co-glycolic acid) based drug delivery devices for tissue engineering and regenerative medicine. Ankem Derg, 26(2), 86–98. doi:10.5222/ankem.2012.086

Keshavamurthy, Tambrallimath, Prabhakar, & Sekhar. (2019)., Additive Manufacturing Processes and their Applications for Green Technology. In Handbook of Research on Green Engineering Techniques for Modern Manufacturing. IGI Global. doi:10.4018/978-1-5225-5445-5

Khadilkar, A., Wang, J., & Rai, R. (2019). Deep learning-based stress prediction for bottom-up SLA 3D printing process. Int J Adv Manuf Technol, 102.

Khan, I., & Shaikh, A. A. (2014). A Review of FDM Based Parts to Act as Rapid Tooling. International Journal of Modern Engineering Research, 4, 59–65.

Khan, M.F., Alam, A., Siddiqui, M.A., Alam, M.S., Rafat, Y., & Salik, N. (2020). Real-time defect detection in 3D printing using machine learning. Mater. Today: Proc.

Khan, W., Muthupandian, S., Farah, S., Kumar, N., & Domb, A. J. (2011). Biodegradable polymers derived from amino acids. Macromolecular Bioscience, 11(12), 1625–1636. doi:10.1002/mabi.201100324 PMID:22052719

Khanzadeh, M., Chowdhury, S., Marufuzzaman, M., Tschop, M.A., & Bian, L. (2018). Porosity prediction: supervised-learning of thermal history for direct laser deposition. J Manuf Syst, 47.

Khanzadeh, M., Chwodhury, S., Tschopp, M.A., Doude, H.R., Marufuzzaman, M., & Bian, L. (2019). In-situ monitoring of melt pool images for porosity prediction in directed energy deposition processes. IISE Trans, 51.

Khanzadeh, M., Rao, P., Jafari Mardani, R., Smith, B.K., Tschopp, M.A., & Bian, L. (2018). Quantifying geometric accuracy with unsupervised machine learning: using self-organizing map on fused filament fabrication additive manufacturing parts. J Manuf Sci Eng, 140.

Khanzadeh, M., Tian, W., Yadollahi, A., Doude, H.R., Tschop, M.A., & Bian, L. (2018). Dual process monitoring of metal-based additive manufacturing using tensorde composition of thermal image streams. Addit. Manuf., 23.

Khoo, Z. X. (2015). 3D printing of smart materials: A review on recent progresses in 4D printing. Virtual and Physical Prototyping, 10(3), 103–122.

Khosravani, M. R., & Reinicke, T. (2020). 3D-printed sensors: Current progress and future challenges. Sensors and Actuators. A, Physical, 305, 111916. doi:10.1016/j.sna.2020.111916

Khosravani, M.R., & Reinicke, T. (2020). Effects of raster layup and printing speed on strength of 3D-printed structural components. Procedia Struct. Integrity, 28.

Khosravani, M. R., & Reinicke, T. (2020). On the environmental impacts of 3D printing technology. Appl. Mater. Today, 20, 100689.

Khosravani, M. R., Berto, F., Ayatollahi, M. R., & Reinicke, T. (2020). Fracture behavior of additively manufactured components: A review. Theoretical and Applied Fracture Mechanics, 109, 102763.

Kim, G. D., & Oh, Y. T. (2008). A benchmark study on rapid prototyping processes and machines: Quantitative comparisons of mechanical properties, accuracy, roughness, speed, and material cost. Proceedings - Institution of Mechanical Engineers, 22, 201–215.

Kim, H. (2017). Increased piezoelectric response in functional nanocomposites through multiwall carbon nanotube interface and fused-deposition modeling three-dimensional printing. MRS Communications, 7(4), 960–966.

Kim, H., Fernando, T., Li, M., Lin, Y., & Tseng, T.-L. B. (2018*). Fabrication and characterization of 3D printed* BaTiO3/ PVDF nanocomposites. Journal of Composite Materials, 52(2), 197–206.

Kim, H., Torres, F., Villagran, D., Stewart, C., & Lin, Y. (2017). 3D printing of BaTiO3/PVDF composites with electric in *situ poling for pressure sensor applications. Macromolecular* Materials and Engineering, 302(11), 1700229.

Kim, J., & Creasy, T. (2004). Selective laser sintering characteristics of nylon 6/clay-reinforced nanocomposite. Polymer Testing, 23, 629–636.

Kim, J. A., Kim, H. N., Im, S. K., *Chung, S., Kang, J. Y., & Choi, N. (2015). Collagen-based* brain microvasculature model in vitro using three-dimensional printed template. Biomicrofluidics, 9(2), 24115. doi:10.1063/1.4917508 PMID:25945141

Kim, S., Kim, H. J., & Park, J. C. (2009). Application of recycled paper sludge and biomass materials in manufacture of green composite pallet. Resources, Conservation and Recycling, 12(53), 674–679.

Kim, S. Y., Shin, Y. S., Jung, H. D., Hwang, C. J., Baik, H. S., & Cha, J. Y. (2018*). Precision and trueness of dental models m*anufactured with different 3-dimensional printing techniques. American Journal of Orthodontics and Dentofacial Orthopedics, 153(1), 144–153. doi:10.1016/j.ajodo.2017.05.025 PMID:29287640

Kim, T. H., Lee, J. H., *Ahn, C. B., Hong,* J. *H.*, Son, K. H., & Lee, J. W. (2019). Development of a 3D-printed drug-eluting stent for treating obstructive salivary gland disease. ACS Biomaterials Science & Engineering, 5(7), 3572–3581. doi:10.1021/acsbiomaterials.9b00636 PMID:33405739

Kishore, S., *Sampathkumaran,* P., Seetharamu, S., Vynatheya, S., Murali, A., & Kumar, R. K. (2000). SEM observations of the effects *of velocity and load on the sliding wear characteristics o*f glass fabric-epoxy composites with different fillers. Wear, 237(1), 20–27. doi:10.1016/S0043-1648(99)00300-2

Klein, *G. T., Lu, Y., & Wang, M. Y. (*2013). 3D printing and neurosurgery—Ready for prime time? World Neurosurgery, 80(3–4), 233–235. doi:10.1016/j.wneu.2013.07.009 PMID:23871811

Koenig, J. L., & Angoo*d, A. C. (1970). Raman spectra of pol*y (ethylene glycols) in solution. Journal of Polymer Science. Part A-2, Polymer Physics, 8(10), 1787–1796. doi:10.1002/pol.1970.160081013

Kola, M., Rabah, A., Raghav, D., Sabra, S., Harby, N. *M., Khalil, H., & Shah, A. (2015).* Surgical templates for dental implant positioning; current knowledge and clinical perspectives. Nigerian Journal of Surgery: Official Publication of the Nigerian Surgical Research Society, 21(1), 1. doi:10.4103/1117-6806.152720 PMID:25838757

Kong, L., Ambrosi, *A., Nasir, M.Z.M., Guan, J., &* Pumera, M. (2019). Selfpropelled 3D-printed "Aircraft Carrier" of light-powered smart micromachines for large-volume nitroaromatic explosives removal. Adv Funct Mater, 19, 1-9.

Koronis, G., Silva, A., & Fontul, M. (2013). Green composites: A review of adequate *materials for automotive applica-*tions. Composites. Part B, Engineering, 44(1), 120–127. doi:10.1016/j.compositesb.2012.07.004

Koshut, W. J., Smoot, D., Rummel, C., Kirillova, A., & Gall, K. (2020). Tensil*e Fatigue of Poly (Vinyl Alcohol) Hydro-gels with* B*io*-Friendly Toughening Agents. Macromolecular Materials and Engineering, 305(3), 1900784. doi:10.1002/mame.201900784

Kousiatza, C., & Karalekas, D. (2016). In-situ monitoring of strain and temperature distributions during fused deposition modeling pr*ocess. Materials & Design*, 97, 400–406. doi:10.1016/j.matdes.2016.02.099

Krajangsawasdi, N., Blok, L. G., Hamerton, I., & Longana, M. L. (2021). Fused Deposition *Modelling of Fibre Reinforced Polym*er Composites: A Parametric Review. Academic Press.

Kroeze, R. J., Helder, M. N., Govaert, L. E., & Smit, T. H. (2009). Biodegradable polymers in bone tissue engineering. Materials (Basel), 2(3), 833–856. doi:10.3390/ma2030833

Kroll, E., & Artzi, D. (2011). Enhancing aerospace engineering students' learning with 3D printing wind-tunnel models. Rapid Prototyping Journal, 17(5), 393–402. doi:10.1108/13552541111156522

Kuang, X., Chen, K., Dunn, C. K., Wu, J., Li, V. C. F., & Qi, H. J. (2018). 3D Printing of Highly Stretchable, Shape-Memory, and Self-Healing Elastomer toward Novel 4D Printing. ACS Applied Materials & Interfaces, 10(8), 7381–7388. doi:10.1021/acsami.7b18265 PMID:29400445

Kubalak, J. R., Wicks, A. L., & Williams, C. B. (2018). Using multi-axis material extrusion to improve mechanical properties through surface reinforcement. Virtual and Physical Prototyping, 13(1), 32–38. doi:10.1080/17452759.2017.1392686

Kumar, A. L., & Prakash, M. (2021). The effect of fiber orientation on mechanical properties and machinability of GFRP composites by end milling using cutting force analysis. Polymers & Polymer Composites, 29(9), 178–187.

Kumar, D., Sharma, M., Verma, S., & Saroha, K. (2016). Natural polymers and herbal medicine based therapy for colonic diseases. International Journal of Herbal Medicine, 4(3), 49–56.

Kumar, P., Sampath, B., Kumar, S., Babu, B. H., & Ahalya, N. (2023). Hydroponics, Aeroponics, and Aquaponics Technologies in Modern Agricultural Cultivation. In Trends, Paradigms, and Advances in Mechatronics Engineering (pp. 223–241). IGI Global.

Kumar, Singh, Singh, & Batish. (n.d.). Fused filament fabrication: A comprehensive review. J. Thermoplast. Compos. Mater. doi:10.1177/0892705720970629

Kumbar, S., Laurencin, C., & Deng, M. (Eds.). (2014). Natural and synthetic biomedical polymers. Newnes.

Kumekawa, N., Mori, Y., Tanaka, H., & Matsuzaki, R. (2022). Experimental evaluation of variable thickness 3D printing of continuous carbon fiber-reinforced composites. Composite Structures, 288, 115391. doi:10.1016/j.compstruct.2022.115391

Kunanopparat, T., Menut, P., Morel, M. H., & Guilbert, S. (2008). Plasticized wheat gluten reinforcement with natural fibers: Effect of thermal treatment on the fiber/matrix adhesion. Composites. Part A, Applied Science and Manufacturing, 39(12), 1787–1792.

Kurimoto, M., Yamashita, Y., Ozaki, H., Kato, T., & Funabashi, T. (2015). 3D printing of conical insulating spacer using alumina/ UV-cured-resin composite. In Electrical insulation and dielectric phenomena (CEIDP), IEEE Conference on, 2015. IEEE.

Kwon, G. S., & Furgeson, D. Y. (2007). Biodegradable polymers for drug delivery systems. In Biomedical polymers (pp. 83–110). Woodhead Publishing. doi:10.1533/9781845693640.83

Lagoda, A., & Gabor, R. (2018, October). Strength parameters of the ABS materials used in 3D printing. AIP Conference Proceedings, 2029, 020036. Advance online publication. doi:10.1063/1.5066498

Lan, X., Liu, Y., Lv, H., Wang, X., Leng, J., & Du, S. (2009). Fiber reinforced shape-memory polymer composite and its application in a deployable hinge. Smart Materials and Structures, 18(2), 024002.

Landers, R., & Mülhaupt, R. (2000). Desktop manufacturing of complex objects, prototypes and biomedical scaffolds by means of computer-assisted design combined with computer-guided 3D plotting of polymers and reactive oligomers. Macromolecular Materials and Engineering, 282(1), 17–21. doi:10.1002/1439-2054(20001001)282:1<17::AID-MAME17>3.0.CO;2-8

Lasikun, Ariawan, D., Surojo, E., & Triono, J. (2018). Effect of fiber orientation on tensile and im*pact* properties of Zalacca Midrib fiber-HDPE composites by compression molding. AIP Conference Proceedings, (1), 1–5.

Laske, S. (2015). Polymer Clay Nanocomosite (1st ed.). Els*evier.*

Lau, K., Hung, P., Zhu, M.-H., & Hui, D. (2018). Properties of natural fibre composites for struc*tural engineering applications. Composit*es. Part B, Engineering, 136(1), 222–233. doi:10.1016/j.compositesb.2017.10.038

Laurencin, C. T., Norman, M. E., Elgendy, H. M., El-Amin, S. F., Allcock, H. R., Pucher, S. R., & Ambros*io, A. A. (1993). Use of* polyphosphazenes for skeletal tissue regeneration. Journal of Biomedical Materials Research, 27(7), 963–973. doi:10.1002/jbm.820270716 PMID:8360223

Lawlor, K. T., Vanslambrouck, J. M., Higgins, J. W., Chambon, A., Bishard, *K., Arndt, D., Er, P. X., Wilson, S. B., How*d*en*, S. E., Tan, K. S., Li, F., Hale, L. J., Shepherd, B., Pentoney, S., Presnell, S. C., Chen, A. E., & Little, M. H. (2021). Cellular extrusion bioprinting improves kidney organoid reproducibility and conformation. Nature Materials, 20(2), 260–271. doi*:10.103841563-020-00853-9 PMID:33230326*

*Le, H. P. (1998). Progress and trends in ink-jet print technology. The Journal of Ima*ging Science and Technology, 42, 49–62.

Leal, R., Barreiros, F.M., Alves, L., Romeiro, F., Vasco, J.C., *& Santos, M. (2017). Addi*tive manufacturing tool*ing for the automotive industry. Int J Ad*v Manuf Technol, 92, 1671-6.

Lee, C., & Tarbutton, J. A. (2014). Electric poling-assisted additive manufacturing process for PVDF polymer-based piezoelectric device applications. Smart Materials and Structures, 23(9), 095044.

Lee, *J., Lee, H., Cheon, K. H., P*ark, C., Jang, T. S., Kim, H. E., & Jung, H. D. (2019). Fabrication of poly (lactic acid)/ Ti composite scaffolds with enhanced mechan*ical properties and biocompatibilit*y *v*ia fused filament fabrication (FFF)–based 3D printing. Additive Manufacturing, 30, 100883. doi:10.1016/j.addma.2019.100883

*Lee, J.-Y., An, J., & Chua, C. K. (20*17). Fundamentals and applications of 3D printing for novel materials. Appl. Mater. Today, 7, 120–133.

Lee, J. S., Chae, G. S., Kim, M. S., Cho, S. H., Lee, H. B., & Khang, G. (2004). *Degradation behaviour in vitro for poly (D, L-lac*tide-co-glycolide) as drug carrier. Bio-Medical Materials and Engineering, 14(2), 185–192. PMID:15156109

Lee, J. S., Hong, J. M., Jung, J. W., Shim, J. H., Oh, J. H., & Cho, D. W. (2014). 3D *printing of com*po*s*ite tissue with complex shape applied to ear regeneration. Biofabrication, 6(2), 24103. doi:10.1088/1758-*5082/6/2/024103 PMID:24464765*

Lee, K., & Jee, H. (2015). Slicing algorithms for multi-axis 3-D metal printing of overhangs. Journal of Mechanical Science and Technology, 29(12), 5139–5144. doi:10.100712206-015-1113-y

Lee, K. S., Kim, R. H., Yang, D. Y., & Park, S. H. (2008). Advances in *3D nano/*microfabrication using two-photon initiated polymerization. Progress in Polymer Science, 33, 631–681.

Lee, K. Y., & Mooney, D. J. (2012). Alginate: Properties and bio*medical ap*plic*a*tions. Progress in Polymer Science, 37(1), 106–126. doi:10.1016/j.progpolymsci.2011.06.003 PMID:22125349

Lee, L., & Jain, R. (2012). Fiber Reinforced Polymer (FRP) Composites for Infrastructure Applications. Springer Science+Business Media. www.springer.com

Lee, Y.-S., Collins, G., & Arinzeh, T. L. (2011). Neurite extension of primary neurons on electrospun piezoelectric scaffolds. Acta Biomaterialia, 7(11), 3877–38*86.*

Leigh, S. J., Bradley, R. J., Purssell, C. P., Billson, D. R., & Hutchins, D. A. (2012). A simple, low-cost conductive composite material for 3D printing of electronic sensor*s. PLoS One, 7(11)*, e49365.

Lendlein, A., & Kelch, S. (2002). Shape-memory polymers. Angewandte Chemie Int*ernational Edition, 41(12), 2034–2057.*

*Leng, J., La*n, X., Liu, Y., & Du, S. (2011). Shape-memory polymers and their composites: Stimulus methods and applications. Progress in Materials *Science, 56(7), 1077–1135.*

*L*i, H., Liu, Y., Liu, Y., Zeng, Q., Hu, K., & Lu, Z. (2020). Effect of debinding temperature under an argon atmosphere on the microstructure and properties of 3D-printed alumina ceramics. Materials Characterization, 168, 110548.

Li, L., Haghighi, A., & Yang, Y. (2018). A novel 6-axis hybrid additive-subtractive manufacturing process: Design and case studies. Journal of Manufacturing Processes, 33(March), 150–160. doi:10.1016/j.jmapro.2018.05.*008*

Li, N., Li, Y., & Liu, S. (2016). Rapid prototyping of continuous carbon fiber reinforced polylactic acid composites by 3D printing. Journal of Materials Processing Technology, 238(2000), 218–225. doi:10.1016/j.jmatprotec.2016.07.025

Li, X., Liu, B., Pei, *B., Chen, J., Zhou,* D., Peng, J., Zhang, X., Jia, W., & Xu, T. (2020). Inkjet Bioprinting of Biomaterials. Chemical Reviews, 120(19), 10793–10833. doi:10.1021/acs.chemrev.0c00008 PMID:32902959

Li, Y., Li, X., Zhang, G., Hor*vath, I., & Han,* Q. (2021). Interlayer closed-loop control of forming geometries for wire and arc additive manufacturi*ng based on fuzzy-logic inferenc*e. *J* Manuf Process, 63.

Li, Y., Sun, Y., Han, Q., Zhang, G., & Horvath, I. (2018). Enhanced beads overlapping model for wire and arc additive manufactu*ring of multi-layer multi-bead* metallic parts. J Mater Process Technol, 252.

Li, Z., Liu, R., & Wu, D. (2019). Data-driven smart manufacturing: tool wear monitoring with a*udio signals and machine learning. J Manuf* Proc*ess, 48.

Li, Z., Zhang, Z., Shi, J., & Wu, D. (2019). Prediction of surface roughness in extrusion-based additiv*e manufacturing with machine learn*ing. Robot Comput Integrated Manuf, 57.

Liao, E., Yaszemski, M., Krebsbach, P., & Hollister, S. (2007). Tissue-engineered cartilage constructs *using composite hy*al*ur*onic acid/collagen I hydrogels and designed poly (propylene fumarate) scaffolds. Tissue Engineering, 13(3), 537–550. doi:10.1089/ten.2006.0117 PMID:17319795

Ligon, S. C., Liska, R., Stampfl, J., Gurr, M.*, & Mülhaupt, R. (2017). Polymers for 3D Printing and* Customized Additive Manufacturing. Chemical Reviews, 117(15), 10212–10290. doi:10.1021/acs.chemrev.7b00074 PMID:28756658

Ligon, S. C., Liska, R., Stampfl, J. R., Gurr, M., & Mülhaupt, R. (20*17). Polymers for* 3D printing and customized additive manufacturing. Chemical Reviews, 117(15), 10212–10290.

Lim, C. W. J., Le, K. Q., Lu, Q., & Wong, C. H. (2016). An overview *of 3-D printing in manufacturi*ng, aerospace, and automotive industries. IEEE Potentials, 35(4), 18–22. doi:10.1109/MPOT.2016.2540098

Lim, S., Buswell, R. A., Valentine, P. J., Piker, D., Austin, S. A., & De Kestelie*r, X. (2016). Modelling curved-layered print*ing paths for fabricating large-scale construction components. Additive Manufacturing, 12, 216–230. doi:10.1016/j.addma.2016.06.004

Lin, D., Jin, S., Zhang, F., Wang, C., Wang, Y., Zhou, C.*, & Cheng, G. J.* (2015). 3D stereolithography printing of graphene oxide reinforced complex architectures. Nanotechnology, 26(43), 434003.

Lin, P. P. A., Willis, K., & Wilhelm, E. J. *(2018). Intel*ligent 3D printing through optimization of 3D print parameters. US Patent 10,073,424, 1(19). https://patents.google.com/patent/US10073424B2/en%0Ahttps://patentimages.storage.googleapis.com/28/1d/07/a0fdac*f1cf3db7/ US10073424.pdf*

Lin, W., Tong, T., Gao, Q., Guo, D., Du, X., Yang, Y., Guo, G., Xiao, M., Du, M., & Qu, X. (2018). Convolutional neural networks-based MRI image analysis for the Alzheimer's disease prediction from mild cognitive impairment. Frontiers in Neuro*science, 12, 777. doi:10.3389/fnins.20*18.*0*0777 PMID:30455622

Liu, C., Qin, H., & Mather, P. (2007). Review of progress in shape-memory polymers. Journal of Materials Chemistry, 17(16), 1543–1558.

Liu, H., He, H., Peng, X., Huang, B., & Li, J. (2019). Three dimensional printing of poly (lactic acid) b*io based* composites with sugarcane bagasse fiber: Effect of printing orientation on tensile performance. Polymers for Advanced Technologies, 30(4), 910–*922.*

Liu, H., Li, X., Zhou, G., Fan, H., & Fan, Y. (2011). Electrospun sulfated silk fibroin nanofibrous scaffolds for vascular tissue engineering. Biomaterials, 32(15), 3784*–3793. doi:10.1016/j.biomaterials.2011.02.002 PMID:213763*91

Liu, P., Chen, W., Liu, C., Tian, M., & Liu, P. (2019). A novel poly (vinyl alcohol)/poly (ethylene glycol) scaffold for tissue engineering with a unique bimodal open-celled structure fabricated using supercritical fluid foaming. *Scientific Reports, 9(1), 1–12. doi:10.103841598-019-46061-7 PMID:31267014*

Liu, Q., Leu, M. C., & Schmitt, S. M. (2006). Rapid prototyping in dentistry: Technology and application. International Journal of Advanced Manufacturing Technology, 29(3–4), 317–335. doi:10.100700170-005-2523-2

Liu, Y., Niu, C., Wang, Z., Gan, Y., Zhu, Y., & Sun, S. (202*0). Machine learni*ng in materials genome initiative: a review. J Mater Sci Technol, 57.

Liu, Y., Zhao, T., Ju, W., & Shi, S. (2017). Materials disco*very and design using machine l*earning. J Mater, 3.

Liu, Z., Li, M., Weng, Y., Qian, Y., Wong, T. N., & Tan, M. J. (2020). Modelling and parameter optimization for filament deformation in 3D cementitious material printing using *support vector* mac*hin*e. Composites. Part B, Engineering, 193, 108018.

Liu, Z., Wang, Y., Wu, B., Cui, C., Guo, Y., & Yan, C. (2019). A cr*itical review of fused deposition modeling 3D printing technology in manufacturing poly*lactic acid parts. International Journal of Advanced Manufacturing Technology, 102(9), 2877–2889.

Lodish, H., Berk, A., Zipursky, S. L., *Matsudaira, P.,* Ba*l*timore, D., & Darnell, J. (2000). Collagen: The fibrous proteins of the matrix. Molecular and Cellular Biology, 4.

LogithKumar, R., KeshavNarayan, A., Dhiv*ya, S., Chawla, A., Sarav*anan, S., & Selvamurugan, N. (2016). A review of chitosan and its derivatives in bone tissue engineering. Carbohydrate Polymers, 151, 172–188. doi:10.1016/j.carbpol.2016.05.049 PMID:27474556

Lohfeld, S., Tyndyk, M. A., Cahill, *S., Flaherty, N., Barron, V., & M*cH*ug*h, P. E. (2010). A method to fabricate small features on scaffolds for tissue engineering via selective laser sintering. Journal of Biomedical Science and Engineering, 03(02), 138–147. doi:10.4236/jbise.2010.32019

*Lu, L., Zheng, J., & Mishra, S. (20*15). A layer-to-layer model and feedback control of ink-jet 3-D printing. IEEE/ASME Transactions on Mechatronics, 20(3), 1056–1068. doi:10.1109/TMECH.2014.2366123

Lu, Q. Y., & Wong, C. H. (2018). Additive manufacturing process monitoring and control by non-destructive testing techniques: Challenges and in-process monitoring. Virtual and Physical Prototyping, 13(2), 39–48. doi:10.1080/17452759.2017.1351201

Lu, X., & Chen, Z. (2005). Curved pi-conjugation, curved pi-conjugation, aromaticity, and the related chemistry of small fullerenes (bC60) and single-walled carbon nanotubes. Chemical Reviews, 105, 3643–3696.

Lu, Z.L., Li, D.C., Lu, B.H., Zhang, A.F., Zhu, G.X., & Pi, G. (2010). The prediction of the building precision in the laser engineered net shaping process using advanced networks. Opt Laser Eng, 48.

Luo, Y., Le Fer, G., Dean, D., & Becker, M. L. (2019). 3D printing of poly (propylene fumarate) oligomers: Evaluation of resin viscosity, printing characteristics and mechanical properties. Biomacromolecules, 20(4), 1699–1708. doi:10.1021/acs.biomac.9b00076 PMID:30807696

Lužanin, O., Movrin, D., & Plan, M. (2013). Experimental Investigation of Extrusion Speed and Temperature Effects on Arithmetic Mean Surface Roughness in Fdm-. Journal of Technology of Plasticity, 38(2), 179–190.

Lyman, D. J. (1974). Polymers in medicine. Angewandte Chemie International Edition in English, 13(2), 108–112. doi:10.1002/anie.197401081 PMID:4207434

Lüchtenborg, J., Burkhardt, F., Nold, J., Rothlauf, S., Wesemann, C., Pieralli, S., Wemken, G., Witkowski, S., & Spies, B. C. (2021). Implementation of fused filament fabrication in dentistry. Applied Sciences (Switzerland), 11(14), 6444. Advance online publication. doi:10.3390/app11146444

Ma, G., Yang, D., Li, Q., Wang, K., Chen, B., Kennedy, J. F., & Nie, J. (2010). Injectable hydrogels based on chitosan derivative/polyethylene glycol dimethacrylate/N, N-dimethylacrylamide as bone tissue engineering matrix. Carbohydrate Polymers, 79(3), 620–627. doi:10.1016/j.carbpol.2009.09.015

Machment, T., Sanjayan, J., & Xia, M. (2019). Method of enhancing interlayer bond strength in construction scale 3D printing with mortar by effective bond area amplification. Materials & Design, 169, 107684. doi:10.1016/j.matdes.2019.107684

Macocinschi, D., Filip, D., Zaltariov, M. F., & Varganici, C. D. (2015). Thermal and hydrolytic stability of silver nanoparticle polyurethane biocomposites for medical applications. Polymer Degradation & Stability, 121, 238–246. doi:10.1016/j.polymdegradstab.2015.09.017

Madhavadas, V., Srivastava, D., Chadha, U., Raj, S. A., Sultan, M. T. H., Shahar, F. S., & Shah, A. U. M. (2022). A review on metal additive manufacturing for intricately shaped aerospace components. CIRP Journal of Manufacturing Science and Technology, 39, 18–36. doi:10.1016/j.cirpj.2022.07.005

Madsen, B., & Lilholt, H. (2003). Physical and mechanical properties of unidirectional plant fibre composites-an evaluation of the influence of porosity. Composites Science and Technology, 63(9), 1265–1272. doi:10.1016/S0266-3538(03)00097-6

Magalhães, L. S. S. M., Santos, F. E. P., de Maria Vaz Elias, C., Afewerki, S., Sousa, G. F., Furtado, A. S. A., Marciano, F. R., & Lobo, A. O. (2020). Printing 3D hydrogel structures employing low-cost stereolithography technology. Journal of Functional Biomaterials, 11(1), 12. doi:10.3390/jfb11010012 PMID:32098431

Magdum, Y., Pandey, D., Bankar, A., Harshe, S., Parab, V., Mahesh, M., & Kadam, S. (2008). Process Parameter Optimization for FDM 3D Printer. International Research Journal of Engineering and Technology, 1472.

Mahjoub, R., Yatim, J. M., Sam, A. R. M., & Hashemi, S. H. (2014). Tensile properties of kenaf fiber due to various conditions of chemical fiber surface modifications. Construction & Building Materials, 55, 103–113.

Mahmood, M. A., Visan, A. I., Ristoscu, C., & Mihailescu, I. N. (2021). Artificial neural network algorithms for 3D printing. In Materials (Vol. 14, Issue 1, pp. 1–29). doi:10.3390/ma14010163

Mahmood, S., Qureshi, A., Goh, K. L., & Talamona, D. (2017). Tensile strength of partially filled FFF printed parts: Experimental results. Rapid Prototyping Journal.

Majid, R. A., Ismail, H., & Taib, R. M. (2010). Effects of polyethylene-g-maleic anhydride on properties of low density polyethylene/thermoplastic sago starch reinforced kenaf fibre composites. Iranian Polymer Journal, 19, 501–510.

Manavitehrani, I., Fathi, A., Wang, Y., Maitz, P. K., Mirmohseni, F., Cheng, T. L., & Dehghani, F. (2017). Fabrication of a biodegradable implant with tunable characteristics for bone implant applications. Biomacromolecules, 18(6), 1736–1746. doi:10.1021/acs.biomac.7b00078 PMID:28535038

Manoukian, O. S., Sardashti, N., Stedman, T., Gailiunas, K., Ojha, A., Penalosa, A., ... Kumbar, S. G. (2019). Biomaterials for tissue engineering and regenerative medicine. Academic Press.

Marandi, M., & Tarbutton, J. (2019). Additive manufacturing of single-and double-layer piezoelectric PVDF-TrFE copolymer sensors. Procedia Manufacturing, 34, 666–671.

Martinschitz, K. J., Boesecke, P., Garvey, C. J., Gindl, W., & Keckes, J. (2008). Changes in microfibril angle in cyclically deformed dry coir fibers studied by in-situ synchrotron X-ray diffraction. Journal of Materials Science, 43(1), 350–356. doi:10.100710853-006-1237-7

Martorelli, M., Gerbino, S., Giudice, M., & Ausiello, P. (2013). A comparison between customized clear and removable orthodontic appliances manufactured using RP and CNC techniques. Dental Materials, 29(2), e1–e10. doi:10.1016/j.dental.2012.10.011 PMID:23140842

Mateos-Timoneda, M. A. (2009). Polymers for bone repair. In Bone repair biomaterials (pp. 231–251). Woodhead Publishing. doi:10.1533/9781845696610.2.231

Materials and filaments for industrial 3D printing. (2015). Retrieved January 20, 2023, from https://www.stratasys.com/en/materials/materials-catalog/

Mathew, J., Das, J. P., Tp, M., & Kumar, S. (2022). Development of poly (butylene adipate-co-terephthalate) PBAT toughened poly (lactic acid) blends 3D printing filament. Journal of Polymer Research, 29(11), 1–15. doi:10.100710965-022-03320-y

Mathew, M. T., Padaki, N. V., Rocha, L. A., Gomes, J. R., Alagirusamy, R., Deopura, B. L., & Fangueiro, R. (2007). Tribological properties of the directionally oriented warp knit GFRP composites. Wear, 263(7-12), 930–938. doi:10.1016/j.wear.2006.12.001

Maxey, K. (2013, September 3). Infographic: The history of 3D printing. Retrieved January 20, 2023, from https://www.engineering.com/story/infographic-the-history-of-3d-printing

Mazurchevici, A. D., Nedelcu, D., & Popa, R. (2020). Additive manufacturing of composite materials by FDM technology: A review. Indian Journal of Engineering and Materials Sciences, 27(2), 179–192. doi:10.56042/ijems.v27i2.45920

Mazzanti, V., & Mollica, F. (2017). In-process measurements of flow characteristics of wood plastic composites. Journal of Polymers and the Environment, 25, 1044–1050.

Mazzanti, V., de Luna, S. M., Pariante, R., Mollica, F., & Filippone, G. (2020). Natural fiber-induced degradation in PLA-hemp biocomposites in the molten state. Composites. Part A, Applied Science and Manufacturing, 137, 105990.

Mazzanti, V., Malagutti, L., & Mollica, F. (2019). FDM 3D printing of polymers containing natural fillers: A review of their mechanical properties. Polymers, 11(7), 1094.

McHale, M. K., Bergmann, N. M., & West, J. L. (2013). Histogenesis in Three-Dimensional Scaf*folds. Handbook* of *St*em Cells (2nd ed.). Academic Press.

Mehrali, M., Bagherifard, S., Akbari, M., Thakur, A., Mirani, B., Mehrali, M., & Dolatshahi-Pirouz, A. (2018). Blending electronics with the human body: A pathway toward a cybernetic future. Advancement of Science, 5*(10), 1700931. doi:10.1002/*ad*vs*.201700931 PMID:30356969

Meijer, H. E. H., & Govaert, L. E. (2005). Mechanical performance of polymer systems: The relation between structure a*nd properties. Progress in Polymer* Sci*ence (Oxford), 30(8–9), 915–938. doi:10.1016/j.progpolymsci.2005.06.009

Melchels, F. P. W., Feijen, J., & Grijpma, D. W. (2010). A review on stereolithography and its applications in biomedical engineering. Biomaterials, 31, 6121–6130.

Mel*nikova, R., & Ehrmann,* A. (2014). Finsterbusch K 3D printing of textile-based structures by Fused Deposition Modelling (FDM) with different polymer materials. In IOP Conference Series: Materi*als Science and Enginee*ring (vol. 1). IOP Publishing.

Melnikova, R., Ehrmann, A., & Finsterbusch, K. (2014, August). 3D printing of textile-based structures by Fused Deposition Modelling (FDM) with different polymer materials. IOP Conference Series. Materials Science and Engineering, 62(1), 012018. doi:10.1088/1757-899X/62/1/012*018*

Menachem, L. (2007). Handbook of Fiber Chemistry. In Encyclopedic Dictionary of Polymers (3rd ed.). CRC Press. doi:10.1007/978-0-387-30160-0_2928

Meng, H., & Li, G. (2013). A review of stimuli-responsive shape memory polymer composites. Polymer, 54(9), *2199–2221.*

Meng, Q., & Hu, J. (2019). A review of shape memory polymer composites and blends. Compos. A Appl. Sci. Manuf., 40(11), 1661–1672. doi:10.1016/j.compositesa.2009.08.011

Merlini, C., Soldi, *V., & Barra, G. M. O. (*2011). Influence of fiber surface treatment and length on physico-chemical properties of short random b*anana fibe*r-reinforced castor oil polyurethane composites. Polymer Testing, 30(8), 833–840. doi:1*0.1016/j.polymertesting.2011.08.008*

Messaoud, S., Bradai, A., Bukhari, S. H. R., Quang, P. T. A., Ahmed, O. B., & Atri, M. (2020). A survey on machine learning in Internet of Things: Algorithms, strategies, and applications. Internet Things, 12, 100314.

Mihaly, *J., Sterkel, S., Ortner, H. M.,* Koc*s*is, L., Hajba, L., Furdyga, E., & Mink, J. (2006). FTIR and FT-Raman spectroscopic study on polymer based high pressure d*igestion vessels. Croatica Chemica Acta, 79(3), 497–501.*

Mi*lo*sevic, M., Stoof, D., & Pickering, K. L. (2017). Characterizing the mechanical properties of fused deposition modelling natural fiber recycled polypropylene composites. Journal of Composites Science, 1(7), 1–14.

Mironov, V., Kasyanov, V., & Mar*kwald, R. R. (2011). Organ prin*ting*:* From bioprinter to organ biofabrication line. Current Opinion in Biotechnology, 22(5), 667–673. doi:10.1016/j.copbio.2011.02.006 PMID:21419621

Moby, V., Dupagne, L., Fouquet, V., Att*al, J. P., François, P., & Dursun, E. (2022). Mechanical Pro*per*ties of Fused Deposition Modeling of Polyetheretherketone (PEEK) and Interest for Dental Restorations: A Systematic Review. Materials (Basel), 15*(19), 6801. Advan*ce *on*line publication. doi:10.3390/ma15196801 PMID:36234139

Mohamed, O.A., Masood, S.H., & Bhowmik, J.L. (2016). Investigation of dynamic elastic deformation of parts processed by fused deposition modeling ad*ditive manufacturing. Adv. Pro*d. Eng. Manage., 13.

Mohamed, O.A., Masood, S.H., & Bhowmik, J.L. (2017). Influence of processing parameters on creep and recovery behavior of FDM manufactured part using definitive screening design *and ANN. Rapid Prototyp J, 23.*

Mohammed, L., Ansari, M. N. M., Pua, G., Jawaid, M., & Islam, M. S. (2015). A review on natural fiber reinforced polymer composite and its applications. International Journal of Polymer Science, 2015, 1–15. doi:10.1155/2015/243947

Mohan, N., Natarajan, S., Kumaresh Babu, S. P., & Siddaramaiah. (2011). Investigation on sliding wear behaviour and mechanical properties of jatropha oil cake-filled glass-epoxy composites. Journal of the American Oil Chemists' Society, 88(1), 111–117. doi:10.100711746-010-1657-z

Mohanraj, K., Sethuraman, S., & Krishnan, U. M. (2013). Development of poly (butylene succinate) microspheres for delivery of levodopa in the treatment of parkinson's disease. Journal of Biomedical Materials Research. Part B, Applied Biomaterials, 101(5), 840–847. doi:10.1002/jbm.b.32888 PMID:23401377

Mohanty, A. K. (2005). Natural Fibers, Biopolymers and Biocomposites. Taylor & Francis Group.

Mohanty, A. K., Misra, M., & Drzal, L. T. (2002). Sustainable Bio-Composites from renewable resources: Opportunities and challenges in the green materials world. Journal of Polymers and the Environment, 10, 19–26.

Mohanty, A. K., Misra, M., & Hinrichsen, G. (2000). Biofibres, biodegradable polymers and biocomposites: An overview. Macromolecular Materials and Engineering, 276/277(1), 1–24. doi:10.1002/(SICI)1439-2054(20000301)276:1<1::AID-MAME1>3.0.CO;2-W

Momenzadeh, N., Miyanaji, H., Porter, D. A., & Berfield, T. A. (2020). Polyvinylidene fluoride (PVDF) as a feedstock for material extrusion additive manufacturing. Rapid Prototyping Journal.

Montazeri, M., & Rao, P. (2018). Sensor-based build condition monitoring in laser powder bed fusion additive manufacturing process using a spectral graph theoretic approach. Journal of Manufacturing Science and Engineering, 140, 091002.

Monz'on, M., Ortega, Z., Martínez, A., & Ortega, F. (2015). Standardization in additive manufacturing: Activities carried out by international organizations and projects. International Journal of Advanced Manufacturing Technology, 76(5–8), 1111–1121.

Moon, S., Hasan, S. K., Song, Y. S., Xu, F., Keles, H. O., Manzur, F., Mikkilineni, S., Hong, J. W., Nagatomi, J., Haeggstrom, E., Khademhosseini, A., & Demirci, U. (2010). Layer by layer three-dimensional tissue epitaxy by cell-laden hydrogel droplets. Tissue Engineering. Part C, Methods, 16(1), 157–166. doi:10.1089/ten.tec.2009.0179 PMID:19586367

Morales Rodríguez. (2017). Printed Sensors for Energy Applications: Challenges and Implications. Academic Press.

Moran, J., Alvarez, V., Petrucci, R., Kenny, J., & Vazquez, A. (2007). Mechanical properties of polypropylene composites based on natural fibers subjected to multiple extrusion cycles. Journal of Applied Polymer Science, 103, 228–237.

Morelli, A., Puppi, D., & Chiellini, F. (2013). Polymers from renewable resources. Journal of Renewable Materials, 1(2), 83–112. doi:10.7569/JRM.2012.634106

Morris, B. A. (2017). 4-Commonly Used Resins and Substrates in Flexible Packaging. The Science and Technology of Flexible Packaging. William Andrew Publishing.

Mota, C., Puppi, D., Dinucci, D., Gazzarri, M., & Chiellini, F. (2013). Additive manufacturing of star poly(ε-caprolactone) wet-spun scaffolds for bone tissue engineering applications. Journal of Bioactive and Compatible Polymers, 28, 320–340.

Mozaffar, M., Paul, A., Al Bahrani, R., Wolff, S., Choudhary, A., & Agrawal, A. (2018). = Data-driven prediction of the highdimensional thermal history indirected energy deposition processes via recurrent neural networks. Manuf. Lett., 18.

Mudge, R. P., & Wald, N. R. (2007). Laser engineered net shaping advances additive manufacturing and repair. Weld. J. N. Y., 86, 44–48.

Mulhaupt, R. (1999). Essay: International Historic Chemical Landmark: The Foundation *of Polymer Science by Hermann Staudinger (1881-1965). Macromo*lecular Chemistry and Physics, 200(8).

Munprom, R., & Limtasiri, S. (2019). Optimization of stereolithographic 3D printing parameters using Taguchi *method for improvement in* mechanical properties. Mater. Today: Proc., 17.

Murphy, S. V., & Atala, A. (2014). 3D bioprinting of tissues and organs. Nature Biotechnology, 32(8), 773–785. doi:10.1038/nbt.2958 PMID:25093879

Murr, L. E. (201*6). Frontiers of 3D printing/addi*tive manufacturing: From human organs to aircraft fabrication. Journal of Materials Science and Technology, 32(10), 987–995. doi:10.1016/j.jmst.2016.08.011

Murr, L.E. (2019). Metallurgy principles applied to powder bed fusi*on 3D printing/additi*ve manufacturing of personalized and optimized metal and alloy biomedical implants: An overview. J. Mater. Res. Technol.*, 9.*

*Murr, L. E., Gayta*n, S. M., Ramirez, D. A., Martinez, E., Hernandez, J., Amato, K. N., Shindo, P. W., Medina, F. R., & Wicker, R. B. (2012). Metal fabrication by additive manufacturing using laser and electron beam melting technologies. Journal of Materials Science and Technology, 28, 1–14.

Mussig, J., Fischer, H., Graupner, N., & Drieling, A. (2010). Testing methods for measuring physical and mechanical fibre properties (plant and *animal f*ibres). In J. Mussig (Ed.), Industrial Applications of Natural Fibres: Structure, *Properties and Technical Applications* (pp. 269–310). John Wiley & Sons, Ltd. doi:10.1002/9780470660324.ch13

Mustapha, K. B., Saad, M., & Abakr, Y. A. (2020). Flatwise Compression and Buckling Characterizations of Adhesive-Free Additively Manufactured Defected Architected St*ructures. In Proceedings of International Conference of* Aerospace and Mechanical Engineering 2019. Springer.

Muth, J. T., Vogt, D. M., Truby, R. L., Mengüç, Y., Kolesky, D. B., Wood, R. J., & Lewis, J. A. (2014). Embedded 3D printing of strain sensors within high*ly stretchable elastomers. Advanced Materials, 26, 63*07–*63*12.

Muthuraj, R., Misra, M., & Mohanty, A. K. (2015). Studies on mechanical, thermal, and morphological characteristics of biocomposites from biodegradable po*lymer blends* and natural fibers. In M. Misra & J. Pandey (Eds.), Biocomposites design and mechanical Performance. doi:10.1016/B978-1-78*242-373-7.00014-7*

*Myilsamy, S., & Sampa*th, B. (2021). Experimental comparison of near-dry and cryogenically coo*led near-dry machining in wire-cut electrical discharge machining p*rocesses. Surface Topography: Metrology and Properties, 9(3), 035015. doi:10.1088/2051-672X/ac15e0

Myilsamy, S., Boopathi, S., & Yuvaraj, D. (2021). A study on c*ryogenically treated molybdenum wire el*ectrode. Materials Today: Proceedings, 45(9), 8130–8135. doi:10.1016/j.matpr.2021.02.049

Nabi Saheb, D., & Jog, J. (1999). Natural fiber polymer composites: A review. Advances in Polymer Tech*nology, 18(4),* 351–363. doi:10.1002/(SICI)1098-2329(199924)18:4<351::AID-ADV6>3.0.CO;2-X

Nagarajan, V., Mohanty, A. K., & Misra, M. (2013). Sustainable green composites: Value addition to agricu*ltural residues and perennial grasses. ACS Sustainable C*hemistry & Engineering, 1(3), 325–333. doi:10.1021c300084z

Nair, N. R., Sekhar, V. C., Nampoothiri, K. M., & Pandey, A. (2017). Biodegradation of biopolymers. In Current Developments in Biotechnology *and Bioengineering (pp. 739–755). Elsevier. doi:10.1016/*B978-0-444-63662-1.00032-4

Narayanan, L. K., Huebner, P., Fisher, M. B., Spang, J. T., Starly, B., & Shirwaiker, R. A. (2016). 3D-bioprinting of polylactic acid (PLA) nanofiber–alginate hydrogel bioink containing human adipose-derived stem cells. ACS Biomaterials Science & Engine*ering, 2(10), 1732–1742. doi:10.*1021/acsbiomaterials.6b00196 PMID:33440471

Nasiri, S., & Khosravani, M. R. (2020). Progress and challenges in fabrication of wearable sensors for *health* mon*i*toring. Sensors and Actuators. A, Physical, 312, 112105. doi:10.1016/j.sna.2020.112105

Navarro-Baena, I., Sessini, V., Dominici, F., Torre, L., Kenny, J. M., & Peponi, L. (2016). Design of biodegradable blends based on PL*A and PCL: From* mo*r*phological, thermal and mechanical studies to shape memory behavior. Polymer Degradation & Stability, 132, 97–108.

Naveed, N. (2021). Investigate the effects of process parameters on material properties and microstructural changes of 3D-printed *specimens using fu*sed deposition modelling (FDM). Materials Technology, 36(5), 317–330. doi:10.1080/10667857.2020.1758475

Nayar, S., Bhuminathan, S., & Bhat, W. (2015). Rapid prototyping and stereolithography in *dentistry. Journal o*f Pharmacy & Bioallied Sciences, 7(April), S216–S219. doi:10.4103/0975-7406.155913 PMID:26015715

Neto, A. H. (2009). The electronic properties of graphene. Reviews of Modern Physics, 81, 109–*162.*

*Neumann, L., & Matas, J. (*2011). A Method for Text Localization and Recognition in Real-World Images. In R. Kimmel, R. Klette, & A. Sugimoto (Eds.), Lect*ure Notes in Computer Science: V*ol. 6494. Computer Vision – ACCV 2010. ACCV 2010. Springer. doi:10.1007/978-3-642-19318-7_60

Neumann, R., Neunzehn, J., Hinueber, C., Flath, T., Schulze, F. P., & Wiesmann, H. P. (2019). 3D-printed poly-ε-caprolactone-CaCO 3-biocompositescaffolds for hard tissue regeneration. Express Polyme*r Letters, 13(1),* 2–*17.* doi:10.3144/expresspolymlett.2019.2

Neuse, E. W. (2008). Synthetic polymers as drug-delivery vehicles in medicine. Metal-Based Drugs, 2008, 2008. doi:10.1155/2008/469531 PMID:18497867

Ngo, T. D., Kashani, A., Imbalzano, G., Nguyen, K. T. Q., & *Hui, D. (2018). Additive m*anufacturing (3D printing): A review of materials, methods, applications and challenges. Composites. Part B, Engineering, 143, 172–196. doi:10.1016/j.compositesb.2018.02.012

Nguyen, N. A., *Bowland, C. C., & Na*skar, A. K. (2018). A general method to improve 3D-printability and inter-layer adhesion in lignin-based composites. Applied Materials Today, 12, 138–152.

Nguyen, T. P., Nguyen, Q. V., Nguyen, V.-H., *L*e, T.-H., Huynh, V. Q. N., Vo, D.-V. N., Trinh, Q. T., Kim, S. Y., & Le, Q. V. (2019). Silk Fibroin-Based Biomaterials for Biomedical Applications: A Review. Polymers, 11(12), 1933. doi:10.3390/polym11121933 PMID:31771251

*Nielsen, A. S., Batchelder, D. N., & Pyrz, R. (*2002). Estimation of crystallinity of isotactic polypropylene using Raman spectroscopy. Polymer, 43(9), 2671–2676. doi:10.1016/S0032-3861(02)00053-8

Ning, F., Cong, W., Qiu, J., Wei, J., & Wang, S. (2015). Additive manuf*acturing of carbon fiber reinforced thermoplastic composites using fu*sed *de*position modeling. Composites. Part B, Engineering, 80(1), 369–378. doi:10.1*016/j.compositesb.2015.06.013*

*Noorani, R. (*2006). Rapid Prototyping—Principles and Applications. John Wiley & Sons, Inc.

Norani, M. N. M., Abdollah, M. F. Bin, Abdullah, M. I. H. C., Amiruddin, H., Ramli, F. R., & Tamaldin, N. (2021*). 3D printing parameters of acrylonitr*ile butadiene styrene polymer for friction and wear analysis using response surface methodology. Proceedings of the Institution of M*echanical Engineers, Part J: Journal of* Engineer*i*ng Tribology, 235(2), 468–477. 10.1177/1350650120925601

Okaro, I.A., Jayasinghe, S., Stucliffe, C., Black, K., & Paoletti, P. (2019). Automatic fault detection for laser powder-bed fusion using semi-supervised machine learning. Addit. Manuf., *27.*

*Oladapo, B. I., Zah*edi, S. A., & Adeoye, A. O. M. (2019). 3D printing of bone scaffolds with hybrid biomaterials. Composites. Part B, Engineering, 158, 428–436. doi:10.1016/j.compositesb.2018.09.*065*

*Oliveira, J., Correia, V., Castro, H., Mart*ins, *P.*, & Lanceros-Mendez, S. (2018). Polymer-based smart materials by printing technologies: Improving application and integration. Addit. Manuf., 21, 269–283. .addma.2018.03.012 doi:10.*1016/j*

Olson, E. (2011). AprilTag: A robust and flexible visual fiducial system. 2011 IEEE International Conference on Robotics and Automation, 3400-3407. 10.1109/ICRA.2011.5979561

Onwubolu, G. C., & Rayegani, F. (2014). Characterization and Optimization of Mechanical Properties of ABS Parts Manufactured by the Fused Deposition Mod*elling Process. International Journ*al *of* Manufacturing Engineering, 2014, 1–13. doi:10.1155/2014/598531

Ormiston, J. A., & S*erruys, P. W. (2009). Bioabsorbable coronary stents. Circulation: Ca*rdiovascular Interventions, 2(3), 255–260. doi:10.1161/CIRCINTERVENTIONS.109.859173 PMID:20031723

Oryan, A., Kamali, A., Moshirib, A., & Eslaminejad, M. B. (2017). Role of Mesenchymal Stem Cells in Bone *Regenerative Medicine: What Is the* Evidence? Cells, Tissues, Organs, 204(*2), 59–83. doi:10.1159/000469704 PMID:28647733*

*Osada, Y., & Matsuda, A. (1995). Shape memory in hydrogels. N*ature, 376(6537), 219–219.

Osidak, E. O., Kozhukhov, V. I., Osidak, M. S., & Domogatsky, S. P. (2020). Collagen as bioink for bioprinting: A comprehensive review. International Journal of Bioprinting, 6(3), 1–10. d*oi:10.18063/ijb.v6i3.270 PMID:33088985*

*Osma*n, *M.* A., & Atia, M. R. (2018). Investigation of ABS-rice straw composite feedstock filament for FDM. Rapid Prototyping Journal, 24(6), 1067–1075.

Ottenbrite, R. M., Kuus, K., & Kaplan, A. M. (1983). Characteristic biological effects of *anionic polymers*. In Polymers in Medicine (pp. 3–22). Springer. doi:10.1007/978-1-4615-7643-3_1

Ozsoy, Aksoy, & Salman. (2021). *Investigation of* the dimensional accuracy using image processing techniques in powder bed fusion. Proc IMechE Part E: J Process Mechanical Engineering.

Pal, A. K., *Mohanty, A. K., & Misra, M. (2021).* Addi*ti*ve manufacturing technology of polymeric materials for customized products: Recent developments and future prospective. RSC Advances, 11(58), 36398–36438. doi:10.1*039/ D1RA04060J PMID:354*94368

Palanikumar, K., Mudhukrishnan, M., & Prahba, P.S. (2020). Technologi*es in additive manuf*acturing for fiber reinforced composite materials: a review. Curr. Opin. Chem. Eng., 28, 51-9.

Pallikari, F., Chondrokoukis, G., Rebelakis, M., & Kotsalas, Y. (2001). Raman spectroscopy: A technique for estimating extent of polymerization *in PMMA. Materials Resear*ch Innovations, 4(2-3), 89–92. doi:10.1007100190000076

Panda, B. P., Krishnamoorthy, R., Bhattamisra, S. K., Shiva*shekaregowda, N. K. H., Seng, L. B., & Patn*aik, S. (2019). Fabrication of second generation smarter PLGA based nanocrystal carriers for improvement of drug delivery and therapeutic efficacy of gliclazide in type-2 diabetes rat model. Scientific Reports, 9(1), 1–15. doi:10.103841598-019-53996-4 PMID:31758056

Panthapulakkal, S., L*aw, S., & Sain, M. (2005). Enhancement of* processability of rice husk filled high-density polyethylene composite profiles. Journal of Thermoplastic Composite Materials., 18, 445–458.

Paolini, A., Kollmannsberger, S., & Rank, E. (2019). *Additive manufacturing in construction: A review on processes, applications, and digital planning method*s. Additive Manufacturing, 30, 100894. doi:10.1016/j.addma.2019.100894

Papke, B. L., Ratner, M. A., & Shriver, D. F. (1981). Vibrational spectroscopy and structure of polymer electrolytes, poly (ethylene oxide) complexes of alkali metal *salts. Journal of Physics and Chemistry of Solids, 42(6), 493–500. doi:10.1016/0022-*3697(81)90030-5

Papon, E. A., & Haque, A. (2020). Review on process model, structure-property relationship of composites and future needs in fused filament fabrication. J. Reinf. Plast. Compos., *39(19–20), 758–789.* 0731684420929757 doi:10.1177/

Paradkar, R. P., Leugers, A., Patel, R., Chen, X., & Meyers, G. (2009). Raman Spectroscopy of Polymers. In Encyclopedia of Analytical Chemistry: Applications, Theory and Instrumentation. John Wiley & S*ons, Ltd. doi:10.1002/9780470027318. a9032*

*Para*ndoush, P., & Lin, D. (2017). A review on additive manufacturing of polymer-fiber composites. Composite Structures, 182, 36–53.

Parandoush, P., & Lin, *D. (2017). A review on additi*ve manufacturing of polymer-fiber composites. Composite Structures, 182, 36–53. doi:10.1016/j.compstruct.2017.08.088

Park, S. H., Park, S. A., Kang, Y. G., Shin, J. W., Park, Y. S., Gu, S. R., & Shin, J. W. (2017). PCL/β-*TCP composite scaffolds exhibit positive ost*eogenic differentiation with mechanical stimulation. Tissue Engineering and Regenerative Medicine, 14(4), 349–358. doi:10.10*0713770-017-0022-9 PMID:30603*491

Parker, G. (2001). Encyclopedia of materials: Science and technology. Academic Press.

Parker, M., Inthavong, A., Law, E., Waddell, S., Ezeokeke, N., Matsuzaki, R., & Arola, D. (2022). 3D printing of continuous carbon fiber *reinforced polyphenylene sulfide: Explo*ring printability and importance of fiber volume fraction. Additive Manufacturing, 54, 102763. doi:10.1016/j.addma.2022.102763

Pascual-Gonza*lez, C., Martı´n, P.S., Lizarralde, I., Fernandez, A.,* Leon, A., & Lopes, C.S. (2021). Post-processing effects on microstructure, interlaminar and thermal properties of 3D printed continuous carbon fibre composites. Composites, Part B, 210.

Pa*tel, D. K., Dutta, S. D., Shin, W.* C., Ganguly, K., & Lim, K. T. (2021). Fabrication and characterization of 3D printable nanocellulose-based hydrogels for tissue engineering. RSC Advances, 11(13), 7466–7478. doi:10.1039/D0RA09620B PMID:35423276

Patnaik, A., & Satapathy, A. (2010). Investigations on *Three-Body Abrasive Wear and Mechani*cal Properties of Particulate Filled Glass Epoxy Composites Composites for wind turbine blade View project Development and selection of biomaterial for hip joint replacement View project. Res*earchgate. Net, 5(2), 37–*48.

Patricio, T., Domingos, M., Gloria, A., & Bártolo, P. (2013). Characterisation of PCL and PCL/PLA scaffolds for tissue engineering. Procedia CIR*P, 5, 110–114. doi:10.1016/j.procir.2013.0*1.022

Pattanayak, D. K., Fukuda, A., Matsushita, T., Takemoto, M., Fujibayashi, S., Sasaki, K., Nishida, N., Nakamura, T., & *Kokubo, T. (2011). Bioa*ctive Ti metal analogous to human cancellous bone: Fabrication by selective laser melting and chemical treatments. Acta Biomaterialia, 7(3), 1398–1406. doi:10.1016/j.actbio.2010.09.034 PMID:20883832

Paul, D., & *Robenson, L. (2008). Polymer nanotech*nology: Nanocomposites. Polymer, 49, 3187–3204.

Paul, V., Kanny, K., & Redhi, G. G. (2015). Mechanical, thermal and morpholo*gical properties* of *a b*io-based composite derived from banana plant source. Composites. Part A, Applied Science and Manufacturing, 68, 90–100. doi:10.1016/j.compositesa.2014.08.032

Paxton, N. C., Allenby, M. C., Lewis, P. M., & Woodruff, M. A. *(2019). Biomedical applications of* polyethylene. European Polymer Journal, 118, 412–428. doi:10.1016/j.eurpolymj.2019.05.037

Pazhamannil, R.V., Govindan, P., & Sooraj, P. (2020). Prediction of the tensile strength of polylactic acid fuse*d deposition models usi*ng *a*rtificial neural network technique. Mater. Today: Proc.

Pei, S., Wang, K., Chen, C.B., Li, J., Li, Y., & Zeng, D. (2021). Process structure- property analysis of short carbon fiber reinforced polymer composite via f*used filamen*t fabrication. J Manuf Process, 64.

Penumakala, P. K., Santo, J., & Thomas, A. (2020). A critical review on the fused deposition modeling of thermoplastic polymer composites. Compos. B Eng*., 201. do*i:10.1016/j.compositesb.2020.108336

Pereira, R. F., & Bártolo, P. J. (2015). 3D bioprinting of photocrosslinkable hydrogel constructs. Journal of Applied Polymer Science, 132(48). Advance online publication. doi:10.*1002/app.42458*

Perez, R. A., & Me*s*tres, G. (2016). Role of pore size and morphology in musculo-skeletal tissue regeneration. Materials Science and Engineering C, 61, 922–939. doi:10.1016/j.msec.2015.12.087 PMID:26838923

Pervaiz, M., & Sain, M. (2003). Carbon storage potential in natural fiber composites. Resources, Conservation and Recycling, 39(4), 325–340.

Petchwattana, N., Channuan, W., Naknaen, P., & Narupai, B. (201*9). 3D printing filaments prepared from modified poly(lactic acid)/teak wood flour composites: An inv*estigation on the particle size effects and silane coupling agent compatibilisation. The Journal of Physiological Sciences; JPS, 30(2), 169–188.

Pham, D. T., & Ji, C. (2000). Design for stereolithography. Proce*edings - Institutio*n of Mechanical Engineers, 214, 635–640.

Phatharapeetranun, N., Ksapabutr, B., Marani, D., & Bowen, J. R. (2017). 3*D-printed barium titanate/poly-*(vi*ny*lidene fluoride) nano-hybrids with anisotropic dielectric properties. Journal of Materials Chemistry. C, Materials for Optical and Electronic Devices, 5(47), 12430–12440.

Pihlehvar, S., Arnhof, M., Erichsen, A., Valentini, L., *& Kjoniksen, A.* (2021). Investigation of severe lunar environmental conditions on the physical and mechanical properties of lunar rego*lith geopolymers. J. Mater. Res. Technol., 12.*

*Pillai, S., Up*adhyay, A., Khayambashi, P., Farooq, I., Sabri, H., Tarar, M., Lee, K. T., Harb, I., Zhou, S., Wang, Y., & Tran, S. D. (2021). Dental 3d-printing: Transferring art from the laboratories to the clinics. Polyme*rs, 13(1), 1–25. doi:10.3390/polym13010157 PMID:33406617*

Pivec, T., Smole, M. S., Gasparic, P., & Kleinschek, K. S. (2017). Polyurethanes for Medical Us. Tekstilec, 60(4).

Porter, D. A., Hoang, T. V., & Berfield, T. A. (2017). Effects of *in-situ poling and process* parameters on fused filament fabrication printed PVDF sheet mechanical and electrical properties. Addit. Man*uf., 1*3, *81*–92.

Pritchard, A. (1969). Statistical bibliography or bibliometrics. The Journal of Documentation, 25(4), 348–349.

Promper, E. (2010). Natural fibre-re*inforced polymers in automotive inte*rio*r* applications. In J. Mussig (Ed.), Industrial Applications of Natural Fibres: Structure, Properties and Technical Applications (pp. 423–436). John Wiley & So*ns, Ltd.*

*Puppi, D., & Chiell*ini, F. (2020). Biodegradable polymers for biomedical additive manufacturing. Applied Materials Today, 20, 100700. doi:10.1016/j.*apmt.2020.100700*

*P*uppi, D., Chiellini, F., Piras, A. M., & Chiellini, E. (2010). Polymeric materials for bone and cartilage repair. Progress in Polymer Science, 35(4), 403–440. doi:10.1016/j.progpolymsci.2010.01.006

Qaiss, A. E. K., Bouhfi, D. R., *& Essabir, H. (2014). Natural f*ibers reinforced polymeric matrix: Thermal, mechanical and interfacial properties. Biomass and Bioenergy Processing and Properties, 225–245.

Qi, X., Chen, G., Li, Y., *Cheng, X., & Li, C. (2019). Applying neural-networkbased machine le*arning to additive manufacturing: current applications, challenges, and future perspectives. Engineering, 5.

Quintelier, J., De Baets, P., Samyn, P., & Van Hemelrijck, D. (2006). *On the SEM features of glass-polyester composite s*ystem subjected to dry sliding wear. Wear, 261(7–8), 703–714. doi:10.1016/j.wear.2006.01.006

Raffic, N. M., Babu, K. G., & Madhan, P. (2019). Application *of taguchi's experimental* design and range analysis in optimization of FDM printing parameters for PET-G, PLA and HIPS. International Journal of Scientific and Technology Research, 8(9), 8*91–902.*

*Rafiee, Farahani, & T*herriault. (2020). Multi-Material 3D and 4D Printing: A Survey. Academic Press.

Rahman, A., Deshpande, P., Raude, M., Odegard, G. M., Gowtham, S., & Ghosj, S. (2021). A machine learning *framework for predicting the shear strength* of carbonnanotube polymerr interfaces based on molecular dynamics simulation data. Composites Science and Technology, 207, 108627.

Ramalingam, P.S., Mayandi, K., Balasubramani*an, V., Chandrasekar,* K., Stalany, V.M., & Munaf, A. (2020). Effect of 3D printing process parameters on the impact strength of onyx-Glass fiber reinforced composites. Mater. Today: Proc.

Ramesh, M., Palanikumar, K., & Reddy, K. H. (2017). P*lant fibre based bio-composites: Sustainab*le *a*nd renewable green materials. Renewable and Sustainable Energy Reviews, Elsevier, 79, 558–584. doi:10.1016/j.rser.2017.05.094

Ramirez-Hernandez, A., Aguilar-Flores, C., & Aparicio-Saguilan, A. (2019). *Fingerprint analysis of FTI*R spectra of polymers containing vinyl acetate. Dyna, 86(209), 198–205. doi:10.15446/dyna.v86n209.77513

Rastogi, P., Njuguna, J., & Kandasubramanian, B. (2019). Exploration of elastomeric and polymeric liquid crystals with photothermal actuation: A review. European Polymer Journal, 121, 109287.

Ravi, M. (2015). 3D Printing - A Current Opinion. Academic Press.

Rayappa Shrinivas Mahale, *V., Shamanth, K., He*m*anth*, S. K. N., Sharath, P. C., Shashanka, R., Patil, A., & Shetty, D. (2022). Processes and applications of met*al additive manufact*u*ring*. Materials Today: Proceedings, 54(2), 228–233. doi:10.1016/j.matpr.2021.08.298

Raza, Z. A., Abid, S., & Banat, I. M. (2018). Polyhydroxyalkanoates: Characteristics, production, recent developments and applications. International Biodeterioration & Biodegradation, 126, 45–56. doi:*10.1016/j.ibiod.2017.10.00*1

Reddy, M. M., Misra, M., & Mohanty, A. K. (2012). Bio-based materials in the new bio-economy. Chemical Engineering Progress, 108(5), 37–42.

Redwood, Sch¨offer, & Garret. (2017). The 3D Printing Handbook: Technologies, Design and Applications. 3D Hubs B.V.

Reis, R. L.*, & San Román, J. (200*4). Biodegradable systems in tissue engineering and regenerative medicine. CRC Press. doi:10.1201/9780203491232

Remya, K. R., Joseph, J., Mani, S., John, A., Varma, H. K., & Ramesh, *P. (2013).* N*an*ohydroxyapatite incorporated electrospun polycaprolactone/polycaprolactone–polyethyleneglycol–polycaprolactone blend scaffold for bone tissue engineering applications. Journal of Biomedical Nanotechnology, 9(9), 1483–1494. doi:10.1166/jbn.2013.1640 PMID:23980497

Ren, D., Li, S., Wang, H., Hou, W., Hao, Y., & Jin, W. (2019). Fatigue *behavior of* T*i-6*Al-4V cellular structures fabricated by additive manufacturing technique. J Mater Sci Technol, 35, 285-94.

Rengier, F., Mehndiratta, A., Von Tengg-Kobligk, H., Zechmann, C. M., Unterhinninghofen, R., Kauczor, H. U., & Giesel, F. L. (2010). 3D printing based on imaging dat*a: Review of medic*al applications. International Journal of Computer Assisted Radiology and Surgery, 5(4), 335–341. doi:10.100711548-010-0476-x PMID:20467825

Report, P. T. (2016, February 26). Retrieved January 20, 2023, from https://www.hubs.com/get/trends/

Reports on polymer nanocomposites drive opportunities in the automotive se*ctor. (2016, February 16). Retrieved January 20, 2023, fr*om https://www.nanowerk.com/category-spotlight.php

Ricard-Blum, S. (2011). The Collagen Family. Cold Spring Harbor Perspectives in Biology, 3(1), a004978. doi:1*0.1101/cshperspect.a0049*78 *PM*ID:21421911

Roberts, J. J., & Bryant, S. J. (2013). Comparison of photopolymerizable thiol-ene PEG and acrylate-based PEG hydrogels for cartilage development. Biomaterials, 34(38), 9969–9979. doi:10.1016/j.biomaterials.201*3.09.020 PMID:2406*0418

Robinson, J., Saint Louis, P. M., & Padmaraju, A. (2001). Polymer in medicine. Introduciton to Polymers.

Rohman, A., Windarsih, A., Lukitaningsih, E., Rafi, M., Betania, K., & Fadzil*lah, N. A. (2019).* The use of FTIR and Raman spectroscopy in combination with chemometrics for analysis of biomolecules in biomedical fluids: A review. Biomedica*l Spectroscopy and* Imaging, 8(3-4), 55–71. doi:10.3233/BSI-200189

Roopaa, T. S., Narasimha Murthy, H. N., Praveen Kumar, V. V., & Krishna, M. (2018). Developm*ent and Characterization of PVDF Th*in *F*ilms for pressure sensors. Materials Today: Proceedings, 5(10, Part 1), 21082–21090. https://doi.org/ 10.1016/j.matpr.2018.06.503

Roopavath, U. *K., & Kalaskar, D. M. (2017). Intro*duct*i*on to 3D printing in medicine. In 3D Printing in Medicine (pp. 1–20). Elsevier. doi:10.1016/B978-0-08-100717-4.00001-6

Rosa, S. M. L., Santos, E. F., Ferreira, C. A., & Nachtigall, S. M. B. (2009). Studies on the properties of rice-husk-filled-PP composites - E*ffect of maleated PP. Materials Research, 1*2(*3*), 333–338.

Royan, N. R. R., Sulong, A. B., Yuhana, N. Y., Chen, R. S., Ab *Ghani, M. H., & Ahmad, S. (2018). UV/O3 treatment* as *a* surface modification of rice husk towards preparation of novel biocomposites. PLoS One, 13(5).

Ruberu, K., Senadeera, M., Rana, S., Gupta, S., Chung, J., & Yue, Z. (2021). Coupling machine l*earning with 3D bioprinting to fast track optimisation of extrusion printing.* App*l.* Mater. Today, 22, 100914.

Rueden, L., Mayer, S., Sifa, R., Bauckhage, C., & Garcke, J. (2020). Combining machine learning and simulation to a hybrid modelling approach: current and future directions. Proceedings of the 18th i*nternational symposium on* intelligent data analysis.

Russ. (2016). Feature-Specific Measurements. doi:10.1201/b10720-14

Sachs, E., Cima, M., & Cornie, J. (1990). Three-dimensional printing: Rapid tooling and prototypes directly from a CAD model. *CIRP* Ann. Manuf. Technol., 39, 201–204.

Saeed, K., McIlhagger, A., Harkin Jones, E., Kelly, J., & Archer, E. (2021). Predication of the in-plane mechanical pr*operties of* continuous carbon fibre reinforced 3D printed polymer composites using classical laminated-plate theory. Composite Structures, 259, 113226.

Saekhor, K., Udomsinprasert, W., Honsawek, S., & Tachaboo*nyakiat, W. (*20*19*). Preparation of an injectable modified chitosan-based hydrogel approaching *for bone tissue engineering. I*nternational Journal of Biological Macromolecules, 123, 167–173. doi:10.1016/j.ijbiomac.2018.11.041 PMID:30423397

Sahoo, S., *Chakraborti, C. K., Behera, P. K., & Mishra, S. C. (2012). FTIR and Raman spectroscopic inv*estigations of a Norfloxacin/Carbopol934 Polymerie suspension. Journal of Young Pharmacists, 4(3), 138–145. doi:10.4103/0975-1483.100017 PMID:23112532

Sampath, B., & Myilsamy, S. (2021). *Experimental investigation of a cryogenically* cooled oxygen-mist near-dry wire-cut electrical discharge machining process. Strojniski Vestnik. Jixie Gongcheng Xuebao, 67(6), 322–330. doi:10.5545v-jme.2021.7161

Sampath, B., Naveenkumar, N., Sam*pathkumar, P.,* Silambarasan, P., Venkadesh, A., & Sakthivel, M. (2021). Experimental comparative study of banana fiber composite with glass fiber composite material using Tag*uchi* met*h*od. Materials Today: Proceedings, 49(5), 1475–1480. doi:10.1016/j.matpr.2021.07.232

Sampath, B., Sasikumar, C., & Myilsamy, S. (2023). Application of TOPSIS Optimization Technique in the Micro-Machining Process. In Trends, Parad*igms, and Advances in Mechatronics Engineering (pp. 162–187*). IGI Global.

Sampson, R., Lancaster, R., Stucliie, M., *Carswell, D., Hauser, C., & Barras, J. (20*20). An improved methodology of melt pool monitoring of direct energy deposition processes. Optics & Laser Technology, 127, 106194.

Samtas & Gulesin. (2011). Digital image processing and its applications in different fields. Academic Press.

Samuel, E. *J. J., & Mohan, S. (2004). FTIR* and *F*T Raman spectra and analysis of poly (4-methyl-1-pentene). Spectrochimica Acta. Part A: Molecular and Biomolecular Spectroscopy, 60(1-2), 19–24. doi:10.1016/S1386-1425(03)00212-9 PMID:14670457

Sangiorgi, G., Melzi, G., Ago*stoni, P., Cola, C*., Clementi, F., Romitelli, P., & Colombo, A. (2007). Engineering aspects of stents design and their translation into clinical pract*ice. Annali dell'Istituto Superiore di Sanita, 43(*1*),* 89–100. PMID:17536159

Saniei, H., & Mousavi, S. (2020). Surface modification of PLA 3D-printed implants by electrospinning with enhanced bioactivity and cell affinity. Polymer, 196, 122467. do*i:10.101*6/j.polymer.2020.122467

Santiago, D., Ferrando, F., & De la Flor, S. (2014). Influence of Holding Time on Shape Recovery in a Polyurethane Shape-Memory Polymer. Journal of Materials Engineering and Per*formance, 23(7), 2567–257*3. *h*ttps://doi.org/10.1007/s1166*5-014-0997-5*

*Santoni, A., Bonf*iglio, P., Fausti, P., Marescotti, C., Mazzanti, V., Mollica, F., & Pompoli, F. (2019). Improving the sound absorption performance of sustainable thermal insulation materials: Natural hemp fibres. Applied *Acoustics, 150, 279–289.*

*Santos*h, Nithyanandh, Ashwath, & Lalith Kishore. (2022, November). Comparison of internal f*riction measurements on Ni-Ti reinforced smart composites prepar*ed by additive manufacturing. Journal of Alloys and Compounds, 924(30), 166027.

Saravanan, M., Vasanth, M., Boopathi, S., Sureshkumar, M., & Haribalaji, V. (2022). Optimization *of Quench Polish Quenc*h (QPQ) Coating Process Using Taguchi Method. Key Engineering Materials, 935, 83–91. doi:10.4028/p-z569vy

Satyanarayana, K. G., Arizaga, G. G. C., & Wypych, F. (2009). Biodegradable composites based on lignocellulosic fibers-An overview. *Progress in Polymer Science, 34(9), 982–10*21. doi:10.1016/j.progpolymsci.2008.12.002

Schmidt, C., Chowdhury, A. M. S., Neuking, K., & Eggeler, G. (2011). Studies o*n the cycling, pro*cessing and programming of an industrially applicable shape memory polymer Tecoflex® (or TFX EG 72D). High Performance Polymers, 23(4), 300–307. https://doi.org/10.1177/0954008311405245

Schroeter, K. (2008). Printed sensors: Enabling new applications. Sensor *Review.*

Schubert, U. S., & Zechel, S. (2020). The Year of Polymers–100 Years of Macromolecular Chemistry. Macromolecular Chemistry and Physics, 221(1), 1900530. doi:10.1002/macp.201900530

Schwentenwein, M., & Homa, J. (2015). Additive manufacturing of dense alumina ceramics. International Journal of Applied Ceramic Technology, 12(1), 1–7.

Scime, L., & Beuth, J. (2018a). A multi-scale convolutional neural network for autonomous anomaly detection and classification in a laser powder bed fusion additive manufacturing process. Addit. Manuf., 24.

Scime, L., & Beuth, J. (2018b). Anomaly detection and classification in a laser powder bed additive manufacturing process using a trained computer vision algorithm. Addit. Manuf., 19.

Scime, L., & Beuth, J. (2019). Using machine learning to identify in-situ melt pool signatures indicative of flaw formation in a laser powder bed fusion additive manufacturing process. Addit. Manuf., 25.

Scribante, A., Gallo, S., Pascadopoli, M., Canzi, P., Marconi, S., Montasser, M. A., Bressani, D., Gandini, P., & Sfondrini, M. F. (2022). Properties of CAD/CAM 3D Printing Dental Materials and Their Clinical Applications in Orthodontics: Where Are We Now? Applied Sciences (Switzerland), 12(2), 551. Advance online publication. doi:10.3390/app12020551

Shah, D. U. (2013). Developing plant fibre composites for structural applications by optimising composite parameters: A critical review. Journal of Materials Science, 48(18), 6083–6107. doi:10.100710853-013-7458-7

Shah, T. V., & Vasava, D. V. (2019). A glimpse of biodegradable polymers and their biomedical applications. e-Polymers, 19(1), 385-410.

Shahrubudin, N., Lee, T. C., & Ramlan, R. (2019). An overview on 3D printing technology: Technological, materials, and applications. Procedia Manufacturing, 35, 1286–1296. doi:10.1016/j.promfg.2019.06.089

Sharma, N., & Honigmann, P. (2020). Dimensional characteristics of FDM 3D printed PEEK implant for craniofacial reconstructions. Journals.Infinite-Science.De, 2(1), 2–3. doi:10.18416/AMMM.2020.2009011

Sharma, T., Je, S.-S., Gill, B., & Zhang, J. X. (2012). Patterning piezoelectric thin film PVDF–TrFE based pressure sensor for catheter application. Sensors and Actuators. A, Physical, 177, 87–92.

Shen, H., Du, W., Sun, W., Xu, Y., & Fu, J. (2020). Visual Detection of Surface Defects Based on Self-Feature Comparison in Robot 3-D Printing. Applied Sciences (Basel, Switzerland), 10(1), 235. doi:10.3390/app10010235

Shen, L., Haufe, J., & Patel, M.K. (2009). Product overview and market projection of emerging bio-based plastics. PRO-BIP 2009, Final Report.

Shen, X., Yao, J., Wang, Y., & Yang, J. (2004). Density prediction of selective laser sintering parts based on artificial neural network. Proceedings of the international symposium on neural networks.

Sheoran, A. J., Kumar, H., Arora, P. K., & Moona, G. (2020). Bio-medical applications of additive manufacturing: A review. Procedia Manufacturing, 51, 663–670. doi:10.1016/j.promfg.2020.10.093

Shepelin, N. A. (2020). Printed recyclable and self-poled polymer piezoelectric generators through single-walled carbon nanotube templating. Energy & Environmental Science, 13(3), 868–883. https://doi.org/10.1039/C9EE03059J

Sherman, L. (2015, February 25). Additive manufacturing: Materials for 'real-world' parts. Retrieved January 20, 2023, from https://www.ptonline.com/articles/additive-manufacturing-materials-for-real-world-parts

Shesan, O. J., Stephen, A. C., Chioma, A. G., Neerish, R., & Rotimi, S. E. (2019). Fiber- Matrix Relationship for Composites Preparation. In Renewable and Sustainable Composites (pp. 1-30). INTECH Open Access Publisher.

Shi, B., Bai, X., & Yao, C. (2016). An end-to-end trainable neural network for image-based sequence recognition and its application to scene text recognition. IEEE Trans Pattern Anal Mach Intell, 39.

Shofner, M. L., Lozano, K., & Rodrı, F. J. (2002). Nanofiber-Reinforced *Polymers Prepared by* Fused Deposition Modeling. Journal of Applied Polymer Science, 89(11), 3081–3090. doi:10.1002/app.12496

Short, D. B. (2015). 3D. Print Addit Manuf, 2(4), 209–215.

Short, D. B. (2015). Use of 3D *printing by museums: educational exhibits*, artifact education, and artifact restoration. 3D Printing and Additive Manufacturing, 2(4), 209-215.

Siesler, H. W. (1980). Fourier transform infrared (ftir) spectroscopy *in polymer research. Journal of* Molecular Structure, 59, 15–37. doi:10.1016/0022-2860(80)85063-0

Silva, M.A. (2012). Coating of polyethylene medical devices. Academic Press.

Sing, S. L., Huang*, S., & Yeong, W. Y. (2020).* Effect of solution heat treatment on microstructure and mechanical properties of laser powder bed fusion produced cobalt-28cromium-6 molybdenum. Materials Science and Engineering A, 769, 138511.

Sing, S. L., Huang, S., Goh, G. D., Goh, G. L., Tey, C. F.*, & Tan, J. H. K. (2021).* Emerging metallic systems for additive manufacturing: Insitu alloying and multi-metal processing in laser powder bed fusion. *Progress in M*aterials Science, 119, 100795.

Singh, A., Cooper, D., Blundell, N., Gibbons, G., & Pratihar, D. (2012*). Modelling of direct metal laser* sintering of EOS DM20 bronze using neural networks and genetic algorithms. Proceedings of the 37th international MATADOR conference.

S*ingh, A. V., Ansari, M. H. D., Wang, S., Laux, P.,* Luch, A., Kumar, A., & Nussberger, S. (2019). The adoption of three-dimensional additive manufacturing from biomedical material design to 3D organ printing. Applied Sciences-Basel, 9(4).

Singh, R. (2011*). Process capabi*lity study of poly jet printing for plastic components. Journal of Mechanical Science and Technology, 25, 1011–1015.

Singh, R., Singh, S., Singh, I. P., Fabbrocino, F.*, & Fraternali, F.* (2017). Investigation for surface finish improvement of FDM parts by vapor smoothing process. Composites. Part B, Engineering, 111, 228–234. doi:10.1016/j.compositesb.2016.11.062

Sizemor*e, N.E., Nogueir*a, M.L., Greis, N.P., & Davies, M.A. (2020). Application of machine learning to the prediction of surface roughness in diamond machining. Procedia Manuf, 48.

Slootmaker, T., & Müssig, J. (2010). SEM catalogue for animal and plant fibres. In Industrial Applic*ations of Natural Fibres: Struct*ure, Properties and Technical Applications (pp. 311–336). John Wiley & Sons, Ltd. doi:10.1002/9780470660324.ch14

Smiti, A. (2020). When machine learning meets medical world: Current status and future ch*allenges. Computer Science Revi*ew, 37, 100280.

Snell, R., Tammas-Williams, S., Chechik, L., Lyle, A., Hernandez-Nava, E., & Boig, C. (2019). Methods for rapid pore classification in metal additive manufacturing. *J Occup Med, 72.*

*Sn*ively, C. M., & Koenig, J. L. (2017). IR and Raman Spectroscopies, Polymer Applications. In Encyclopedia of Spectroscopy and Spectrometry (3rd ed., pp. 365-371). Elsevier.

Sodeifi*an, G., Ghaseminejad, S., & You*sefi, A. A. (2019). Preparation of polypropylene/short glass fiber composite as Fused Deposition Modeling (FDM) filament. Results in Physics, 12(August), 205–222. doi:10.1016/j.rinp.2018.11.0*65*

Song, R., Murphy, M., Li, C., Ting, K., Soo, C., & Zheng, Z. (2018). Current development *of biodegradable polymeric materials for biomedical applications. Drug* Design, Development and Therapy, 12, 3117–3145. doi:10.2147/DDDT.S165440 PMID:30288019

Song, X., Pan, Y., & Chen, Y. (2015). Development of a low-cost parallel kinematic *machine for multidirectional additive* manufacturing. Journal of Manufacturing Science and Engineering, 137(2), 021005. Advance online publication. doi:10.1115/1.4028897

Sood, A.K., Equbal, A., Toppo, V., Ohdar, R.K., *& Mahapatra, S.S. (2012). An investigation on sliding wear* of FDM built parts. CIRP J. Manuf. Sci. Technol., 5.

Sood, A. K., Ohdar, R. K., & Mahapatra, S. S. (2010). Parametric appraisal of mechanical *property of fused deposition modelling* processed parts. Materials & Design, 31(1), 287–295. doi:10.1016/j.matdes.2009.06.016

Sood, A.K., Ohdar, R.K., & Mahapatra, S.S. (2012). Experimental investigation and empirical modelling of FDM process for compressive strength improvement. J Adv Res, 3.

Soyler, M. (2019). 3B yazicilarda kullanilan mumlarin üretilmesi ve geri kazanilmasi [Master's thesis]. Kahramanmaras Sutcu Imam University.

Spinelli, G. (2018). Morphological, rheological and electromagnetic properties of nanocarbon/poly (lactic) acid for 3d printing: Solution blending vs Melt mixing. Materials (Basel), 11(11), 2256.

Stamboulis, A., Baillie, C. A., & Peijs, T. *(2001). Effects of environmental conditions* on mechanical and physical properties of flax fibers. Composites. Part A, Applied Science and Manufacturing, 32(8), 1*105–1114. doi:10.1016/S1359-835X(01)00032-X*

Standard Terminology for Additive Manufacturing – General Principles – Terminology (ASTM52900), International Organization for Standardization: Geneva, S*witzerland. (2015). https://www.iso.org/o*bp/ui/#iso:std:iso-astm:52900:ed-1:v1:en

Stansbury, J., & Idacavage, M. (2015). 3D printing with polymers: Challenges among expanding options and opportunities. Den*tal Materials, 32(1), 54–64. doi:1*0.1016/j.dental.2015.09.018 PMID:26494268

Steer, D. L., & Nig*am, S. K. (2004). Developmental approach*es to kidney tissue engineering. American Journal of Physiology - Renal Physiology, 286. doi:10.1152/ajprenal.00167.2003

Steiner, M. L., Bartley, T. D., Byers, F. M., & Krovetz, L. J. (1965). Polyet*hylene catheter in the heart: Report of a case with suc*cessful removal. Journal of the American Medical Association, 193(12), 1054–1056. doi:10.100*1/jama.1965.03090120062022 PMID:14338809*

Stoof, D. (2016). Development of Recycled Polypropylene Composite Materials for Applications in 3D Printing [Ph.D. Thesis]. University of Waikato, Hamilton, New Zealand.

Stoof, D., P*ickering, K., & Zhang, Y. (2017). Fu*sed deposition modelling of natural fibre/polylactic acid composites. Journal of Composites Science, 1(8).

Stratasys. (n.d.). https://www.stratasys.com/

Stuart, M. A. C. (2010). Emerging applications of stimuli-responsive polymer materials. Nature Materials, *9(2), 101–113.*

Subramaniam, A., & Sethuraman, S. (2014). Biomedical applications of nondegradable polymers. In Natural and synthetic biomedical polymers (pp. 301–308). Elsevier. doi:10.1016/B978-0-12-396983-5.00019-3

Subramanian, A., *Krishnan, U. M., & Sethuraman, S. (2012).* Fabrication, characterization and in vitro evaluation of aligned PLGA–PCL nanofibers for neural regeneration. Annals of Biomedical Engineering, 40(10), 2098–2110. doi:10.*100710439-012-059*2-6 PMID:22618802

Summerscales, J., Dissanayake, N. P. J., Virk, A. S., & Hall, W. (*2010). A review of bast fibers and their composites.Part 1-Fibers as reinforcements. Composi*tes. Part A, Applied Science and Manufacturing, 41(10), 1329–1335.

Sun, Q., Rizvi, G., Bellehumeur, C., & Gu, P. (2008). FDM polymer filament bonding quality influenced b*y processing conditions. Ra*pid Prototyping Journal, 14(2), 72–80.

Sun, Q., Rizvi, G. M., Bellehumeur, C. T., & Gu, P. (2008). Effect of processing conditions on the bonding quality of FDM polymer filame*nts. Rapid Prot*otyping Journal, 14(2), 72–80. doi:10.1108/13552540810862028

Sun, S., Sun, S., Cao, X., & Sun, R. (2016). The role of pretreatment in improving the enzymatic hydrolysis of lignocellulosic materials. Bioresource Technology, 199, 49–58. doi:10.1016/j.biortech.2015.08.061 PMID:26321216

Sun, W., Starly, B., Darling, A., & Gomez, C. (2004). *Computer-aided tissu*e engineering: Application to biomimetic modelling and design of tissue scaffolds. Biotechnology and Applied Biochemistry, 39(1), 49. doi:10.1042/BA20030109 PMID:14556653

Sun, Yue, Yu, S*hao, Peng, Zhou, Demoly, Zhao, & Qi. (2022). Mach*ine Learning-Evolutionary Algorithm Enabled Design for 4D-Printed Active Composite Structures. Academic Press.

Sunny, B. C., Benedict, S., Rajan, M. P., & Srinivas, M. (2019). Impact of printing *parameters on energy consumpti*on of 3D printers using IoT cloud architecture. 2019 IEEE 16th India Council International Conference, INDICON 2019 - Symposium Proceedings, 1–4. 10.1109/INDI*CON47234.201*9.9029069

Suparyanto Dan Rosad (2015). Accuracy of Stereolithographic Models of Human Anatomy. Suparyanto Dan Rosad, 5(3), 248–253.

Suresha, B., C*handramohan, G., Pra*kash, J. N., Balusamy, V., & Sankaranarayanasamy, K. (2006). The Role *of Fillers on Friction and Slide Wear Characteristics in Glass-E*poxy Composite Systems. Journal of Minerals & Materials Characterization & Engineering, 05(01), 87–101. doi:10.4236/jmmce.2006.51006

Suresha, B., Chandramohan, G., Rao, P. R. S., Sampathkumaran, P., Seetharamu, S., & Ven*kateswarlu, V. (2006).* Friction and slide wear characteristics of glass-epoxy and glass-epoxy filled with SiCp composites. Indian Journal of Engineering and Materials Sciences, *13(6), 535–541.*

*Syrlybayev, D., Zharylkassyn, B., Seise*kulova, A., Akhmetov, M., Perveen, A., & Talamona, D. (2021). Optimisation of strength properties of FDM printed parts—A critical review. Polymers, 13(10), 1587. Advance online publication. doi:10.3390/polym13101587 PMID:34069144

Szycher, M., Siciliano, A. A., & Reed, A. M. (1991). Polyurethanes in medical devices. Medical Design and Material, 1(2), 18–25. PMID:10171592

Taj, S., Munawar, M. A., & Khan, *S. (2007). Natur*al *fi*ber-reinforced polymer composites. Proceedings of the Pakistan Academy of Sciences, 44(2), 129–144.

Takezawa, A., & Kobashi, M. (2017). An optimizatio*n-based multimaterial topology and additive manufactur*ing method for making porous composites with variable thermal expansion. Composites. Part B, Engineering, 131, 21–29.

Tambrallimath, V., & Keshavamurthy, R. (2021). Mechanical Charact*erization of PC-ABS based Graphene reinfor*ced *P*olymer Nanocomposites fabricated by FDM Process. Journal of Polymers, 13(17). Advance on*line publication. doi:10.3390/polym13172951*

Taormina, G., Sciancalepore, C., Messori, M., & Bondioli, F. (2018). 3D printing processes for photocurable polymeric materials: Technologies, materials, and future trends. Journal of Applied Biomaterials & Functional Material*s, 16(3), 151–160. doi:10.1177/*2280800018764770 PMID:29609487

Tapia, G., Elwany, A.H., & Sang, H. (2016). Prediction of porosity in metalbased additive manufacturing using *spatial Gaussian* process models. Addit. Manuf., 12.

Tapia, G., Khairallah, S., Matthews, M., King, W.E., & Elwany, A. (2017). Gau*ssian process-based surrogate modeling f*rame work for process planning in laser powder-bed fusion additive manufacturing of 316L stainless steel. Int J Adv Manuf Technol, 94.

Tarbuttona, J., Leb, T., Helfrichb, G., & Kirkpatrickb, M. (2017). Phase Transform*ation and Shock Sensor Response of Additively Manufac*tured Piezoelectric PVDF. Procedia Manufacturing, 10, 982–989. https://doi.org/10.*1016/j.promfg.2017.07.089*

*Teimouri, A., Azadi, M., Emadi, R., Lari, J., & Chermahini, A. N. (201*5). Preparation, characterization, degradation, and biocompatibility of different silk fibroin based composite scaffolds prepared by freeze-drying method for tissue engineering appl*ication. Polymer Degradation & Stability, 121, 18–29. doi:10.1016/j.polymdegradstab.2015.08.*004

Tekinalp, H. L., Kunc, V., Velez-Garcia, G. M., Duty, C. E., Love, L. J., Naskar, A. K., Blue, C. A., & Ozcan, S. (2014). Highly oriented carbon fiber–polymer composites via add*itive manufacturing. Composit*es Science and Technology, 105, 144–150.

Ten Kate, J., Smit, G., & Breedveld, P. (2017). 3D-printed upper limb prostheses: A review. Disability and Rehabilitation. Assist*ive Technology, 12(3), 30*0*–31*4. doi:10.1080/17483107.2016.1253117 PMID:28152642

Testad, I., Corbett, A., & Aarsland, D. (2013). ORE Open Research Exeter. Journal of Cleaner Produc*tion.*

Therriault, D., Shep*herd, R. F., White, S. R., & Lewis, J. A. (2005). Fugitive inks for direct-write assembly of three-dimensional microvascular networks. Advanced Materials, 17, 395–399.

Therriault, D., White, S. R., & Lewis, J. A. (*2003). Chaotic mixing in three-dimensio*nal microvascular networks fabricated by direct-write assembly. Nature Materials, 2, 347.

The strongest choice for 3D printing. (2014). *Retrieved January* 20, 2023, from https://www.stratasys.com/en/resources

The Third Industrial Revolution. (2012, April 21). Retrieved January 20, 2023, from https://www.economist.com/leaders/2012/04/21/the-third-industrial-revol*ution*

*Tian, H., Tang, Z., Zhuang, X., Chen, X., & Jing, X. (*2012*).* Biodegradable synthetic polymers: Preparation, functionalization and biomedical application. Progress in Polymer Science, 37(2), 237–280. doi:10.1016/j.progpolymsci.2011.06.004

Tian, Y., Chen, C., Xu, X., Wang, J., Hou, X., Li, K., Lu, X., Shi, H., Le*e, E., & Jiang, H. B. (2021). A Review of 3D Printing in* Dentistry: Technologies, Affecting Factors, and Applications. Academic Press.

Tibbits, S. (2014). 4D printing: Multi-material shape change. Architec. Des., 84(1), 116–121.

Tigmeanu, C. V., *Ardelean, L.* C., Rusu, L. C., & Negrutiu, M. L. (2022). Additive Manufactured Polymers in Dentistry, Current State-of-the-Art and Future Perspectives-A Review. Polymers, 14(17), 1*–24. doi:10.3390/polym141736*58 PMID:36080732

Tlegenov, Y., Hong, G. S., & Lu, W. F. (2018). Nozzle condition monitoring in 3D printing. Robotics a*nd Computer-Integrated Manufacturing, 54, 45–55. do*i:10.1016/j.rcim.2018.05.010

Todica, M., Stefan, R., Pop, C. V., & Olar, L. (2015). IR and Raman investigation of some poly (acrylic) acid gels in aqueous and neutralized state. Acta Physica *Polonica A, 128(1), 128–135. doi:10.*12693/APhysPolA.128.128

Todor, M. P., & Kiss, I. (2016). Systematic approach on materials selection in the automotive industry for making vehicles lighter, safer and more *fuel–efficient. Appli*ed Engineering Letters, 1(4), 91–97.

Todor, P., Bulei, C., & Kiss, I. (2017). An Overview on Fiber-Reinforced Composites Used in the automotive industry. International Journal of Engineering, XV(2), 181–184.

Tootooni, M., Dsouza, A., *Donovan, R., Rao, P. K., Kong, Z., & Borgesen, P. (2017).* Classifying the dimensional variation in additive manufactured parts from laser-scanned three journal dimensional point cloud data using machine learning approaches. Journal of Manufacturing Science a*nd Engineering, 1*39, 091005.

Top, N., Şahin, İ., Gökçe, H., & Gökçe, H. (2021). Computer-aided design and additive manufacturing of bone scaffolds for tissue engineering: State of the art. Journal of Materials Research, 36(19), 3725–3745. doi*:10.155743578-021-00156-y*

Torrado Perez, A. R., Roberson, D. A., & Wicker, R. B. (2014). Fracture surface analysis of 3D-printed tensile specimens of novel ABS-based materials. Journal of Fa*ilure Analysis and Pre*vention, 14(3), 343–353. doi:10.100711668-014-9803-9

Tran, P., Ngo, T. D., Ghazlan, A., & Hui, D. (2017). Bimaterial 3D printing and numerical analysis of bio-inspired composite structures under in-plane and transverse loadings. Composites. Part B, Engineeri*ng, 108, 210–223. doi:10.1016/j.* c*ompo*sitesb.2016.09.083

Tran, P., Ngo, T. D., Ghazlan, A., & Hui, D. (2017). computational *study of bio-inspired composi*te structures using bimaterial 3D printing under transverse and in-plane loads. Composites. Part B, Engineering, 108, 210–223.

Tranchard, S., & Rojas, V. (2015, May 5). Manufacturing our 3D future. Retrieved January 20, 2023, from https://www.iso.org/news/2015/05/Ref1956.html

Trindade, W. G., de Paiva, J. M. F., Leao, A. L., & Frollini, E. (2008). Ionized-Air-Treated Curaua Fibers as Reinforcement for Phenolic Matrices. Macromolecular *Materials and Eng*ineering, 293(6), 521–528.

Trivedi, A., & Gurrala, P.K. (2021). Fuzzy logic based expert system for prediction of tensile strength in fused filament fabrication (FF*F) process. Mate*r. Today: Proc., 44.

Trojovsky, P., Dhasarathan, V., & Boopathi, S. (2023). Experimental investigations on cryogen*ic friction-stir welding of similar ZE42 magnesium alloys. Alexandria Engineering Journ*al, 66, 1–14. doi:10.1016/j.aej.2022.12.007

Truong, V., Blakey, I., & Whittaker, *A. K. (2012). Hyd*rophilic and amphiphilic polyethylene glycol-based hydrogels with tunable degradability prepared by "click" chemistry. Biomacromolecules, 13(12), 4012–4021. doi:10.1021/bm3012924 PMID:23134321

*Trusle*r, *G.* A., & Mustard, W. T. (1958). Intravenous polyethylene catheter successfully removed from the heart. Canadian Medical Association Journal, 79(7), 558. PMID:13585269

Tudor, A. M., Melia, C. D., Binns, J. S., Hendra, P. J., Church, S., & Davies, M. C. (1990). The application of Fourier-transform Raman spectroscopy to the analysis of pharmaceuticals and biomaterials. Journal of Phar*maceutical and Biomedic*al *An*alysis, 8(8-12), 717–720. doi:10.1016/0731-7085(90)80109-3 PMID:2100613

Tymrak, B., Kreiger, M., & Pearce, J. M. (2014). by employing open-source 3-D printers, the components of Mechanical characteristics pro*duced in real world environ*ments. Materials & Design, 58, 242–246.

Tymrak, B. M., Kreiger, M., & Pearce, J. M. (2014). Mechanical properties of components fabr*icated with open-source 3-D printers un*der realistic environmental conditions. Materials & Design, 58, 242–246. doi:10.1016/j.matdes.2014.02.038

Ulas, M., Aydur, O., Gurgnec, T., & Ozel, C. (2020). Surface roughness prediction of machined aluminum alloy with wire electrical discharge machining by differ*ent machine learning algorithms. J. Mater. Res.* Techn*o*l., 9.

Ulery, B. D., Nair, L. S., & Laurencin, C. T. (2011). Biomedical applications of biodegradable polymers. Journal of Polymer Science. Part B, Polymer Physics, 49(12*), 832–864. doi:10.1002/polb.*22259 PMID:21769165

Vahabi, E., & Rahmati, S. (2016). Application of an RBF neural network for FDM parts' surface roughness prediction for enhancing surface quality. Int J Precis Eng Manuf, 17.

Valad*ez-Gonzalez, A., Cervantes-Uc, J. M., Olay*o, *R.*, & Herrera-Franco, P. J. (1999). Effect of fiber surface treatment on the fiber–mat*rix bond strength of natura*l fiber reinforced composites. Composites. Part B, Engineering, 30(3), 309–320.

Valino, A. D., Dizon, J. R. C., Espera, A. H. Jr, Chen, Q., Messman, J., & Advincula, R. C. (2019). Advances in 3D printing of thermoplastic polymer composites and nanocomposite*s. Progress in Polymer Science, 98,* 101162. doi:10.1016/j.progpolymsci.2019.101162

Van den Oever, M. J. A., Beck, B., & Müssig, J. (2010). Agrofibre reinforced poly (lactic acid) composites: Effect of moisture on degradation and mechanical properties. Composite*s. Part A, Applied Science and* Man*uf*acturing, 41(11), 1628–1635.

Van Der Klift, F., Koga, Y., Todoroki, A., Ueda, M., Hirano, Y., & Matsuzaki, R. (2015). 3D printing of continuous carbon fibre r*einforced thermo-plastic (CFRTP) tensile* test specimens. Open J Compos Mater, 6(01), 18.

Van Der Klift, F., Koga, Y., Todoroki, A., Ueda, M., Hirano, Y., & Matsuzaki, R. (2016). 3D Printing of Continuous Carb*on Fibre Reinforced The*rmo-Plastic (CFRTP) Tensile Test Specimens. Open Journal of Composite Materials, 06(01), 18–27. doi:10.4236/ojcm.2016.61003

Van Noort, R. (2012). The futu*re of dental devices is digita*l. *D*ental Materials, 28(1), 3–12. doi:10.1016/j.dental.2011.10.014 PMID:22119539

Vapnik, V. N. (2013). The nature of statistical learning theory. Springer.

*Vidakis, N., Petousis, M., Tzouni*s, *L.*, Maniadi, A., Velidakis, E., Mountakis, N., Papageorgiou, D., Liebscher, M., & Mechtcherine, V. (2020). Sustainable additive manufacturing: Mechanical response of polypropylene over multiple recycling processes. Sustainabilit*y, 13(1), 159. doi:10.3390u13010159*

*Von Tengg-Ko*bligk, W. T., Rengier, F., Kotelis, D., Geisbusch, P., Bockler, D., Schumacher, H., & Ley, S. (2008). Imaging modalities for the thoracic aorta. The Journal of Cardiovascular Surgery, 49(4), 429–447. PMID:18665106

Wallace, J., Wang, M. O.*, Thompson, P., B*us*so*, M., Belle, V., Mammoser, N., & Dean, D. (2014). Validating continuous digital light processing (cDLP) additive manufacturing accuracy and tissue engineering utility of a dye-initiator package. Biofabrication, 6(1)*, 015003. doi:10.1*088/1758-5082/6/1/015003 PMID:24429508

Wambua, P., Ivens, J., & Verpoest, I. (2003). Natural fibres: Can they replace glass in fibre reinforced plastics? Composites Science and Technology, 63(9), 1259–1264. *doi:10.1016/S0266-3538(03)*00096-4

Wang, C., Tan, X. P., Tor, S. B., & Lim, C. S. (2020). Machine learning in additive manufacturing: *State-of-the-art and perspectives. Addit. Manuf.,* 36, *1*01538.

Wang, J., Li, H., Liu, R., Li, L., Lin, Y. H., & Nan, C. W. (2018). Thermoelectric and mechanical properties of PLA/Bi0·5Sb1·5Te3 composite wires used for 3D printing. Composites Science and *Technology, 157, 1–9. doi:*10.1016/j.compscitech.2018.01.013

Wang, P., Zou, B., Ding, Sh., Zhuang, Y., Liu, J., & Li, L. (2022). Functionally graded polyetheretherketone-based composites additively manufactured by material *extrusion using a transition inter*face design method. Composites. Part A, Applied Science and Manufacturing, 158, 106977. doi:10.1016/j.compositesa.2022.106977

Wang, Q., Ji, C., Sun, J., Yao, Q., Liu, J., Saeed, R. M. Y., & Zhu, *Q. (2019). Kinetic thermal* beh*a*vior of nanocellulose filled polylactic acid filament for fused filament fabrication 3D printing. Journal of Applied Polymer Science. Advance online publication. doi:10.1002/app.48374

Wang, S., Kempen, D. H., Simha, N. *K., Lewis, J. L., Windebank, A. J., Yaszemski, M. J.,* & Lu, L. (2008). Photo-cross-linked hybrid polymer networks consisting of poly (propylene fumarate) and poly (caprolactone fumarate): Controlled physical properties and regulated bone and nerve cell respons*es. Biomacromolecul*es, 9(4), 1229–1241. doi:10.1021/bm7012313 PMID:18307311

Wang, S., Li, M., Yang, H., Liu, Y., & Sun, Y. (2017). Application of computer aided design and fused deposition modeling technology in the di*gital manufacture of orthodontic st*ud*y* models. Biomedical Research (India), 28(10), 4425–4431.

Wang, T. (2016). Electroactive polymers for sens*ing. Interface F*ocus, 6(4), 20160026.

Wang, X., Jiang, M., Zhou, Z., & Gou, J. D. (2017). 3D printin*g of polymer matrix composites: A review* and prospective. Compos. B Eng., 110, 442–458. doi:10.1016/j.compositesb.2016.11.034

Wang, X., Jiang, M., Zhou, Z., Gou, J., & Hui, D. (2017). 3D printing of polymer matrix composites: A review and prospective. Composites. Part B, Engineering, 110, 442–458.

Wang*, Y., & Wei, Q. (*2020). A review on 3D printed matrix polymer composites: Its potential and future challenges. International Journal of Advanced Manufacturing Technology, 106, 1695–1721. https://doi.org/10.*1007/s00170-019-04534-z*

Wang, Y., Lin, Y., Zhong, R. Y., & Xu, X. (2019). IoT-enabled cloud-based additive manufacturing platform to support rapid product development. International Jou*rnal of Production Research, 57(12),* 3975–3991. doi:10.1080/00207543.2018.1516905

Wang, Y., Zheng, P., Peng, T., Yang, H., & Zou, J. (2020). Smart additive manufacturing: current artificial intelligence-enab*led methods and fut*ure perspectives. Science China Technological Sciences., doi:10.100711431-020-1581-2

Wang, Y.-C., Chen, T., & Yeh, Y.-L. (2018). Advanced 3D printing technologies for the aircraft ind*ustry: A fuzz*y *sys*tematic approach for assessing the critical factors. International Journal of Advanced Manufacturing Technology, 1–11.

Ward, I. M., & Wilding, M. A. (1977). Infra-red and Raman *spectra of poly (m-methylene tere*ph*tha*late) polymers. Polymer, 18(4), 327–335. doi:10.1016/0032-3861(77)90077-5

Wei, L., Wu, S., Kuss, M., Jiang, X., Sun, R., Reid, P., & Duan, B. (2019). 3D printing of silk fibroin-based hybrid scaffold treated with platelet rich plasma for bone tissue engineering. *Bioactive Materials, 4, 256–260. doi:10.1016/j.bioactmat.*2019.09.001 PMID:31667442

Wei, X., Li, D., Jiang, W., Gu, Z., Wang, X., Zhang, Z., & Sun, Z. (2015). 3D printable graphene composite. Scientific Reports, 5, 11181.

Weller, C., Kleer, R., & Piller, F. T. (2015). Economic implications of 3D print*ing: Market structure models in l*ight of additive manufacturing revisited. International Journal of Production Economics, 164, 43–56. https://doi.org/10.1016/j.ijpe.2015.02.020

Wexler, P. (2014). Encyclopedia of Toxicology. Academic Press.

Wickramasinghe, S., Do, T., & Tran, *P. (2020). FDM-based 3D prin*ting of polymer and associated composite: A review on mechanical properti*es, defects an*d treatments. Polymers, 12(7), 1529.

Wickramasinghe, S., Do, T., & Tran, P. (2020). FDM-Based 3D printing of polymer and associated composite: A *review on mechanical properti*es, *de*fects and treatments. Polymers, 12(7), 1–42. doi:10.3390/polym12071529 PMID:32664374

Williams, D. F. (2008). On the mechanisms of biocompatibility. Biomaterials, 29(20), 2941–2953. *doi:10.1016/j.biomaterials.2008.04.023 PMID:18440630*

*Woh*lers, T. (2014). Tracking global growth in industrial-scale additive manufacturing. 3D. Print Addit Manuf, 1(1), 2–3.

Wohlers, T., & Gornet, T. (2015). *History of additive manufacturing. Wohlers R*eport 2015.

Wong, K. V., & Hernandez, A. (2012). A review of additive manufacturing. International Scholarly Research Notices.

Wong, K. V., & Hernandez, A. (2012). A rev*iew of additive manufacturing. ISRN Mechanical Engineering.*

*Wo*odward, D. I., Purssell, C. P., Billson, D. R., Hutchins, D. A., & Leigh, S. J. (2015). Additively-manufactured piez*oelectric devices. Physica Status Solidi. A, Applications and Materials Science, 212(10), 2107–2113.* https://doi.org/10.1002/pssa.201532272

Wright, W. J., Koerner, H., Rapking, D., Abbott, A., & Celik, E. (2022). Rapid fiber alignment quantification in direct write printing of shor*t fiber rei*nforced composites. Composites. Part B, Engineering, 236, 109814. doi:10.1016/j.compositesb.2022.109814

Wu, H., Fahy, W. P., Kim, S., Kim, H., Zhao, N., Pi*lato, L., Kafi, A., B*ateman, S., & Koo, J. H. (2020). Recent developments in polymers/polymer nanocomposites for additive manufacturing. Progress in Materials Science, 111, 100638. doi:1*0.1016/j.pmatsci.2020.100638*

*Wu, H., Yu, Z., & Wan*g, Y. (2016). A new approach for online monitoring of additive manufac*turing based on acoustic em*ission. Proceedings of the ASME 2016 international manufacturing science and engineering conference.

Wu, He, Hu, & Zhao. (2018). Process Monitoring of Fused Deposition Modeling throug*h Profil*e C*o*ntrol. doi:10.1109/CBS.2018.8612192

Wu, M., Song, Z., & Moon, Y.B. (2019). Detecting cyber-physical attacks in cyber manufacturing systems with machine learning methods. J Intell Manuf, 30.

Wu, M., Zhou, H., Lin, L. *L., Sil*va, B., Song, Z., & Cheung, J. (2017). Detecting attacks in cyber manufacturing *systems: Additi*ve manufacturing example. MATEC, 108, 06005.

Xiao, Y., Zhang, J., Fang, B., Zhao, X., & Hao, N. (2022). Acoustics-Actuated Microrobots. Micromachines, 13(3), 481. doi:10.3390/mi13030481 PMID:35334771

Xin, Y. (2016). The use of polyvinylidene fluoride *(PVDF) films as s*en*s*ors for vibration measurement: A brief review. Ferroelectrics, 502(1), 28–42.

Xiong, *J., Zhang, G., Hu, J., & Wu, L. (2014).* Bead geometry prediction for robotic GMAW-based rapid manufacturing throug*h a neural network and a se*cond-orderregression analysis. J Intell Manuf, 25.

Xu, D., Liu, P.F., Li, J.G., & Chen, Z.P. (2019). Damage mode identification of adhesive *composite joints under hygrothermal environment using acous*tic emission and machine learning. Compos Struct, 211.

Xu, F., Celli, J., Rizvi, I., Moon, S., Hasan, T., & Demirci, U. (2011). A three-dimensional in vitro ovarian cancer co-culture model using a high-throughput cell patterning platfo*rm. Biotechnology Journal, 6(2), 204*–212. doi:10.1002/biot.201000340 PMID:21298805

Xu, H., Cui, W., & Chang, J. *(2013). Fabrication of patterned PDLLA/PCL composite scaffold by electro*spinning. Journal of Applied Polymer Science, 127(3), 1550–1554. doi:10.1002/app.37505

Yakacki, C. M., Griffis, J., Poukalova, M., & Gall, K. (2009). Bearing area: *A new indication f*or suture anchor pullout strength? Journal of Orthopaedic Research, 27(8), 1048–1054. https:// doi.org/10.1002/jor.20856

Yan, Q., Dong, H., Su, J., Han, J., Song, B., Wei, Q., & Shi, Y. (201*8). A Review of 3D Printing Techno*logy for Medical Applications. Engineering, 4(5), 729–742. doi:10.1016/j.eng.2018.07.021

Yan, X., & Gu, P. (1996). A review of rapid prototyping technologies and systems. *CAD Computer Aided Design, 28(4), 307–318. doi:10.1016/0010-4485(95)00035-6*

*Yang, J., Che*n, Y., Huang, W., & Li, Y. (2017). Survey on artificial intelligence for additive manufacturing. Proceedings of the 23rd international conference on automati*on an*d *com*puting.

Yao, X., Moon, S.K., & Bi, G. (2017). A hybrid machine learning approach for additive manufa*cturing design fe*ature recommendation. Rapid Prototyp J, 23.

Yaszemski, M. J., Payne, R. G., Hayes, W. C., Langer, R. S., Aufdemorte, T. B., & Mikos, A. G. (1995). The ingrowth of ne*w bone tissue* an*d i*nitial mechanical properties of a degrading polymeric composite scaffold. Tissue Engineering, 1(1), 41–52. doi:10.1089/ten.1995.1.41 PMID:19877914

Yaxin, Z. F. H. E. Z. L. Z. (2023, January). *Effect of the sup*port bath on embedded 3D printing of soft elastomeric composites. Materials (Basel), 331(15), 133475.

Ye, D., Fuh, J.Y.H., Zhang, Y., Hong, G.S., & Zhu, K. (2018). In situ monitoring of *selective laser m*elting using plume and spatter signatures by deep belief networks. ISA Trans, 81.

Ye, D., Hong, G.S., Zhang, Y., Zhu, K., & Fuh, J.Y.H. (2018). Defect detection in selective lasermelting *technology by acous*tic signalswith deep belief networks. Int J AdvManuf Technol, 96.

Yeo, T., Ko, Y. G., Kim, E. J., Kwon, O. K., Chung, H. Y., & Kwon, O. H. (2021). Promoting bone regeneration by 3D-printed *poly (glycolic acid)/hydroxyap*ati*te* composite scaffolds. Journal of Industrial and Engineering Chemistry, 94, 343–351. doi:10.1016/j.jiec.2020.11.004

Yin, J., Zhu, G. Z., & Deng, B. (2013). Multi-*walled carbon nanotubes (MWNTs)/polysulfone (PSU) mixed matrix hollow fiber* membranes for enhanced water treatment. Journal of Membrane Science, 437, 237–248.

Yousif, B. F., & El-Tayeb, N. S. M. (2008). Wear and *friction c*haracteristics of CGRP composite under wet contact condition using two different test techniques. Wear, 265(5–6), 856–864. doi:10*.1016/j.wear.2008.01.016*

Yu*an*, P. F., Meng, H., Yu, L., & Zhang, L. (2016). Robotic Fabrication in Architecture, Art and Design 2016. Robotic Fabrication in Architecture, Art and Design. Advance online public*ation. doi:10.1007/9*78-3-319-26378-6

Yuanjian, T., & Isaac, D. H. (2008). Combined impact and fatigue of glass fiber reinforced composites. Composites. *Part B, Enginee*ring, 39(3), 505–512. doi:10.1016/j.compositesb.2007.03.005

Yupapin, P., Trabelsi, Y., Nattappan, A., & Boopathi, S. (2022). Performance Improvement of Wire-Cut Electrical Discharge Machi*ning Process* Using Cryogenically Treated Super-Conductive State of Monel-K500 Alloy. Iranian Journal of Science and Technology. Transaction of Mechanical Engineering. Advance onl*ine publication. doi:10.10*0740997-022-00513-0

Zaheri, Fenner, Russell, Restrepo, Daly, Wang, Hayashi, Meyers, Zavattieri, & Espinosa. (2018). Revealing the Mechanics of Helicoidal Composites *through Additive Manufactu*ring and Beetle Developmental Stage Analysis. Advanced Functional Materials, 28(33).

Zaldivar, R.J., Mclouth, T.D., Ferrelli, G.L., Patel, D.N., Hopkins, A.R., & Witkin, D. (201*8).* Ef*fec*t of initial filament moisture content on the microstructure and mechanical performance of ULTEM 9085 3D printed parts. Addit. Manuf., 24.

Zanchetta, *E., Cattaldo, M., Franchin, G.,* Sch*w*entenwein, M., Homa, J., Brusatin, G., & Colombo, P. (2016). Stereolithography of SiOC ceramic microcomponents. Advanced Materials, 28(2), 370–376.

Zarrintaj, P.*, Saeb, M. R., Jafari, S. H., & Mozafari, M. (2020*). Application of compatibilized polymer blends in biomedical fields. In Compatibilization of Polymer Blends (pp. 511–537). Elsevier. doi:10.1016/B978-0-12-816006-0.00018-9

Zein, I., Hutmacher, D. W., Tan, K. C., & Teoh, S. H. (2002). Fused deposition modeling of novel sc*affold architectures for tissue engineering applications. Biomaterials, 23(4), 1*169–1185.

Zhang,, M., Sun,, C.N., Zhang,, X., Goh,, P.C., & Wei,, J. (2019). High cycle fatigue life prediction of laser additive manufactured stainless steel: A machine learning approach. International Journal of Fatigue, 128, 105194.

Zhang, B., Liu, S., & Shin, Y.C. (2019). *In-process monitoring of porosi*ty during laser additive manufacturing process. Addit. Manuf., 28.

Zhang, Q., Cai, H., Zhang, A., Lin, X., Yi, W., & Zhang, J. (2018). Effects of lubricant and toughening agent on the fluidity and toughness of poplar powder-rei*nforced polylacti*c acid 3D printing materials. Polymers, 10(932), 1–11.

Zhang, S. U. (2019). Classifying Degraded Three-Dimensionally Printed Polylactic Acid Specimens Using Arti*ficial Neural Netw*orks based on Fourier Transform Infrared Spectroscopy. Applied Sciences (Basel, Switzerland), 9(13), 2772. doi:10.3390/app9132772

Zhang, W., Chen, B., Zha*o, H., Yu, P., Fu, D., Wen, J., & P*eng, X. (2013). Processing and characterization of supercritical CO2 batch foamed poly (lactic acid)/poly (ethylene glycol) scaffold for tissue engineering application. Journal of Applied Polymer Science, 130(5), 3066–3073. *doi:10.1002/*app.39523

Zhang, W., Mehta, A., Deasi, P. S., & Higgs, C. (2017). Machine learning enabled powder spreading process map for metal additive *manufacturing (AM*). Proceedings of the 28th annual international solid freeform fabrication symposium.

Zhang, Y., & Atala, A. (2019). Regenerative medicine of the bladder. In Principles of Regener*ative Medicine* (pp. 1263–1279). Academic Press. doi:10.1016/B978-0-12-809880-6.00072-2

Zhang, Y., Hong, G.S., Ye, D., Zhu, K., & Fuh, J.Y.H. (2018). Extraction and evaluation of melt pool, plume *and spatter information for powd*er-b*e*d fusion AM process monitoring. Mater Des, 156.

Zhang, Z., Liu, Z., & Wu, D. (2021). Prediction of melt pool temperature in directed energy deposition using machine learning. Addit. Manuf., 37, 101692.

Zhang, Z.-X., *He, Z.-Z., Y*ang, J.-H., Huang, T., Zhang, N., & Wang, Y. (2016). Crystallization controlled shape memory behaviors of dynamically vulcanized poly (l-lactide)/ poly (ethylene vinyl acetate) blends. Polymer *Testing, 51, 82–92.*

Zhao, D. X., Cai, X., Shou, G. Z., Gu, Y. Q., & Wang, P. X. (2016). Study on the preparation of bamboo plastic composite intend for additive manufacturing. Key Engineering Materials, 667, 250–258.

Zhao, *H., He, Y., Fu, J., & Qiu, J. (2017, November). Inclined layer printing for fused* deposition modeling without assisted supporting structure. Robotics and C*omputer-integrated Manufacturing, 51*, 1–13. doi:10.1016/j.rcim.2017.11.011

Zhao, H., Yang, F., Fu, J., Gao, Q., Liu, A., Sun, M., & He, Y. (2017). email protected]: From Medical Models to Organ Implants. ACS Biomaterials Science & Engin*eering, 3(12), 30*83–3097. doi:10.1021/acsbiomaterials.7b00542 PMID:33445353

Zhao, Q., Qi, H. J., & Xie, T. (2015). Recent progress in shape memory polymer: New behavior, enabling materials, and mechanistic understanding. Progress in Polym*er Science, 49,* 79–120.

Zhong, W., Li, F., Zhang, Z., Song, L., & Li, Z. (2001). Short fiber reinforced composites for fused deposition modeling. Materials Science and Engineering A, 3*01(2), 125–130. doi:10.10*16/*S0*921-5093(00)01810-4

Zhou, Z., Cunningham, E., Lennon, A., McCarthy, H. O., Buchanan, F., & Dunne, N. (2018). Development of three-dimensional printing polymer-*ceramic scaffolds with enhanced compressive prope*rties and tuneable resorption. Materials Science and Engineering C, 93, 975–986.

Zhu, J., Hu, Y., Tang, Y., & Wang, B. (2017). Effects of styrene–acrylonitrile contents on the properti*es of ABS/SAN blends for fu*sed deposition modeling. Journal of Applied Polymer Science, 134(7), 1–5. doi:10.1002/app.44477

Zhu, Z., Answer, N., Huang, Q., & Mathie*u, L. (2018). Machine learning in t*oleran*cing for additive manufacturing. CIRP Ann Manuf Technol, 67.

Zhu, Z. H., Zhang, N., Wang, T., & Hao, M. Y. (2020, February). Short review of polymer composites for 3D Printing. IOP Conference Series. Materials Science and Engineering, 758(1), 012046. doi:10.1088/175*7-899X/758/1/012046*

*Zini, E., & Sc*and*o*la, M. (2011). Green composites: An overview. Polymer Composites, 32(12), 1905–1915.

Zohdi, T.I. (2018). Dynamic thermomechan*ical modeling and simulati*on of the design of rapid free-form 3D printing processes with evolutionary machine learning. Comput Methods Appl Mech Eng, 331.

Zolfagharian, A., Durran, L., *Gharaie, S., Rolfe, B., Kaynak,* A., & Bodaghi, M. (2021). 4D printing soft robots guided by machine learning and finite element models. Sensors and Actuators. A, Physical, 328, 112774.

*Zuniga, J. M. (2018). 3D printed antibacterial prosthese*s. *Ap*plied Sciences (Basel, Switzerland), 8(9), 1651. doi:10.3390/app8091651

Özcan, M., Hotza, D., Fredel, M. C., Cru*z, A., & Volpato,* C. A. M. (2021). Materials and manufacturing techniques for polymeric and ceramic scaffolds used in implant dentistry. Journal of Composites Science, 5(3), 78. doi:10.3390/jcs5030078

*Özpozan, T., Schrader, B., & Ke*ller, S. (1997). Monitoring of the polymerization of vinylacetate by near IR FT Raman spectroscopy. Spectrochimica Acta. Part A: Molecular and Biomolecular Spectroscopy, *53(1), 1–7. doi:10.1016/S1386-1425(97)*83003-X

About the Contributors

R. Keshavamurthy is full time professor and Head of Department of Mechanical Engineering at Dayananda Sagar College of Engineering, Bangalore, India. He has completed Bachelor of Engineering in Mechanical Engineering followed by Masters of Technology in Manufacturing Science and Engineering and PhD in Mechanical Engineering. He started his career as Research Associate in 2007, followed by Scientist at Central Manufacturing Technology Institute (CMTI) (GOI) Bangalore, in Rapid manufacturing division and presently working at Dayananda Sagar College of Engineering. He has been a recipient of many sponsored research projects as Principal Investigator which was mainly sponsored by SERB, Dept. of Science & Technology, GOI, VGST Govt. Of Karnataka and AICTE, New Delhi. Under his guidance more than 10 batches of students have obtained sponsored research grant from Karnataka State Council for Science and Technology. His patents are also numerous and one being in the field of Additive Manufacturing. He has guided numerous PhD, postgraduate and graduate students. He is life time member of numerous professional bodies. He is recipient of numerous awards and also Gold medalist for securing highest percent in Master of Technology. He is editorial member and reviewer of numerous reputed journals. He has more than 150 research articles publication in journals and conferences and authored more than 20 book chapters. He has lifelong citations of 2224 and has h-index of 27.

Vijay Tambrallimath completed his Bachelor of Engineering from VTU, Karnataka, India in 2011 and Master of Technology from VTU, Karnataka, India in 2013. PhD in synthesis and characterization of polymer composites developed by 3D printing technology was completed in Bangalore, India in 2020. He is recipient of numerous best research paper awards in international conferences. He has been a reviewer at Scopus journals in the field of Additive Manufacturing and composites. He has published more than 25 research papers and 8 book chapters till date. To his credit he has a granted Australian Patent in the field of Additive Manufacturing and one published patent in Indian Patent Office. He has a research grant in the field of Additive Manufacturing. Also he has been an invited speaker in national level faculty development program in the field of Additive Manufacturing. He has extensive international research collaborations at his very young age.

J. Paulo Davim received his Ph.D. degree in Mechanical Engineering in 1997, M.Sc. degree in Mechanical Engineering (materials and manufacturing processes) in 1991, Mechanical Engineering degree (5 years) in 1986, from the University of Porto (FEUP), the Aggregate title (Full Habilitation) from the University of Coimbra in 2005 and the D.Sc. from London Metropolitan University in 2013. He is Eur Ing by FEANI-Brussels and Senior Chartered Engineer by the Portuguese Institution of Engineers with an MBA and Specialist title in Engineering and Industrial Management. Currently, he is Professor at the Department of Mechanical Engineering of the University of Aveiro, Portugal. He has more than 30 years of teaching and research experience in Manufacturing, Materials, Mechanical and Industrial Engineering, with special emphasis in Machining & Tribology. He has also interest in Management, Engineering Education and Higher Education for Sustainability. He has guided large numbers of postdoc, Ph.D. and master's students as well as coordinated & participated in several financed research projects. He has received several scientific awards. He has worked as evaluator of projects for international research agencies as well as examiner of Ph.D. thesis for many universities. He is the Editor in Chief of several international journals, Guest Editor of journals, books Editor, book Series Editor and Scientific Advisory for many international journals and conferences. Presently, he is an Editorial Board member of 25 international journals and acts as reviewer for more than 80 prestigious Web of Science journals. In addition, he has also published as editor (and co-editor) more than 100 books and as author (and co-author) more than 10 books, 80 book chapters and 400 articles in journals and conferences (more than 200 articles in journals indexed in Web of Science core collection/h-index 45+/6000+ citations and SCOPUS/h-index 53+/8500+ citations).

* * *

Sevim Akyuz is a professor of physics, currently working at the Physics Department of Istanbul Kultur University (Istanbul-Turkey). Her research interest is on experimental and theoretical molecular spectroscopy and mainly focused on the study of a relationship between the structure and spectral properties of macromolecules using combined theoretical and experimental methods. She has 202 scientific papers as SCI record and total number of citations according to Web of Knowledge is 2253. Her H-index according to Web of Knowledge is 25.

Abhishek B. is working as an Assistant Professor in the Department of Electronics and communication engineering at Dayananda Sagar College of engineering.

Praveena B. A. received his Bachelor of Engineering in Mechanical Engineering from Visvesvaraya Technological University, India, Post-graduation in Computer Integrated Manufacturing from Visvesvaraya Technological University, India and Ph.D in Mechanical Engineering Sciences from Visvesvaraya Technological University, India in the year 2008, 2012 and 2021 respectively. He has worked in industry for about 2 years. Since 2012 he is serving in the Department of Mechanical Engineering, Nitte Meenakshi Institute of Technology Bangalore, India as a teaching faculty. He has published number of research papers in international journals and conferences.

Sampath Boopathi () completed his undergraduate in Mechanical Engineering and postgraduate in the field of Computer-Aided Design. He completed his Ph.D. from Anna University and his field of research includes Manufacturing and optimization. He published 60 more research articles in Internationally Peer-reviewed journals, one Patent grant, and three published patents.He has 16 more years of academic and research experiences in the various Engineering Colleges in Tamilnadu, India.

Prakash C. P. S. completed Bachelor of Engineering in Mechanical Stream in 1987. He completed Master of Technology in Production Engineering System Technology in 1990 and completed Doctoral research in the field of Mechanical engineering in 2009. He has worked in various industries. Presently, he is working in the capacity of Principal and Professor in Dayananda Sagar College of Engineering, Bangalore, India. He has published number of research papers in international journals and conferences.

Juan Carlos Campos Rubio has a Bachelor's in Mechanical Engineering from Federal University of Santa Maria (1988), Master in Mechanical Engineering from Federal University of Santa Catarina (1992) and a Ph.D. in Mechanical Engineering from University of São Paulo (2000). In the 2006-2007 period, he participated in a post-doctoral internship at the University of Aveiro (Portugal). He is currently a full professor at the Federal University of Minas Gerais. He is part of the editorial board of several specialized international journals. He has experience in the areas of Mechanical and Production Engineering, with an emphasis on Manufacturing and Automation Processes, working mainly on the following topics: precision engineering, design, automation, machining, and composite materials.

Sefa Celik is an Associate Professor at the Physics Department of Istanbul University (Turkey). His research is focused on the elucidation of the structure of biomacromolecules using combined experimental (molecular spectroscopy) and theoretical methods (ab initio, molecular dynamics), particularly on the molecular docking studies for shedding light on new drug design. He has 39 scientific papers as SCI record.

Jaya Christiyan completed his B.E degree in Mechanical Engineering from C.S.I. Institute of Technology, Nagercoil in 2000. Later he worked as Production Engineer for a period of one year. In 2001 he joined for M.E in Computer Aided Design from Hindustan College of Engineering, Chennai. After he joined as CAD/CAM Engineer in Eco Pulp products India (Pvt) Ltd, Hosur for the period of 4 years. From 2006 to 2007 he worked as Lecturer in P.E.S. School of Engineering, Bangalore. Later he joined M S Ramaiah Institute of Technology as Assistant Professor. Dr. Christiyan, has a vast experience of 5 years in Industry and 16 Years in academics. From 2011 onwards he has actively participated in various activities on Additive Manufacturing Research and published more than 35 papers in International conference and Journal.

Wanderson de Oliveira Leite is a PhD in Production Engineering in UFMG (Brazil) with studies in the Universidad de Castilla-la Mancha (Spain). He has a Master's degree in Production Engineering, a Specialization in Teaching in Basic Education, a degree in Production Engineering and a degree in Industrial Design and Graduation.

Jalal El Bahaoui is an Associate Professor of mechanical engineering, Physics department of Faculty of Sciences of Tetouan, Abdelmalek Essaadi University, Morocco.

Vibha G. is working as an Assistant Professor in the Department of Electronics and Communication Engineering at Dayananda Sagar College of engineering.

Bindu H. M. is working as an Assistant Professor in the Department of Electronics and Communication Engineering at Dayananda Sagar College of Engineering.

Issam Hanafi is professor at National School of Applied Sciences - Al Huceima (ENSAH), Abdelmalek Essaâdi University, Morocco. He holds a Ph.D. in mechanical engineering from the University of Abdelmalek Essaâdi, Morocco. His main interest field is devoted to the development of novel machining processes for composite materials. His graduate degrees include a master's degree in civil engineering from Faculty of Sciences and Techniques of Tangier, Morocco, and a Master of advanced studies (DESA) from Faculty of Sciences at Tétouan, Morocco. His present research interests are in the areas of quality engineering, intelligent control as it relates to neural networks and fuzzy control, also genetic algorithms in industrial applications.

Karibasappa K. is working as Professor in the Department of Computer science engineering from Graphics Era (Deemed to be University), Dehradun, Uttarkhand.

Sukhila Krishnan completed her Ph.D in Polymer Science and Technology at Laboratory for Advanced Research in Polymeric Materials (LARPM), Central Institute of Plastic Engineering and Technology, Bhubaneswar and had done his M.Tech degree in Polymer science from Cochin University and is currently working at Sahrdaya College of Engineering and Technology, Kodakara. Her main research interests focus on biopolymers from natural resources.

Sudheer Kumar is working as a Polymer Scientist at JC OrthoHeal Pvt. Ltd. Vadodara Gujarat. He completed his MTech degree in Polymer Nanotechnology from the Biju Patnaik University of Technology Rourkela Odisha. Dr. Sudheer Kumar did his Ph.D. in Polymer Science and Technology at CIPET Government of India affiliated to Utkal University Bhubaneswar. His main research interests are thermoset epoxy resins, photocurable resins, biodegradable polymers, 3D filament fabrication, endcap oligomers, and polymeric coatings. He has published more than 17 research publications in peer-reviewed international journals and 07 book chapters.

Harish Kumar M. is Assistant Professor, Department of Mechanical and Automobile Engineering, CHRIST (Deemed to be university), Bangalore. Diploma in Mechanical Engineering from P.V.P Polytechnic, DTE. Bachelor of Automobile Engineering from PDA college of Engineering, Gulbarga, Visveswaraya Technological University and Master's Degree from Visveswaraya Technological University centre for Post-graduation Studies, Bangalore, Visveswaraya Technological University. His research interests include impact characterization and machining effects on composite laminates, FDM, Material coatings, Stress Analysis, and Biofuels. Has published 14 papers in reputed peer-reviewed journals and international conferences.

Aysen E. Ozel is a Professor at the Physics Department of Istanbul University (Turkey). Her research is focused on the molecular spectroscopy and on the molecular modeling. She has 64 scientific papers as SCI record.

Sharath P. C. has obtained his doctorate degree in Metallurgical and Materials Engineering from NITK, Surathkal. He completed his post-graduation (Process Metallurgy) from NITK, Surathkal and carried out major project in Deakin University, Australia and pursued graduation (Mechanical) from Malnad College of Engineering, Hassan under VTU, Belgaum. He has previously served in various reputed organization and currently working in Metallurgical and Materials Engineering department,

Jain University and has 10 years of teaching and research experience. His research interests are metal forming, severe plastic deformation, composites and its characterization, electro-chemical sensors and applications. He has awarded first prize for presentation in processing of automotive materials at the 70th annual technical meeting of the Indian Institute of Metals (NMD-ATM 2016) held at IIT Kanpur. He has published more than 25 international Journal and chair for many conferences. He is reviewer for many reputed international journals and received excellent reviewer award by ASM International - Journal on materials engineering and performance on March 2019.

K. Prabakaran is working as assistant professor in department of physics, KPR Institute of Engineering and Technology, Coimbatore, Tamil Nadu. Prior to his present position, he was working as postdoctoral researcher in Shenzhen University, China and SRM Institute of Science and Technology, Chennai. Dr. K. Prabakaran did his PhD in CIPET (Govt of India) Bhubaneswar, Odisha. He published 27 research articles and 3 book chapters in reputed international journals. His research focuses on polymer nanocomposites for biomedical and energy conversion applications.

M. Puviyarasan, a Professor in the Department of Mechanical Engineering at Panimalar Engineering College in Chennai, believes that science and technology should be accessible to everyone and should solve the community's specific needs. He has more than 18 years of Teaching and Research experience in the field of Mechanical Engineering. Currently, he is working on additive manufacturing and machinability investigations. He has published over 13 papers in Peer- reviewed reputed Scientific Journals and 24 papers in various international journals, international and national conferences.

Manasa R. is working as an Assistant Professor in the Department of Electronics and communication engineering at Dayananda Sagar college of engineering.

Sachit S. is working as an Assistant professor in the department of Robotics and Automation at Symbiosis Institute of Technology, Symbiosis International Deemed University Pune, Maharastra. Research areas are Additive Manufacturing, Smart Manufacturing, Machine Learning, and Industrial Automation.

Vijay Muni T. has over 13 years of experience in teaching courses in both undergraduate and postgraduate courses. Authored 5 textbooks on Electrical Engineering. Published over 55 articles in peer-reviewed journals. Filed 5 patents with one granted. Completed 5 global certifications in the field of AI, Multi Cloud Computing, Outcome Based Education and Foundations of Google Data Analytics.

Satyanarayana Tirlangi is working as an Associate Professor in the Department of Mechanical Engineering VIET College of Engineering, has about 10 years of teaching experience. He received his B.Tech degree in Mechanical Engineering from JNTUK and M.E. degree in Machine Design with distinction and PhD in the area of Metal matrix composites from Andhra University, A.P. He has published 15 research papers in refereed international journals and various international conferences. He has received several best paper awards for his research papers at various international conferences. His areas of research include Composite Materials, and Weldings, Renewable Energy Sources and CFD. He is an active member of IEEE and IEI

Srinivasan Vidyavasthu received his Bachelor of Engineering in Mechanical Engineering from Madras University, India, Post-graduation in Manufacturing Engineering from Anna University, India and Ph.D in Mechanical Engineering Sciences from Visvesvaraya Technological University, India in the year 2003, 2007 and 2019 respectively. He has worked in industry for about 5 years. Since 2007 he is serving in the Department of Mechanical Engineering, Dayananda Sagar College of Engineering, India as a teaching faculty. He has published number of research papers in international journals and conferences.

Index

www.ingramcontent.com/pod-product-compliance
Lightning Source LLC
LaVergne TN
LVHW060422300926
849162LV00006B/1499